The political culture of planning

The Political Culture of Planning is written for two distinct readerships. The main body of the book synthesizes a mass of information to provide an overview of a complex and amorphous field. This material is designed to meet the needs of students who require a succinct account of the American system of land use planning. These readers can ignore the notes. For those who are embarking upon a much wider and deeper study of land use planning in the United States the notes are crucial: they provide the guideposts to an immensely rich literature.

The first four parts of the text present the main issues of land use planning in the United States. Part I assesses the United States zoning system. The introductory chapter discusses the meaning of zoning (and its difference from planning), the primacy of local governments, the constitutional framework and the role of the courts. Chapter 2 provides the historical background to zoning and an outline of the classic Euclid case. Chapter 3 discusses the objectives and nature of zoning and the use which local governments have made of its inherently inflexible character. Chapter 4 acts as a corrective to this view, describing how lawyers and planners have shown remarkable ingenuity in adapting zoning to the demands of a changing society. Part II deals with the perennial issues of discrimination, financing infrastructure for new development and the process for negotiating zoning matters. Part III presents a discussion of two overlapping issues of increasing significance – aesthetics and historic preservation. Part IV focuses on the main issue facing land use planners: attempting to channel the forces of development into spatial forms held to be socially desirable.

Part V consists of a series of broad-ranging essays which discuss land use planning in the United States, its institutional and cultural framework and the reasons for its differences from the British and Canadian systems. Part VI discusses the limited possibilities for land use reform in the United States – drawing on the author's considerable experience in both Britain and Canada – in order to interpret the limitations and potentialities of land use planning in the United States.

The political culture of planning

American land use planning in comparative perspective

J. Barry Cullingworth

ROUTLEDGE
New York and London

First published in 1993 by

Routledge Inc.
29 West 35 Street
New York, NY 10001

Published in Great Britain by

Routledge
11 New Fetter Lane
London EC4P 4EE

Library of Congress Cataloging in Publication Data

Cullingworth, J.B.
        The political culture of planning : American land use planning in
    comparative perspective / J. Barry Cullingworth.
            p.   cm.
    Includes bibliographical references (p.   ) and index.
    ISBN 0-415-08812-7
    1. Land use—United States—Planning. 2. Land use—Law and
legislation—United States. 3. Land use—Canada—Planning. 4. Land
use—Great Britain—Planning. I. Title.
HD205.C85   1993   92-18452
33.73'0973–dc20   CIP

British Library Cataloguing in Publication Data also available.

This book is dedicated with sincere thanks to
Susan Brynteson, Director of Libraries,
and the staff of the Morris Library
who helped me even more than they knew.
A good library service is the heart of a University.
Delaware is fortunate in having such an excellent one.

Political culture comprises
the orientations and beliefs that people have toward
the structure and practice of their political institutions.
(Mercer 1982)

Contents

Tables

Preface

The genesis of this book, and its growth from a modest narrowly focused introductory account of American land use planning to a broad discussion of the political culture of planning in three countries, is outlined in the Introduction. Here, the reader is informed of the purpose of the book and the debts I owe to so many who have helped me along the way.

This book is written for two quite different readerships. First, the main parts of the narrative are intended to meet the needs of those who require a reasonably succinct account of the American system of land use planning, with a minimum of technical and legal encumbrances. Readers who fall into this class can ignore the notes: the text, together with supplementary material from the course instructor, should suffice. At the least, the account should be comprehensible, though not, of course, comprehensive.

The second class of readers are those who are embarking upon a much wider and deeper study of land use planning in the United States. For these, the notes are crucial: they provide the guideposts to an immensely rich literature. Few will want to follow all the guideposts; but those wanting to know where to explore further the innumerable themes touched upon in the text should find them a useful starting point.

It will be immediately apparent to any reader that the author is not a legal scholar. No apology is necessary on this account. There are many excellent texts on the law of land use regulation, though all present difficulties to students who are not in law school. Herein lies a major problem for anyone who attempts to write about land use planning in America: the field is dominated by legal experts. There is nothing strange about this: the primacy of a written constitution guarantees that lawyers will be the leading actors in any matter of dispute: an all-embracing category. But it is a major objective of this volume to present an account which is broadly accurate, and also comprehensible to the non-lawyer. It is not directed at readers who wish to become expert in the law of land use planning.

The first part of the book consists of four chapters which give a rapid survey of the U.S. zoning system. The introductory chapter discusses the meaning of zoning (and its difference from planning), the primacy of local governments, the constitutional framework, and the role of the courts. Chapter 2 provides some of the historical background to zoning, and an outline of the classic *Euclid* case. The third chapter discusses the objectives and nature of zoning and the use which local governments have made of its inherently inflexible character. The fourth chapter acts as a corrective to this view (that zoning is inflexible). It shows the ways in which lawyers and planners have exercised a remarkable ingenuity in adapting the procrustean bed of zoning to the demands of a continually changing society. Much of this first part of the book involves a discussion of, or at least a reference to, a host of legal cases now largely forgotten except by those involved in the never-ending succession of court cases.

Part II consists of three chapters which deal with the perennial issues of discrimination, financing infrastructure for new development, and the process of negotiating zoning matters. This is followed, in Part III, by a discussion of two overlapping issues which have achieved increased salience in recent years – aesthetics and historic preservation. The term "environmental quality" might have been used instead, but this suggests a broader field than is covered here.

In Part IV there are three chapters which focus on the main issue with which land use planners are faced: attempting to channel the forces of development into spatial forms which are thought to be socially desirable. (Whether this is, or is not, feasible is a matter which – like many similar ones – is not directly addressed in this book, though the reader is given some pointers to further reading.)

These four parts form the major body of the book, and attempt to discuss the main issues of land use planning in the United States. The coverage, of course, is not complete. In particular, there is no discussion of environmental policies: to have included those would have extended the book to inordinate length. (Other important subjects which are omitted include capital improvement programs and local economic development.) But sufficient has been reviewed to provide the reader with an introduction to American land use planning.

The last two parts are rather different. Part V consists of a series of broad-ranging essays which discuss land use planning in the United States, its institutional and cultural framework, and the reasons for its particular character, while Part VI contains a single chapter in which I discuss the limited possibilities for land use reform in the United States. These chapters attempt to draw upon my experience elsewhere, particularly in Britain and, to a lesser extent, in Canada, in order to identify the nature of land use planning in the United States.

The choice of Britain and Canada as the two countries to provide a comparative perspective is personal, but not idiosyncratic. It was determined partly by the fact that I have had personal experience of planning in these three countries, and partly because, despite their differences, they have sufficient in common to make comparison interesting and useful. After some twenty years' experience (largely, though not entirely, academic) of British planning, I spent six years in Canada, followed by (at the time of the main revision of this text) eight years in the United States. This has prompted a series of questions relating to the nature of planning in the three countries. An adequate treatment, of course, would involve many byways of history, culture, psychology, law, and so forth. Perhaps I may attempt that one day; but in the meantime, this personal – and provisional – interpretation may have some value in stimulating thought about the potentialities and limitations of United States land use planning.

What I have attempted to do in this volume is to synthesize a mass of material to provide an overview of an amorphous field. In so doing, I have incorporated my personal interpretation: I believe that the land use planning system in the United States operates in an intolerably regressive and inefficient way. It is a servant of private interests, not the public good. As the reader of the final chapter will note, I find myself with some strange bedfellows on this; but at least the argument might make the book somewhat livelier than it might otherwise have been.

Authors of a book such as this, which straddles so many fields, know how indebted they are to a very large number of people – academics, politicians, administrators, legislators, and so on. Some would be glad to see their contributions recorded; others have requested anonymity. The recorded list is necessarily incomplete. And so I offer thanks to many anonyms (particularly from planning and governmental agencies) who have helped in varying ways. But I must make special mention of a number of colleagues who have helped me greatly, sometimes without knowing it. Among these are Richard Babcock, David Callies and Robert Freilich, John DeGrove, Charles Haar, Dan Mandelker, Urlan Wannop, William H. Whyte, Norman Williams, and the late Don Hagman. The staff of the Morris Library at the University of Delaware have been extremely helpful to me. I would like to make particular mention of the staff in the Inter-Library Loan Section who, over a period of more than three years, have responded so readily to my innumerable requests for assistance. And finally my academic colleagues: David Ames, Tim Barnekov, Marvin Gilman, Bernie Herman, Caroline Torma, and, above all, Dan Rich have provided me with the intellectual stimulus and support which a work of this nature requires. The errors and blemishes are, of course, my own.

J. Barry Cullingworth

College of Urban Affairs and Public Policy
University of Delaware

Acronyms

ACHP	Advisory Council on Historic Preservation
ACIR	Advisory Commission on Intergovernmental Relations
ACSC	Areas of Critical State Concern (Florida)
AHOP	Areawide Housing Opportunities Plan
ALI	American Law Institute
CAC	Citizen Advisory Committee (Oregon)
CDBG	Community Development Block Grant
CMHC	Canada (formerly Central) Mortgage and Housing Corporation
COG	Council of Government
DLCD	Department of Land Conservation and Development (Oregon)
DRI	Developments of Regional Impact (Florida)
EDA	Economic Development Administration
EIS	Environmental Impact Statement
FHA	Federal Housing Administration
GAO	General Accounting Office
GB	Great Britain (England, Scotland, and Wales)
GPO	Government Printing Office (United States)
HMSO	Her Majesty's Stationery Office (UK)
HUD	Department of Housing and Urban Development
LCDC	Land Conservation and Development Commission (Oregon)
LCP	Local Coastal Plan (California)
LUBA	Land Use Board of Appeals (Oregon)
MSUA	Ministry of State for Urban Affairs (Canada)
NAACP	National Association for the Advancement of Colored People
NEPA	National Environmental Policy Act
NLC	National League of Cities

NTHP	National Trust for Historic Preservation
OAHPP	Office-Affordable Housing Production Program (San Francisco)
OHPP	Office-Housing Production Program (San Francisco)
POS	*Plans d'occupation des sols* (France)
PUD	Planned Unit Development
SDGP	State Development Guide Plan (New Jersey)
SDRDP	State Development and Redevelopment Plan (New Jersey)
SHPO	State Historic Preservation Officer
SSZEA	Standard State Zoning Enabling Act
TDC	Transfer of Development Credits
TDR	Transfer of Development Rights
UDAG	Urban Development Action Grant
UDC	Urban Development Corporation
UGB	Urban Growth Boundary (Oregon)
ULURP	Uniform Land Use Review Process (New York City)
USHA	United States Housing Authority
VA	Veterans Administration

Introduction

This book originated in a simple desire to understand the system of land use planning in America. Having spent some time studying and writing about the British and Canadian planning systems, it seemed appropriate, after a few years in the United States, to inquire into the nature of American planning. The inclination was spurred by the large gap which seemed to exist between what was written about planning and what happened in practice. I could see little that was common between the standard descriptions of American planning and what was happening in my own area. It did not take any brilliant detective work to establish that it was my area that was typical of reality, not the normative discussions of "planning." But herein lay the problem: much more so than was the case in either Britain or Canada,[1] the system proved to be resistant to description, let alone explanation. Indeed, the term "system" is a misnomer. On reflection, perhaps it could not have been otherwise. There are some 40,000 local governments in the United States which administer zoning, and most of them do so autonomously. They operate within fifty states, each with its own history, culture, and constitution. The social scientist who attempts to impose a scheme of order upon this abundant variety must surely fail.

But there is much more to the matter than that. There are features of American planning which are quite distinctive. As the relative size and scope of legal textbooks indicate, the courts have a very much larger role to play than in Britain or Canada. On the other hand, most local governments have little contact with the courts. The great majority of decisions are taken on the basis of local issues without any concern for, or by, the courts, and only rarely is there a higher level of government involved in what is generally seen as "local land use planning." There could not be a greater contrast with the situation in Britain, with its comparatively elaborate system of centrally approved structure plans with which local plans and local development controls must broadly comply.[2] American local governments may have few powers (and planning is perhaps the most important) but their autonomy is indeed striking.

Strangely, to my mind, though there is a

substantial library of works on land use planning, there is very little about "planning" in these: their concern is with zoning. "Whatever happened to planning" is the title of a critique of British planning (Ambrose 1986), but it is much more appropriate to the American scene. There is little mystery here, and the matter is explained in a few paragraphs. More problematic and baffling is the strangely inconclusive character of the legal tomes. Though they provide an indispensable resource which is comprehensive, and sometimes encyclopedic, they differ greatly from their British and, less so, from their Canadian counterparts. This is not perversity on the part of academic lawyers: it is a reflection of the real world of zoning law. Cases contradict each other, even though the salient facts may differ in ways which it is impossible for the nonlawyer to detect. (Perhaps this is the objective? If so, it is no different from the obfuscations of other professions and disciplines.) As the number of cases has grown, so has the complexity. Multiplication rather than elucidation is the result.

However, as my progress through the legal thickets continued, I appreciated simple points that escaped me at first. One of these was the importance, in the legal jousts, of distinguishing the very special characteristics of the case under consideration in contrast to all the other cases that have a bearing on the essential issues. The skills involved here serve to benefit the party concerned in that his or her case is identified and treated as being unique. And the one thing that courts are averse to doing is to lay down general principles which can act as guidelines to planners. Sometimes (as with "the taking issue" – discussed at length in various chapters) they seem to go out of their way in refusing to address the issues about which planning practitioners are so concerned. They thus operate in quite the opposite manner to a British or Canadian government department, which provides policy guidance to local governments.

The result is a surprising degree of uncertainty. For example, to attempt to draw up a summary of the law concerning charges (exactions, imposts, etc.) levied on developers is not an exercise in synthesis but a battle with conflicting arguments, cases, and laws. The "conclusion" can be no more than a schedule of types of action which may be legally permissible in differing degrees in different parts of the country. Even the competence of the courts to judge a case is a matter of debate – a matter which, at the time of writing, is still unsettled. A brief reference to a 1973 Oregon case is useful in illustrating the points being made here.

The case at issue is known as *Fasano*, and involved the rezoning of thirty-two acres in Washington County, Oregon, to allow for increased densities and a mobile home park. The change was made at the request of a builder and opposed by local homeowners. Though unsupported by the local planning commission (who presumably were much influenced by the homeowners) the rezoning was approved by the board of county commissioners (presumably on the ground of regional needs). In considering the matter, the court focused on the question as to whether the change was a legislative act (and therefore not generally reviewable by the courts – except in the case of arbitrary or capricious acts) or a quasi-judicial act which could be reviewed. The court decided that the rezoning was quasi-judicial and, moreover, that there was insufficient justification for it.

The decision "challenged nearly a half-century of judicial noninterference" and, not surprisingly, "received a mixed reception" (Haar and Wolf 1989: 336). A number of other courts have followed the *Fasano* lead, but the majority have preferred the traditional view that "a legislative body can best determine which zoning classifications best serve the public welfare."[3]

Though the reported case does not give background particulars (infuriatingly they seldom do), it is replete with legal niceties which can be ignored here. However, several points arise. Firstly, the case was considered in purely legal terms: there was no examination of the planning issues involved (housing need, affordability, market factors, land suitability etc.). Secondly, the power of local homeowners (and electors) is demonstrated, as it frequently is throughout this volume. Thirdly, as already indicated, there is now no national agreement on whether a rezoning can be reviewed by the courts. It will become apparent that all these are common features of American land use planning.

It is notable, in reading legal textbooks, that it is common for the editors (typically there is more than one) to take delight in setting the reader a series of questions, rather than a summary of conclusions. Generally, these seem to be of pedagogic character, designed to stress the variety of circumstances and issues that can inform a legal opinion, and to stretch the law student's mind. (How does one case compare with another? Why did a court facing a similar issue come to a different decision? How much importance should be attached to the dissent? and so on.) But at other times, the underlying politics shines through the legal mask, as in the following quotation from the text by Haar and Wolf, which discusses "inclusionary zoning" – the requirement that a development include a certain proportion of low or moderate income houses:

> As a legislator how would you balance the equities of an inclusionary zoning proposal? Would your answer depend on whether you represented an urban, suburban, or rural constituency? Might there be a disincentive for the inner-city lawmaker to support such efforts to relocate the voters responsible for his or her (re-)election? Or do notions of "fair share" supersede the legislative district and political party lines? Upon which principle should the conservative lawmaker from the white collar suburb act: the protection of the landowner/developer's private property rights or the opposition to

regulating even further the real estate market? What about his or her liberal counterpart, forced to choose between the desire to aid members of minority groups excluded by zoning and planning barriers, and the instinct to resist large scale development in environmentally sensitive regions? As a judge, how would your responses differ?

> (Haar and Wolf 1989: 435)

It is seldom that political matters are so explicitly raised in the legal arena (though Haar and Wolf follow the tantalizing lead of other legal writers in avoiding any answers to the questions which they pose). Indeed, it almost seems as if there is conspiracy of silence among lawyers: their concern is solely with the law. Other matters may be discussed in the newspapers or in the writings of political scientists, but these are distinctly out of bounds, except to iconoclasts such as Richard Babcock, who delights in pulling away the mask and revealing both the underlying politics and the incredible absurdity of the act which is so often played out before the courts.[4]

However, whether or not there are absurdities (and which country can boast that it has none?), it is obvious that any account of American land use planning has to devote a considerable amount of attention to the role of the courts, the law, and the Constitution. Lawyers and the law are the dominating influence in planning, and even though there may be some criticism of their role, the system as a whole attracts (to my mind) a surprising degree of support. (This may well be an indication that I have further learning to do!) In particular, Americans look on their Constitution with an almost religious reverence. It is remarkably easy to step from saying that some action is unfair to declaring that it is unconstitutional. And, of course, it is the courts that are the guardians of the Constitution. There is an almost obsessive belief that disputes are best settled by the legal system. This is true generally, as well as in the particular area of land use regulation. Thus, as my exploration continued, it broadened to cover far more legal issues than I had anticipated.

Some of the cases take us far afield into, for example, the activism of the New Jersey courts where, for a time, they actually took over the zoning functions of some local governments. While stressing that it was not their role to administer zoning, they had no hesitation in so doing when local governments insisted on blatantly exclusionary practices. Some readers may feel that I have devoted too much space to this, particularly since its practical importance has been relatively local. Nevertheless, apart from its intrinsic interest, it provides a case study in judicial activism; and the lesson is that this does not provide an easy – or any – solution to local discrimination. If there is an answer, it lies elsewhere: the subject of the final chapter.

Despite the prominence given to the courts in my account, there is no question that, generally speaking, local governments reign supreme. Though they are occasionally challenged by irate developers who are prepared to risk the cost of delay (and possible failure) in contesting a zoning decision, most appear able to proceed within their own little kingdoms ("tight little islands" as Sager has called them) without much fear of challenge. The problems to which this gives rise would not be so acute if local governments were larger in area. A wider based selfishness would at least enable conflicts to be resolved in a single political arena. However, perhaps too much should not be made of the point, since the problems certainly are not absent from the very large local governments such as New York and Los Angeles.

This, however, is only a part of the story. The more one examines the operation of zoning, the clearer it is that it has been transformed from a rigid, certain method of land use control to a highly flexible instrument which allows negotiation, bonusing, imposts, and "agreements" of all kinds. In fact, the reality is precisely the opposite of the theory. That there

are few plans to which zoning ordinances should be in conformance is, of course, well known; and what is called a "comprehensive plan" is frequently nothing more than a zoning scheme for a wide area. More important, zoning ordinances, whether narrow or wide, are often simply the basis for initial discussions between a developer and the zoning body. "Zoning with a difference" (to use one of the chapter titles) is the norm, not the exception.

It should not be inferred that the dice are loaded against developers. This may be the case in some areas, but more generally development is welcomed – even if developers themselves may be unpopular. This needs some explanation. The first point to make is that, historically, the United States has been growth-minded. Indeed, until relatively recently, few questioned the conventional wisdom that growth was good for everybody. Today this is not as true as it used to be. Many suburban localities have put up formidable barriers to development which might attract low income households or increase tax burdens. So prevalent is this in parts of the country (and in the minds of academic and other critics of zoning) that the 1991 report of the Advisory Commission on Regulatory Barriers to Affordable Housing bears the title *Not in My Back Yard*.

But while blacks and other minorities are unwanted, there is no barrier to the traditional white middle-class household wanting a single-family house on its own lot. Nor is the developer of an office complex, of a shopping mall, or indeed of *any* such attractive development likely to find it difficult to find a good site for his investment. Controls have increased in many areas, but as long as the siting of a development does not detract from the amenities enjoyed by an articulate minority, it is likely to go ahead.

That this should be so prompted me to extend my inquiry much further than I had originally envisaged. What was intended to be a succinct conclusion turned into four lengthy chapters; and still there remain many relevant issues which are not fully analyzed. Indeed, as may be only too apparent, once embarked on this broader quest, it was difficult to decide where to stop. The basic issues covered, however, are fairly clear. First, as already indicated, there is the American love affair with the protean Constitution, and its interpreter, the courts. These sometimes reflect public opinion, but at other times, lead it. They have certainly been a stalwart defender of the rights of those whom exclusionary local governments would prevent from settling within their boundaries, though the practical effectiveness is open to considerable doubt.

Less immediately obvious to a newcomer is the deep-seated sense of the rightness of a system that allows an owner of land to turn it to a profitable use. In determining the use to which land may be put, the typical local government will think in terms of the owners, not the public interest. With this "mind set," it becomes virtually impossible to allow development of one field of agricultural land while refusing development of its neighbor. In the county where I live (New Castle County, Delaware) there has been (to my mind) an extraordinary, long battle over development along a crowded major road (Concord Pike) leading from the sales tax-free State of Delaware to the State of Pennsylvania, which has a sales tax. The road has recently been widened to cater for current use, thereby making it even more attractive to developers. Site after site has been granted permission for development which everyone knows will create traffic havoc in the future; but the seven members (or to be precise, five of the seven) argue that the developers have the right to extract more profit from their land. Though there has been some proven bribery in the sad history of Concord Pike, this is only a sideshow to the main drama.

There are several curious features about the whole matter which I have found baffling, and which I have attempted to account for in the final chapters. Firstly, there is a strong belief that zoning decisions should be "fair." This is interpreted to mean that if one developer is allowed to build a shopping mall on Concord Pike, a second, and a third, should be given similar permission. To do otherwise is to unfairly benefit one developer while penalizing another. Little consideration is given to the wider public interest even when, as in this case, it is (at least to me) abundantly clear that the long term public interest is being seriously harmed. Secondly, it is obviously unacceptable to even raise the question as to whether there are already too many shopping malls in the area: this is felt to be a market decision which can be made only by individual developers. It remains to be seen who will deal with the increasing number of older existing shopping malls which are failing and which mar the suburban landscape. Thirdly, there is a strangely fatalistic attitude taken by many to the development: "it had to come in time" is a common opinion. And so it is with many sites in this county, and elsewhere. Why this should be so takes us far from the narrow aspects of a zoning ordinance; and it explains why I enter into a long discussion of attitudes to government, law, and land.

Despite the length of the book, there are many issues which are touched upon but not developed. Here I will refer to one which I found particularly troublesome. This is the complex of issues that fall under the label "the urban problem." Though there are many references to aspects of this in the book, the subject is not directly addressed. The main reason for this is that much has already been written in this field, that it is a huge subject in its own right, and that my main focus is on land use. But the omission is a serious one for the reader who wishes to obtain a comprehensive view of American planning in the broader sense of the term. Moreover, much of "the urban problem" is a direct result of, or at least is greatly exacerbated by, suburban land policies. Tragically, the United States has become (to use Disraeli's term) "two nations," increasingly divided according to access to opportunity. On the one hand is an affluent suburbia which has been, and continues to be, significantly assisted by public policies (such as preferential tax treatment, and the heavily subsidized building of commuter roads). On the other hand, there is the growing underclass of the inner city, infested by ever-deepening urban ills of crime, drugs, homelessness, poverty, and inadequate public services. The situation may not at present be as explosive as it was for a temporary period in the 1960's, but by any sane criterion it ought to be. The differences between the two nations are intolerable and, in any properly constructed political system would be the subject of intense debate, experimentation, and action at all levels of government. Yet this is not so. The problems are too long term to be of attraction to Congress, too diverse and costly to the states, and either of total unconcern or overwhelming difficulty to local government. Deep-seated urban problems require long term treatment: this will not be provided by a system whose span of attention rarely extends beyond four years, and is typically much shorter than that.[5]

Though I do not deal with these specific issues in this book, they were very much in my mind as I moved from a modest aim of writing about zoning to the unattainable objective of attempting to fathom the depths of American land use policy. And the issue of short-term governments ignoring long-term problems applies equally to land use regulation as it does to inner city problems.

There is one curious exception to this policy myopia. Both the federal and the state governments have introduced, and are implementing, a wide range of environmental policies which are of a long-term nature and also have serious impacts upon property rights. Why this should be so, and why the policies should apparently be widely acceptable, are extremely interesting questions which are not addressed in this book, because of the decision to exclude environmental planning. It was thought that this would both extend the size of the volume unduly and delay its publication. In any case, environmental planning is, in important ways, different from land use planning. As Downs (1972: 47) has noted, it is not "politically divisive," and there are wealthy and powerful polluting "villains" who can be made to bear the burden of the cost (even if this is passed on to consumers).

The omission can be rectified if a second edition is called for. But though this might improve the book, I have no delusion that it could ever be made "complete." As Laski (1948: ix) commented over forty years ago, "there is so much more in America than any one can know."

Part I
Zoning

Chapter 1

The institutional framework

Introduction

> In the United States government has
> perpetually to justify itself, to overcome a set
> of peculiar resistances, from the
> psychological to the physical, to maintain, as
> it were, its ascendancy over the competing
> elements in the dynamic life of a diverse and
> restless people.
>
> (Nicholas 1986)

The distinction between the ideal of planning
and the reality of zoning is an important one.
Planning is concerned with the long-term
development (or preservation) of an area and
the relationship between local objectives and
overall community and regional goals. Zoning
is a major instrument of this; but in the United
States it is more. Indeed, it has taken the place
of the function to which it is supposedly subservient. One of the reasons for this is that
responsibility for land use controls has been
delegated to small units of local government.
These local authorities have traditionally been
concerned with attracting development to their
areas, but, since the 1970s, there has been
increasing pressure from electors for their
communities to be preserved as they are (or at
least safeguarded from undesirable uses such as
industry, apartments and low income housing).
The powers of zoning (and subdivision
control)[1] provide a very effective tool for this –
a tool which can be wielded with a skill that
thwarts judicial action. The contrast with
planning is a sharp one: a comprehensive plan
would deal not only with the needs of the
existing inhabitants of an area, but also with its
role in meeting the needs for housing
newcomers, whatever their income or color.
Additionally, it would make provision for such
undesirable land uses as power stations, land-
fills, and a host of other uses which have given
rise to the acronym NIMBY: "not in my back
yard," and its more recent progeny NIMTOO:
"not in my term of office." This can be, and is,
done by some local governments; but there are
many more who use zoning as a means of
precluding comprehensive planning.

In other countries, planning systems are
hierarchical in structure, with local planning
agencies accountable to higher levels of govern-

ment. Thus, in Britain, each local authority is required to prepare a development plan along lines laid down by the central government; that plan has to be approved by the central government (perhaps after required amendment); and development control has to be administered in harmony with the provisions of the plan.[2] This central control is especially concerned to ensure that local plans have "due regard" to regional and national needs and policies. By contrast, American land use controls are essentially a local matter. Over most of the United States, local governments are not required to operate any system of land use control and, though most do have a zoning system, some have none at all. Only in a few states are local governments subject to any type of control by the state government.

There is, however, a higher level of authority – the courts – but their function is not to co-ordinate policy and seldom do they consider the regional implications of local policies:[3] their role is to ensure that local governments operate in a correct legal and constitutional manner. Indeed, constitutional issues are always in the background of American land use regulation. Accidents of circumstance or sheer pertinacity on the part of an aggrieved developer (or a group such as the NAACP) can bring the constitutional issues to the foreground. The outcome of a case which gets this far is uncertain, partly because of the narrow way in which courts deal with cases (which is discussed later) and partly because the Constitution is a "living" thing, and thus its interpretation changes as social conditions change.

It is, however, easy to be misled about the primacy of the Constitution. The voluminous legal textbooks on land use planning contribute to this incorrect impression. First, their very size is at least in part due to the fact that the courts frequently differ among themselves or refuse to clarify principles which planning authorities can follow. The reader has to digest the cases (which, thanks to the writers of the text books, are reduced to manageable length) and try to

establish how the particular issues in which she is interested might be treated. If the courts acted more helpfully, law books could be very much thinner.

Legal texts are misleading also in that they give the impression that constitutional and legal matters are all-important in land use planning. In fact, though legal issues are in the background, only a very small proportion of cases reach the courts. The general experience of those who come into contact with land use planning is of a bureaucratic rather than a constitutional nature.

Some flavor of the action at the local level is given in the 1991 report of the Advisory Commission on Regulatory Barriers to Affordable Housing (more popularly known as the NIMBY report):

> In addition to lobbying elected officials, NIMBY groups regularly participate in the regulatory process through vocal input at public forums and hearings dealing with land use and development issues. Unlike the strict rules governing judicial proceedings, many localities have no specific rules regarding who can testify at public hearings or what rules of evidence apply. Participants often represent ad hoc groups that coalesce around a particular development issue. They can be very effective at packing hearing rooms and leaving the impression that public opinion is strongly against whatever project they oppose.
>
> (Advisory Commission on Regulatory Barriers to Affordable Housing 1991)

Thus we have a curious paradox. On the one hand, the local governments operate very much as they please, without regard to any higher authority. Their actions are, in practice, seldom challenged. On the other hand, as the existence of a large and prosperous profession of land use lawyers demonstrates, there is a higher level of judicial processing which deals with the cases that are challenged. Though local governments may not follow legal decisions carefully, they are aware that at any time they may stumble across a case in which passions are sufficiently aroused to lead them to the courts.

Planning and Zoning

> American cities seldom make and never carry out comprehensive plans. Plan making is with us an idle exercise, for we neither agree upon the content of a "public interest" that ought to override private ones nor permit the centralization of authority needed to carry a plan into effect if one were made.
>
> (Banfield 1961)

Much if not most of the land use planning in the United States is not planning but zoning. The former implies comprehensive policies for the use, development, and conservation of land. Zoning – which can be one instrument of this – is the division of a local government area into districts which are subject to different regulations regarding the use of land and the height and bulk of buildings which are allowable. As the New Jersey court said in *Angermeier v Borough of Sea Girt*:

> While municipal planning embraces zoning, the converse does not hold true. They are not convertible terms. Zoning is not devoid of planning, but it does not include the whole of planning. Zoning is a separation of the municipality into districts, and the regulation of buildings and structures, according to their construction, and the nature and extent of their use, and the nature and extent of the uses of land. This is the constitutional sense of the term. . . . Planning has a much broader connotation. It has in view . . . the physical development of the community and its environs in relation to its social and economic well-being for the fulfillment of the rightful common destiny, according to a "master plan" based on

careful and comprehensive surveys and studies of present conditions and the prospects of future growth of the municipality, and embodying scientific teaching and creative experience.

This is a rather colorful description, but the essential point is clear.

Zoning is an exercise of the police power: the inherent power of a sovereign government to legislate for the health, welfare, and safety of the community.[4] The Constitution confers the police power upon the states which in turn delegate it to the local governments. All fifty states have passed legislation enabling municipalities (and often counties) to operate zoning controls (Bassett 1940: 13). Most are based on the Standard State Zoning Enabling Act issued by the Department of Commerce in the mid-1920s.[5] The first two sections of this are as follows:

> 1. For the purpose of promoting health, safety, morals, or the general welfare of the community, the legislative body of cities and incorporated villages is hereby empowered to regulate and restrict the height, number of stories, and size of buildings and other structures, the percentage of lot that may be occupied, the size of yards, courts, and other open spaces, the density of population, and the location and use of buildings, structures, and land for trade, industry, residence or other purposes.

> 2. For any or all of said purposes the local legislative body may divide the municipality into districts of such number, shape and area as may be deemed best suited to carry out the purposes of this act; and within such districts it may regulate and restrict the erection, construction, reconstruction, alteration, repair, or use of buildings, structures, or land. All such regulations shall be uniform for each class or kind of building throughout each district, but the regulations

in one district may differ from those in other districts.

Paradoxically (in view of what was said above about the distinctions between planning and zoning) another section provides that zoning regulations "shall be made in accordance with a comprehensive plan." In fact, zoning was conceived (at least by planners, if not by lawyers) as a tool of planning. But generally the part became the whole, and (with notable exceptions considered later) practice does not follow the text of the Act.

> Most of the zoning ordinances in the United States are not based upon a land use plan. Most of the zoning ordinances in the United States were not adopted to help implement a community development plan. Many of the zoning ordinances in the United States were carelessly prepared or were prepared by people who had no competence to deal with this subject.
>
> (Blucher 1960: 146)

The phrase "in accordance with a comprehensive plan" is incorporated in the zoning legislation of most states, but generally it does not mean what the words suggest; instead, it means that zoning should be carried out comprehensively rather than in a piecemeal manner.[6] Even where the zoning legislation mandates comprehensive planning (as in California, Florida and Oregon)[7] it is the zoning ordinance which carries the force of law, not the comprehensive plan. Comprehensive plans:

> represent only a basic scheme generally outlining planning and zoning objectives in an extensive area, and are in no sense a final plan; they are continually subject to modification in the light of actual land use development and serve as a guide rather than a strait jacket.

> The zoning as recommended or proposed in the master plan may well become incorporated in a comprehensive zoning map

... but this will not be so until it is officially adopted and designated as such by the District Council.[8]

Nevertheless, an increasing use is being made of comprehensive plans (or master plans, or general plans: the terms are used interchangeably) by both local governments and the courts. Zoning decisions are much easier to defend before the courts if a strong planning framework can be demonstrated.

It is important to appreciate (again with exceptions) that American zoning is essentially a *local* matter. Even the decision on whether to operate a zoning system is usually a local one. Some localities have highly sophisticated zoning systems; some have none at all. But however complex a zoning system may be, it typically remains what it always has been: "a process by which the residents of a *local* community examine what people propose to do with their land, and decide whether or not they will permit it" (Garner and Callies 1972: 305). Babcock contrasts the "multiplicity of jurisdictions" in zoning with other areas of control:

> In other significant areas of administrative law – the regulation of utilities, control over the issuance of securities, and the arbitration of disputes between employer and employee – there exist if not national at least statewide forums for the resolution of disputes. In the area of zoning there is no such centralized umpire to provide a sense of belonging to a common administrative practice, and, indeed of sharing a common administrative ethic. Among these scattered groups of lay decision-makers there is an almost total lack of communication despite the efforts of innumerable planning groups, each offering earnest if generally diffused guidance.
>
> (Babcock 1966: 12)

The Local Managers of Zoning

The most important and most common zoning action of any local government is the map amendment or rezoning of land. Given the strong legal, cultural, and economic heritage of private property rights in this country, it is virtually impossible for any public body to "plan" – and then to implement through zoning – a specific future use for a piece of undeveloped land.

(Kelly 1988)

More than in any other area of law – including school administration – in zoning the layman is vested with the heady power of direct participation in decision-making, free of any centralized guidance or regulation.

(Babcock 1966)

Local governments carry out their zoning and planning powers within the framework of powers conferred on them by the individual states, either by constitutional home rule authority or by a specific enabling Act. There are thus fifty different systems of local government – which fortunately it is not necessary to analyze here. What has to be said, however, is that though some states exercise varying degrees of control over local governments, most do not. The striking feature of American local government is its independence. In this, the American system differs dramatically from those of Europe (and Canada). American local government is far more "local" than foreigners usually appreciate. A nice (even if extreme) illustration is given in Wakeford's study of *American Development Control* (1990: 31): in this he cites with obvious astonishment the case of an Oregon rural city whose electorate voted not to allow the city to levy taxes. As a result, the city administration was closed down and the local police force disbanded! "Although the county would provide emergency services, little other outside intervention could be expected;

the people would have to work out their own political solution to the problem they had precipitated."

The fifty states contain 80,000 local governmental units. Over half of these are school districts and other "special districts" for particular functions such as natural resources, fire protection, and housing and community development.[9] In 1987, there were 3,042 counties, 19,200 municipalities, and 16,691 townships.

The variation among states is exemplified by a few statistics from the *Census of Governments*. Of the 19,200 municipal governments, 183 have populations of 100,000 or more, while nearly a half (9,369) have less than 1,000. (In terms of inhabitants, 62 million live in the former, while only 40,000 live in the latter.) Illinois, with 1,279 such governments, has more municipalities than any other state; Texas has 1,156, and Pennsylvania 1,022. At the other extreme are states such as Connecticut and Massachusetts, which have fewer than fifty municipalities. Organized county governments are common, though their power and functions vary. Twenty states have "townships" which have powers similar to those of municipalities, except that their boundaries are defined without regard to the concentration or distribution of population.

Counties play a role in zoning and planning in parts of the country. The situation has been summed up by N. Williams as follows:

A principal complaint about American land use controls is that they exist only at the local level. As a description of existing practice, this is justified, but the enabling legislation is far ahead of the practice. The extent of county zoning is not widely understood. All counties are authorized to zone in some states; only some counties in others. In some states, including several of the northeastern metropolitan states, no counties may zone. In the northeast, only Maryland uses county zoning substantially; in Maryland most zoning is by counties,

Table 1 Government Units 1987

Type of government	Number of units
U.S. Government	1
State governments	50
Local governments	83,186
County	3,042
Municipal	19,200
Township	16,691
School district	14,721
Special district	29,532
Total	83,237

Source: U.S. Bureau of the Census, *1987 Census of Governments.*

except for the cities of Baltimore and Annapolis. It is also common in Georgia. Apart from these examples, county zoning is widespread mostly in the west, where there is a substantial amount of zoning at the county level.

(N. Williams 1985–90: 18.03)

One of the distinctive features of planning in the United States is highlighted by a comparison with Britain. A British municipality both prepares and implements its plans. There is no separation between the legislative and executive functions. Moreover, there is only a very limited right of appeal to the courts. Most appeals are dealt with by the administrative system. The appeal is heard by an "inspector" who is appointed by the central or local government and who reports his findings (based not only on his interpretation of the facts of the case, but also on relevant official planning policies) to the appointing body. The whole system is characterized by a high degree of administrative discretion which would be totally unacceptable in the United States.

By contrast, the American system embraces the doctrine of the separation of powers. In brief (and therefore ignoring deviations from the normal rule) a zoning ordinance is passed by the legislative body (e.g. a municipality); applications for rezoning or variances are reviewed by an independent commission (the planning or zoning commission/board);[10] and appeals are to a board of adjustment (or appeals), and sometimes to the legislative body, and finally – on legal or constitutional grounds – to the normal courts. Furthermore, the role for discretion is severely limited (in theory at least). Indeed, zoning was originally conceived as being virtually "self-executing": the zoning ordinance and the zoning map would spell out the permitted land uses in such clarity and detail that there would be little room for doubt or discretion. Thus "policy" is seen firmly as the responsibility of the legislative body, while the commission deals with its execution through the issuance of permits and the occasional variance or exception.

As previously indicated, "policy" is usually a matter only for the local government: no higher level of government is generally involved. However, the courts do hear appeals against local decisions and therefore, in one sense, act as a type of policy-imposing body. However, "policy" enters into the courts' deliberations only to the extent that they do or do not defer to the legislative judgment of municipalities: what is termed the "presumption of validity" (or "judicial deference"). The courts are concerned, not with planning policy issues, but with legal and constitutional matters. As will become apparent, this neat division does not work in practice.

The Constitutional Framework

However styled, ferment in the land use field is now so pervasive that, like Alice's flamingo, hedgehog, and card soldiers, nothing seems to stay put for very long, least of all the point at which judges will draw the line between police power and the power of eminent domain.

(Costonis 1973)

Both the federal and state constitutions include provisions which are binding on municipalities. One of the most important of these is the protection of property rights. The Fifth Amendment to the Constitution provides "... nor shall private property be taken for public use without just compensation." The "taking issue" (alternatively known as the "just compensation issue") is at the heart of the major problem facing zoning: when does the exercise of the police power over land use constitute such an infringement of the property right as to become a "taking"? The crucial matter, of course, is the definition of a "taking." An enormous amount of thought, effort, and scholarship has been applied to this question, yet the position is not clear. Postponing fuller discussion until later, here it suffices to note that it has long been established that (to quote the famous words of Justice Holmes in the case of *Pennsylvania Coal Company v Mahon*) "the general rule ... is, that while property may be regulated to a certain extent, if regulation goes too far it will be recognized as a taking." However, this is not very helpful since we are still left with the puzzle as to where the dividing line is between zoning decisions which are acceptable and those which go "too far." In truth, there is none: the Supreme Court has taken the view that (as with the question of obscenity)[11] no generally applicable definition is possible, and each case must be decided upon its merits. The classic statement on this was made in the *Penn Central* case:

> The question of what constitutes a "taking" for the purposes of the Fifth Amendment has proved to be a problem of considerable difficulty. While this Court has recognized that the "Fifth Amendment's guarantee ... [is] designed to bar Government from forcing some people alone to bear public burdens which, in all fairness and justice,

should be borne by the public as a whole", this Court, quite simply, has been unable to develop any "set formula" for determining when "justice and fairness" require that economic injuries caused by public action be compensated by the Government, rather than remain disproportionately concentrated on a few persons. Indeed, we have frequently observed that whether a particular restriction will be rendered invalid by the Government's failure to pay for any losses proximately caused by it depends largely "upon the particular circumstances [in that] case".

The Fifth Amendment also includes what is termed "the public use doctrine": that property can be "taken" only for a public use. The interpretation of this doctrine has changed significantly in recent decades (illustrating the changes that can take place in the constitutional framework). Until the early fifties, it was conservatively interpreted as meaning that property which was taken had to be literally used by a public body. It could not be taken for a joint public–private venture, and still less for a private use. Justice Douglas made short shrift of this in the 1954 *Berman v Parker* case. This related to an urban renewal project in Washington, D.C. which was to be carried out under the District of Columbia Redevelopment Act of 1945. The Act provided that "the acquisition and the assembly of real property and the leasing or sale thereof for redevelopment pursuant to a project area redevelopment plan ... is hereby declared to be a public use." The Court upheld the constitutionality of this legislation; indeed, in magisterial terms it declared that the public use requirement of the Constitution was "coterminous with the scope of a sovereign's police powers." It also declared that the concept of public welfare was so broad that it could encompass aesthetic matters: "it is within the power of the legislature to determine that the community should be beautiful as well as healthy, spacious as well as clean, well-balanced as well as carefully patrolled."

Later cases have further extended "the public purpose."[12] One particularly well-known and controversial case was the clearance of the Poletown neighborhood of Detroit for the purpose of accommodating a new General Motors plant. The city, faced with the prospect of General Motors moving out of the area, condemned some 465 acres of land and conveyed it on favorable terms to GM. The Supreme Court of Michigan, in *Poletown Neighborhood Council v Detroit*, ruled that "the power of eminent domain is to be used in this instance primarily to accomplish the essential public purpose of alleviating unemployment and revitalizing the economic base of the community. The benefit to a private interest is merely incidental."[13] The new factory "led to the destruction of 1,021 homes and apartment buildings, 155 businesses, churches and a hospital, displaced 3,500 people, and all but obliterated a more or less stably integrated community embodying a century of Polish cultural life" (Hill 1986: 111).

Other relevant constitutional provisions require that land use controls be operated by "due process": the Fourteenth Amendment states that "... no person ... shall be deprived of life, liberty, or property without due process of law." The due process clause applies both substantively (is the action legitimate?) and procedurally (is it administered fairly?). Substantive due process requires that controls serve a legitimate governmental interest such as the public health, safety and general welfare. (A zoning ordinance which excluded low-income families could be challenged on substantive due process grounds). Procedural due process requires that fair and proper procedures are followed in relation, for example, to public notice of, and hearings on, zoning ordinances. It further requires that an ordinance be clear and specific: a property owner must be able to ascertain what he may or may not do with his property. If the ordinance is not clear, it can be challenged as being "void for vagueness." For example, a provision that authorized a planning commission to permit development on criteria which "include but are not limited to" those set out in the provision would be void since it would allow the commission to consider unspecified, alternative criteria.[14]

The Fourteenth Amendment also provides that no state "shall deny to any person within its jurisdiction the equal protection of the laws." An ordinance that involved racial considerations clearly denies equal protection. However, as so often with zoning matters, cases are often not at all clear. Unequal results can be obtained by devious mechanisms such as the prohibition of multi-family dwellings and the imposition of large minimum lot sizes or large minimum dwelling sizes, or even the regulation of laundries.

The Role of the Courts

There is hardly a political question in the United States which does not sooner or later turn into a judicial one.

(Tocqueville 1848)

Why our judicial system has developed as it has is in part explained by the recognition that historically most Americans, besides advocating popular sovereignty (i.e. "government of the people, by the people, and for the people"), have also wanted a "government of laws, not men".

(Goldman and Jahnige 1985)

Constitutional requirements lead to a major role in the land use planning process for the courts. The role is, moreover, an "active" one: decisions change over time in the light of changing economic and social conditions, and also the political complexion of the court.[15] The Supreme Court of the United States, as its name suggests, is the final arbiter; but it does not stand alone. There are over a hundred federal courts, and each of the fifty states has its own system of courts, including a supreme court.[16] Decisions at the state level stand unless overturned by the United States Supreme Court. There are important implications of this. Firstly, until a matter is settled by the United States Supreme Court (and few cases reach this level), the law can differ among the states. At the extreme, it is theoretically possible for there to be fifty different interpretations of a legal issue. This is particularly important in land use planning since the majority of zoning cases are dealt with at the state level. As a result, judgments vary enormously, from judicial intervention in the administration of zoning (*Mount Laurel*) to a denial of power to question the validity of any zoning legislation (*Vulcan*). The New Jersey and Pennsylvania courts are "the national pacesetters in the review of municipal zoning" while the California courts "are exceeded by none in their insensitivity to the nuances of land development law" (Ellickson 1977: 492 and 476). To complicate matters further, a dissenting opinion can, on occasion, be more influential than the court's judgment: typically when it presages a later judgment. In Babcock's words (1966: 181), "the articulate dissent has an established role in American jurisprudence as a forecast of things to come." One example, which is discussed in Chapter 3, is that of Justice Hall in the *Vickers* case – where he dissented from the majority view that Gloucester Township, New Jersey, could exclude mobile homes from their municipality. This became the majority view in the 1975 *Mount Laurel I* case.

Because of the wide variation in state statutes and judicial attitudes to public control, the case law reflects "sharply varying emphases in various periods and various states." However, the decisions *within* the individual states have generally been consistent and predictable. Thus "in the 1950s the Michigan courts would invalidate practically anything; ... the Maryland court has had a strong pro-

neighbor bias ...; in California anything goes." In his monumental study, *American Land Planning Law* (from which this quotation is taken), Norman Williams classifies the "major zoning states" (i.e. the thirteen states which provide 75 percent of all the case law) into four categories: the "pro-zoning" states – California, New Jersey, Massachusetts, and Maryland; the "highly erratic" states – Michigan, Ohio, Florida, New York, and Pennsylvania; "the good gray middle" – Connecticut and Texas; and the strongly "developer-minded" states – Illinois and Rhode Island.

Secondly, unless a matter has been definitively ruled upon by a court, it may be unclear what the law is. This has been the case in recent years with respect to the limits of the police power in relation to land use controls – a matter we shall need to examine closely. The problem has been compounded by the reluctance of the Supreme Court to lay down clear guidelines (and their rejection of various academic theories).[17] A further complication arises in those states where the majority of appeals in zoning cases is decided, not by the state supreme court, but by intermediate appellate courts sitting in different districts: differences can develop between the districts (Mandelker 1988: 13). Courts at different levels can have widely different views on a matter. The landmark case of *Hawaii Housing Authority v Midkiff* provides a nice illustration. The case concerned the Hawaii Land Reform Act, which provided for the breakup of large landed estates and the sale of plots to lessees. The district court held that certain procedures were unconstitutional, but that the body of the Act was in conformity with the public use clause of the Fifth Amendment, i.e. that property was being taken for a public use. On appeal, a U.S. Court of Appeals reversed and ruled that the Act was essentially unconstitutional (because the property was transferred to private persons, not a public body). The U.S. Supreme Court, however, held that there was a legitimate public use, or purpose, in that the legislation was aimed "to attack certain perceived evils of concentrated property ownership in Hawaii."

A decision of a court can reflect deep divisions between the justices. A famous case is *Metromedia Inc v City of San Diego*. The city had passed a sign and billboard ordinance which prohibited most types of signs. The district court declared the ordinance unconstitutional on the grounds that the city had exceeded its police power and also that the ordinance constituted an abridgment of the appellants' First Amendment rights of freedom of speech. The California Court of Appeal confirmed on the first ground, but the California Supreme Court reversed, and held that the ordinance was "a proper application of municipal authority over zoning and land use for the purpose of promoting the public safety and welfare." The U.S. Supreme Court had great difficulties with the case and ruled that the ordinance "reaches too far into the realm of protected speech" and was "unconstitutional on its face."[18] However, the judgment was by no means unanimous, and there were five separate opinions: Justice White wrote the plurality opinion; Justice Brennan wrote a concurring opinion (joined by Justice Blackmun); and separate dissenting opinions were written by Chief Justice Burger and Justices Stevens and Rehnquist.[19]

As this makes apparent, the courts can speak with many voices. Individual justices can exert a remarkable influence on the course of events – in the future if not immediately. Studies such as those of Justices Brennan (Haar and Kayden 1989a) and Stevens (Biddle 1989) clearly illustrate the point.

One final issue on the role of the courts needs to be made here. Their function is to ensure that municipalities are acting in a constitutional manner. It is not their role to act as a "super board of adjustment" or "planning commission of last resort" (*National Land and Investment Co v Kohn* 1965). There is a "presumption of validity" in the actions of a municipality to which the courts give "judicial deference."[20] This is nicely illustrated by a Missouri case (*City of Ladue v Horn* 1986). In the zoning ordinance of this city, a family is defined as "one or more persons related by blood, marriage or adoption, occupying a dwelling unit as an individual housekeeping organization." The case concerned two unmarried adults who were living, along with their teenage children in a single family zone. Clearly they offended the zoning ordinance, but was the ordinance constitutional? Certainly concluded the court:

> The stated purpose of Ladue's zoning ordinance is the promotion of the health, safety, morals and general welfare in the city. Whether Ladue could have adopted less restrictive means to achieve these goals is not a controlling factor in considering the constitutionality of the zoning ordinance. Rather, our focus is on whether there exists some reasonable basis for the means actually employed. In making such a determination, if any state of facts either known or which could reasonably be assumed is presented in support of the ordinance, we must defer to the legislative judgment. We find that Ladue has not acted arbitrarily in enacting its zoning ordinance which defines family as those related by blood, marriage or adoption. Given the fact that Ladue has so defined family, we defer to its legislative judgment.

> (*City of Ladue v Horn* 1986)

Closely related is the "fairly debateable" concept. This holds that if a decision is a matter of opinion, i.e. open to fair or reasonable debate, the court will not substitute its judgment for that of the responsible legislative body. The role of the courts is not to sit in judgment on the wisdom of a local government's legislative actions: that is the function of the political process. The judicial role is circumscribed. Typically, it can overrule a legislative body only if its actions are shown to be clearly arbitrary, capricious and unreasonable. A good statement is given in the *Arlington Heights* case:

> It is because legislators and administrators are properly concerned with balancing numerous competing considerations, that courts refrain from reviewing the merits of their decisions, absent a showing of arbitrariness or irrationality.

In short, "a court does not sit as a super zoning board with power to act de novo, but rather has, in the absence of alleged racial or economic discrimination, a limited power of review" (Wright and Gitelman 1982: 527).

While this is the traditional view, there is no doubt that the local zoning process is frequently subject to irresistible pressure, and decisions are often taken which serve narrow interests. Rezonings, for example, are often made in defiance of the policy enshrined in the zoning ordinance. Some state courts have held that zoning decisions can be administrative rather than legislative in character, i.e. they constitute the *application* of policy as distinct from the *making* of policy. This is particularly so where ad hoc decisions are taken on rezoning. Where this is held, there is no presumption of validity (which applies only to legislative acts), and the court requires to be satisfied that the rezoning is needed in the public interest. The state that has been particularly aggressive on this matter is Oregon,[21] and a few states have followed its lead – but most have not.

If this seems confusing, that is because it is! Or, at the least, "there is no fully accepted definition of the distinction between legislative and administrative decision-making."[22]

Conclusion

This chapter has outlined the major institutions concerned with land use planning in the United States. The reader should be reminded, however, that environmental planning has been explicitly excluded from this account. Were it to be included, it would be necessary to discuss the role of a number of federal agencies and, indeed, the federal government itself. However, so far as land use planning as defined for the purpose of this book is concerned, the federal government has little role to play. Its limited activities are discussed in appropriate chapters, for example, on fair housing, on urban policy and on historic preservation.

The most striking feature of the American planning process is that it lacks the features of a "system." There is little interlocking relationship between the various institutions that operate in the field. It is common to talk of a "separation of powers"; but American planning powers are more than separated: they exist in different spheres, hardly ever communicating except on matters of controversy. There is thus a multiplicity of local systems with a wide range of differing policies. The student is warned that any generalization is likely to be false. Despite this, some significant changes have taken place in recent years, and these will be chronicled in later chapters.

Having set out the general and institutional background, the next chapter provides a historical sketch of the development of zoning as an instrument of public policy.

Chapter 2

Historical background

Introduction

> Urban America was in something of a zoning crisis in the early 1920s. Like a patient who could endure his fever until he suddenly learned that there was a new remedy for it and who was then impatient to be cured, urban America was now sure that it would perish if it did not have zoning.
>
> (M. Scott 1969)

The history of land use controls is as old as history itself. In this chapter, after a short reference to colonial times, a rapid review is given of the foundations and early development of American controls from the nineteenth century "nuisance" cases up to the time of the classic *Euclid* case which laid upon zoning the imprimatur of the U.S. Supreme Court. The full story is a fascinating one, particularly with the virtually cliff-edge climax of the Supreme Court's deliberations (McCormack 1946; Toll 1969; Haar and Kayden 1989a). By way of introduction it is useful to list some of the more important factors which gave rise to zoning. As in European countries (Unwin 1909; Ashworth 1954; Ladd 1990), a major problem was public health, which increased rapidly as unbelievable numbers of immigrants crowded into cities which were totally unprepared to cater for their basic needs. Technological factors also played a major role: electricity increased the spread of the streetcar suburbs – the escape route of the middle class from the horrors of the insanitary and congested city. But they themselves contributed to this congestion. Even more so did the two technological innovations of the steel frame and the elevator, which made towering skyscrapers both possible and practical (Goldberger 1981: 5). Central city uses intensified as the middle class sought semirural respite by new means of transport. Later the wizardry of Henry Ford escalated problems of traffic congestion to huge dimensions. Other changes were in progress or in the wind: widespread regulation of election procedures (Griffith 1938a: 41), and the reform movement which was aimed at securing sound engineering-type solutions to problems of municipal administration. Even "planning" was debated as a rational solution to the problems of the city. This started as a "City

Beautiful" movement, but soon changed its character into a move for the "City Efficient." Neither got very far: they were, then as now, too long-term ("visionary" was the word) for practical men.

But one problem above all demanded attention: the safeguarding of the new suburbs from the blight which had stimulated their development. The solution was found in the extension of the law of nuisance to land uses, by way of zoning. Zoning provided long-term security against change: industry, garages, apartments, corner shops – indeed, anything which might threaten the sanctity of the single family dwelling suburb – could now be precluded. In Mel Scott's (1969: 192) words: "zoning was the heaven-sent nostrum for sick cities, the wonder drug of the planners, the balm sought by lending institutions and householders alike. City after city worked itself into a state of acute apprehension until it could adopt a zoning ordinance." While few might understand what "planning" involved, the protection of zoning was immediately apparent; and it spread at an incredible speed.

Early Land Use Controls

It is an American fable or myth that a man can use his land any way he pleases without regard of his neighbors.
(Bosselman, Callies, and Banta 1973)

Land use controls have a long history in the United States. In fact, in spite of the abundance of land and a strong belief in independence, the colonists soon found it necessary to impose various forms of regulation in both urban and rural areas.

In 1631 the Virginia House of Burgesses passed an Act requiring each white adult male over 16 to grow two acres of corn, or suffer the penalty by forfeiting an entire

tobacco crop. A 1642 Immigration Act required the growing of at least one pound of flax and hemp, and an Act of 1656 required landowners to cultivate at least ten mulberry bushes per 100 acres in order to stimulate the production of silk. . . .

Regulations in urban areas resembled those in London. The cities enforced strict land use regulations designed to promote health and safety. In the aftermath of the great fire of Boston, restrictions were set up on how a property owner could build his home. A series of laws was passed requiring the use of brick or stone in buildings. No dwelling house could be built otherwise, and the roof had to be of slate or tile upon penalty of a fine equal to double the value of the building.
(Bosselman, Callies, and Banta 1973: 82)

There are many other examples of the extensive amount of regulation in colonial times. "Although the settlers may have cherished their new freedom to use their land they also recognized the need to regulate that use for the good of the community as a whole" (Bosselman, Callies, and Banta 1973: 84).

So far as the taking of land for public purposes was concerned, there were few problems. Land was in abundance: so much so that questions of compensation hardly arose. Undeveloped land was perceived to be in such plentiful supply as to have no significant value. However, where developed, improved or enclosed land was physically acquired, compensation was normally payable. The power of eminent domain was accepted as an inherent power of government for which specific legislation was not required (Stoebuck 1972: 553). The "taking issue" which became of such importance later received scant attention. Indeed, there is a paucity of evidence on the reasons why the taking clause became a part of the Constitution.[1]

Matters changed dramatically with the adoption of the Constitution and the Bill of Rights, particularly when (under John Marshall) the Supreme Court claimed the singular power to determine the constitutionality of legislative acts.[2] So far as the taking issue was concerned, it was accepted by both the federal and state courts that a regulatory action could not involve a taking. The term "taking" was applied only to the physical acquisition of land by government – an approach encapsulated in Stoebuck's (1972: 601) phrase: "no taking without a touching." Where the use of property was restricted by regulatory controls, no compensation was payable. This was so even if landowners were deprived of all use of their land, as an 1826 case illustrates. The case is that of *Brick Presbyterian Church v the City of New York*.[3] Some sixty years after conveying land for the purposes of a church and cemetery, the city of New York passed a bylaw prohibiting the cemetery use. The New York Court of Appeals upheld the city:

> Sixty years ago, when the lease was made, the premises were beyond the inhabited part of the city. They were a common; and bounded on one side by a vinyard (*sic*). Now they are in the very heart of the city. When the defendants covenanted that the lessees might enjoy the premises for the purposes of burying their dead, it never entered into the contemplation of either party, that the health of the city might require the suspension, or the abolition of that right.

Since it was generally believed at the time that burying the dead produced unhealthy vapors, the court held that it would be extremely unreasonable to endanger the public by the cemetery use, despite the terms of the lease. In such cases, since the physical property (as distinct from the property rights) had not been invaded, no compensation was appropriate.

In a much later case, that of the famous (or infamous, depending on one's point of view) *Mugler v Kansas*, decided by the U.S. Supreme Court in 1887, Mugler's brewery was made virtually worthless by a Kansas Act which prohibited the manufacture and sale of intoxicating liquor. Justice Harlan held that:

> there is no justification for holding that the State, under the guise merely of police regulations, is here aiming to deprive the citizen of his constitutional rights; for we cannot shut out of view the fact, within the knowledge of all, that the public health, the public morals, and the public safety, may be endangered by the general use of intoxicating drinks. . . . A prohibition simply upon the use of property for purposes that are declared, by valid legislation, to be injurious to the health, morals, or safety of the community, cannot, in any sense be deemed a taking or an appropriation of property for the public benefit.
>
> (*Mugler v Kansas* 1887)

Mugler still retained his premises and could use them for any legal purpose – that is excluding the formerly legal brewery use! (There must have been numerous Muglers in the United States during the prohibition years.)

There were many such cases of the use of the police power. One further important example can be given here: the 1915 case of *Hadacheck v Sebastian*. Hadacheck had owned and operated a brickworks in the open countryside since 1902; but in the following years residential development took place and the area was annexed by the City of Los Angeles. The brickworks now became a nuisance to the local inhabitants, and the city passed an ordinance which effectively prohibited Hadacheck from continuing to operate his brickworks (which gave the land a value of $800,000), though he could use it for other purposes (value $60,000). The court held that "vested interests" could not be asserted against the ordinance because of conditions which previously existed. "To so hold would preclude development and fix a city forever in its primitive condition. There must be progress, and if in its march private interests are in the way they must yield to the good of the community." In the court's view, the ordinance was a proper exercise of the police power.

The number of such cases of the operation of the police power was legion. A landmark 1920 New York case cited a lengthy list of examples:

> In the exercise of the police power, the uses in a municipality to which property may be put have been limited and also prohibited. Thus, the manufacture of bricks; the maintenance of a livery stable; a dairy; a public laundry; regulating billboards; a garage; the installation of sinks and water closets in tenement houses; the exclusion of certain business; a hay barn, wood yard or laundry; a stone crusher, machine shop or carpet beating establishment; the slaughter of animals; the disposition of garbage; registration of plumbers; prohibiting the erection of a billboard exceeding a certain height; regulating the height of buildings; compelling a street surface railroad corporation to change the location of its tracks; prohibiting the discharge of smoke; the storing of oil; and generally, any business, as well as the height and kind of building, may be regulated by a municipality under power conferred upon it by the legislature.
>
> (*Lincoln Trust Co. v Williams Building Corporation* 1920)

Underlying these regulations was the English common law concept of nuisance which held that no property should be used in such a manner as to injure that of another owner. These were largely "negative" instruments, but gradually land use controls developed into more positive tools of planning. For instance, in 1867 San Francisco passed an ordinance which prohibited the building of slaughterhouses, hog storage facilities, and hide curing plants in certain districts of the city (Gerckens 1988: 26). Though clearly in the tradition of nuisance law, the ordinance was notable because it was "preventive rather than after the fact and restricted land uses by physical areas of the city"; it thus

"set the stage for further evolution of land use zoning in the United States" (ibid).

Judicial approval favoring the trend toward greater governmental intervention in private development was illustrated by the 1877 Supreme Court ruling in *Munn v Illinois*:

> When, therefore, one devotes his property to a use in which the public has an interest, he, in effect, grants to the public an interest in that use, and must submit to be controlled by the public for the common good, to the extent of the interest he has thus created.

The provision for "control by the public" increased as the problem of urbanization escalated at a phenomenal rate.

Immigration and Urbanization

> Give me your tired, your poor, your huddled masses yearning to breathe free, the wretched refuse of your teeming shore. Send these, the homeless, tempest-tossed, to me: I lift my lamp beside the golden door.
>
> (Emma Lazarus)

In the decades from 1851–60 to 1871–80 migration into the United States averaged 2.5 million (U.S. Bureau of the Census 1976). In the following decades it rose to 5.25 million (1881–90), to 3.65 million (1891–1900), and to nearly 9 million in 1901–10. In the single year 1907 it reached the staggering height of 1.25 million. Between 1890 and 1920, the population of the United States rose by over 42 million. Urban areas grew at an incredible rate, quite overpowering the ability of city governments to provide basic public services. During the same three decades, the urban population of the United States increased from 22 million to 54 million; the proportion of the population living in cities rose from 35 percent to 51 percent (Miller and Melvin 1987: 79); and the number of cities with a population of 50,000 or more rose from fifty to 144. The growth of individual cities was even more dramatic. Between 1880 and 1920, New York grew from 1,478,000 to 5,620,000; Philadelphia from 847,000 to 1,823,000; Baltimore from 362,000 to 748,000; and Boston from 332,000 to 733,000. The difficulties created by these huge increases in population were exacerbated by the fact that the newcomers were different from previous immigrants:

> In the thirty years from 1890 to 1920, more than eighteen million immigrants poured into America's cities. These new immigrants were more "foreign" than those who arrived before, coming mainly from Italy, Poland, Russia, Greece, and Eastern Europe. Overwhelmingly Catholic or Jewish, they came to cities that were already industrialized and class conscious. They made up the preponderance of the working force in the iron and steel, meatpacking, mining, and textile industries. They shared no collective memories of the frontier or the Civil War, much less of the American Revolution. Few spoke English, and many were illiterate even in their native language.
> (Judd 1988: 118)

They were therefore perceived as a threat to public health as well as to the sensibilities of middle- and upper-class residents of the outer city who had to pass the ghettoes on their way to work. Even more ominously, they threatened the entrenched urban political systems (ibid: ch.5).

The Movement for Planning

> Between 1900 and 1916, the emerging responses to private land use conflict gradually took shape. Building codes, block

ordinances, city plans, and nuisance regulations abounded. But their most powerful and effective expression was undoubtedly zoning. Zoning lent modern scientific legitimacy to the law's traditional concern with conflict avoidance. The method was simple and appealingly balanced, altogether objective and efficient in appearance.
> (Plotkin 1987)

> With some confidence it may be said that lawyers were midwives to the birth of urban planning and zoning in this country.
> (Mandelker 1971a)

The American city planning movement had had an erratic history (Reps 1965, M. Scott 1969). As in other countries, there were several strands: architecture and civic design, parks and open spaces, fire prevention, and public health and housing – spiced with a constant fear of social unrest. These strands remained largely separate, even when planning interests felt sufficiently bold to establish a new professional organization, the American City Planning Institute in 1917. The membership of this was, in all likelihood, the most diverse any professional body has ever witnessed: the fifty-two charter members included fourteen landscape architects, thirteen engineers, six attorneys, five architects, four realtors, two publishers, two "housers," and an assorted group of writers, tax specialists, land economists, educators, and public officials (M. Scott 1969: 163). A common denominator was a recognition of the fact that cities had increasing problems with which existing institutional structures were unable to deal, or even comprehend. The spirit of reform was still in the air, but it wore a many-colored cloak.

At the end of the nineteenth century, the spirit of reform could be seen in many guises – municipal government, public health, tenement regulations, and employment controls to name but a few. Much of this was a far cry from the

City Beautiful movement which had its origins in the small scale concern for physical and artistic improvement in town and village (Peterson 1976: 53) and also – better known – in the grand planning of the style in Paris (from which the movement got its name). This reached its flowering at the turn of the century, with such major landmarks as the Chicago World's Fair of 1893, the McMillan Plan for Washington of 1902 (a revision of L'Enfant's plan), and Burnham's Chicago Plan of 1909. By the later date, however, the City Beautiful movement was on the wane, and the "age of reform" was in full swing. But what to reform: transportation; housing; building densities and heights; sanitation; the legal powers of cities; the role of the states; the political system; management...? There was broad agreement that much more than a concern for aesthetics was called for, and there was a rallying under the banner of "efficiency."

The City Efficient (or City Functional) took over from the City Beautiful. Though catch phrases oversimplify, in this case they neatly encapsulate the way in which the new century saw a powerful turn from the landscape architecture and beaux arts tradition to a business minded concern for efficiently engineered cities (Arnold 1973: 27). The model was F.W. Taylor and his "scientific management"; but the city planning commissions which were set up to give substance to the strive for efficiency – coupled with the municipal information clearing houses, and a burgeoning of planning courses, conferences, and publications, and the like (Hancock 1967: 294) – were severely constrained in what they could actually do. The forces of privatism were too strong to be contained by public officials. Indeed, urban planners seldom did more than follow residential and commercial developers with transportation and sewer systems. George Ford (1916b: 353), in an essay on socioeconomic factors in planning, could argue that "it is now realized that the city is a complex organism, so complex that no doctor is safe in prescribing for it unless he has made a thorough-going and impartial diagnosis of everything that may have even the remotest bearing on the case"; but, as Hancock has put it, "if there was a definite system in this work by 1917, however, there was as yet more hope than empirical data for making meaningful generalizations for this inexact science" (Hancock 1967: 296).

Even had the planners had the tools to retune the urban system, or the extension of it, they lacked the necessary political clout. Zoning, however, had a particular appeal which extended beyond those whose essential concern was with "planning" (and who saw zoning merely as an instrument of planning). The crucial feature of zoning, then as now, is its utility in excluding unwanted neighbors.

> Nothing appeared so destructive of urban order as garages and machine shops in residential areas, or loft buildings in exclusive shopping districts, or breweries amid small stores and light manufacturing establishments. Nothing caused an investor so much anguish as the sight of a grocery store being erected next door to a single family residence on which he had lent money. Nothing made whole neighborhoods feel so outraged and helpless as the construction of apartment houses when the private deed restrictions expired and there was no zoning to prevent vacant lots from being used for multifamily structures. Zoning was the heaven-sent nostrum for sick cities, the wonder drug of the planners, the balm sought by lending institutions and householders alike.
>
> (M. Scott 1969: 192)

There was one major difficulty: it was unclear whether zoning would be accepted by the courts as constitutional. That the fears were justified was clearly illustrated several years later, after the passing of the New York Ordinance, by the rejection of the Euclid ordinance by the lower court (noted later in this chapter). Considerable effort and skill was employed by planners and lawyers in drafting ordinances which would stand judicial scrutiny. The "battle" – the word is appropriate – was between those who saw zoning as "a protection of the suburban American home against the encroachment of urban blight and danger," and those who saw it as "the unrestrained caprice of village councils claiming unlimited control over private property in derogation of the Constitution" (Brooks 1989: 7). However, the first major zoning ordinance emerged in New York where the forces in favor of zoning were exceptionally strong.

The New York Zoning Ordinance of 1916

> Considering the planning framework within which zoning had but recently arrived in the United States, it may seem remarkable that it fell away so rapidly. Yet its early disappearance was not unusual if we recall that the first American sponsorship of the planning idea came from an extremely small group of reform types whose power in the community bore little relation to the strength of its thinking. When the spirit and grand aims of the new planning movement came to be translated into law, they took the impress of whatever groups were politically competent enough to force the legislation.
> (Toll 1969)

The 1916 New York City zoning ordinance is usually regarded as the first comprehensive zoning ordinance in the United States.[4] It was the successful outcome of an open campaign to stop changes that were taking place on Fifth Avenue. It was a war on two fronts, one between carriage trade merchants and the invading garment industry, the other between wealthy residents and the invading retail trade. In Toll's words: "If this was war of sorts, it was in truth a double war: garment manufacturers fighting retail merchants fighting wealthy residents. The entire conflict was much closer in spirit to social Darwinism than to the Geneva Convention. There were no rules and only one objective, survival by any means" (Toll 1969: 110). But it was the encroachment of the Jewish garment workers and their immigrant workers that formed the central issue. Property values fell by a half in the five years up to 1916 (Feagin 1989: 81).

A Commission on Heights of Buildings reported in 1913, and recommended that height, area, and use should be regulated in the interests of public health and safety, and that the regulations should be adapted to the varying needs of the different districts. The city and the state legislature accepted these proposals, and the city charter was amended to include "districting" provisions. In 1916, a comprehensive zoning code was adopted for the whole city (Makielski 1966). The importance of the provision for different regulations in different districts needs to be stressed. As Bassett noted:

> The novel feature of zoning as distinguished from building code regulations, tenement house laws, and factory laws was that suitable regulations for different districts were established. We have become so accustomed to zoning regulations that it is difficult to understand how fixed the popular notion was that all land should be regulated in the same way throughout a municipality. On this account imposing different regulations on different areas appeared to many to be a discriminatory, arbitrary, and therefore an unlawful invasion of private rights. To counteract this impression it was considered important that the regulations within each district should be uniform for the same kind or class of buildings. A provision to this effect was placed in the original zoning clauses of the charter of New York city, and there can be no doubt that

courts which early passed upon these regulations were to a considerable extent persuaded to favor them on account of this requirement of uniformity. If it had been possible to make different regulations for the same sort of buildings in different parts of the same district, it is unlikely that zoning would have received the court approval that it now has.

(Bassett 1940: 26)

There was another difference between the new zoning controls and the well-established police power regulation of buildings and factories. Whereas the latter was intended to solve existing problems and to promote health and safety, zoning applied only to new development (as was the case also with the New York Tenement Law of 1901). So far as Fifth Avenue was concerned, further incursion by the garment industry was preventable, but nothing could be done about the changes already brought about, at least not through zoning itself: political action was another matter (Toll 1969: 178). Existing uses were hallowed as "nonconforming": "In any building or premises any lawful use existing therein at the time of the passage of this resolution may be continued therein, although not conforming to the regulations of the use district in which it is maintained." Thus, ironically, the very problem which gave rise to the zoning ordinance remained untouched. The reason is not far to seek: there was so much concern that the newfangled zoning system would reduce property values that the commission was most anxious to allay the fears. Indeed, "the city set aside enough land in business and industrial zones to accommodate an eventual population of approximately 340 million persons" (Haar and Kayden 1989a: 280).

In a discussion of nonconforming uses some twenty years later, Bassett (1940: 113) commented that "consideration for investments made in accordance with the earlier years has been one of the strong supports of zoning in New York City."[5] Protection of "investments" was, and remains, a major objective of zoning.

Successful though the campaign for the New York ordinance appeared it was, in one crucial respect, a dismal failure: in contrast to the hopes of the proponents of the planning movement, it lacked any "planning" component. In Gerckens' words, it was a "substitute" for a plan – "generally protective of current land interests ... [but] not based on a forecast of future land use demand" (Gerckens 1988: 33). Nevertheless, it was rapidly copied by numerous cities throughout the United States. Indeed:

The movement in favor of zoning took on aspects of a fervid crusade. Whereas, in 1916, Lawrence Veiller had announced that zoning "sounds like a beautiful dream", by the following year George Ford was able to state that, as a result of the success in New York, zoning was being organized or actually carried on in twenty municipalities. In 1922, Frank B. Williams published *The Law of City Planning and Zoning*, the first comprehensive American work in the field. The same year Theodora Kimball could write that "zoning has taken the country by storm"; she reported twenty enabling Acts, nearly fifty ordinances, and about one hundred zone plans in progress.

(Haar and Wolf 1989: 165)

But the most significant indication of progress was the appointment, by Secretary of State Herbert Hoover, of the Advisory Committee on Building Codes and Zoning.[6] This committee drafted a Standard State Zoning Enabling Act which rapidly became the model for a large number of zoning ordinances.

The Standard State Zoning Enabling Act

The scientific management reform movement swept the country between 1910 and 1919. Although they initially focused on industrial management, scientific management advocates claimed its universal application to other areas such as schools, philanthropic institutions, governments, and cities. In short, it could be applied to any activity that sought to combine the values of efficiency with functional organization.

(Boyer 1983)

Zoning was a part of the scientific management movement which was sweeping America in the first quarter of the twentieth century. Herbert Hoover was an important figure on this stage. He was instrumental in creating a new area of federal responsibility: one which Christine Boyer (1983: 146) has termed the "cooperative state."[7] Central to this was scientific study of the facts (and the collection of scientific data), and the establishment of "a central clearing-house for social and economic reforms." Boyer documents some of the areas in which Hoover applied this philosophy; these included the standardization of industrial parts and of the plans, designs, materials, and structural elements of houses; and the coordination of information concerning the housing market. Zoning clearly fell into this kind of thinking:

In the search for a new order to the American city, the division of land uses and regulations restricting building heights and bulk became tactical rearrangements. . . . Zoning, it was claimed, embodied and exemplified the idea of orderliness in city development; it encouraged the erection of the right building, in the right form, in the right place. "What would we think of a housewife who insisted on keeping her gas range in the parlor and her piano in the kitchen?" Yet these were commonplace anomalies in the American city of the 1920s: gas tanks next to parks, garages next to schools, boiler shops next to hospitals, stables next to churches, and funeral parlors next to dwelling houses.[8]

(Boyer 1983: 155)

Hoover's philosophy was that the role of the state was, not to interfere with market forces, but to make them more efficient by, for example, facilitating the production of better market information, advancing the acceptance of standardization, and (in the area of housing and urban development) assisting with the introduction of a system for orderly development which would be safe as an investment for both lenders and borrowers. Zoning was seen as the instrument for providing the necessary security against both unwanted development and legal challenge. In particular, it provided protection to homeowners from uncongenial neighboring uses which would affect both amenity and market value.[9]

The SSZEA gave state legislatures "a procedure, based upon an accepted concept of property rights and careful legal precedent, for each community to follow" (Boyer 1983: 164). A crucial element in the rationale here was the belief that a single legal code would pass legal muster in a way which a multiplicity of individual local ordinances would not. A carefully crafted ordinance, based on this universal model and embodying the fruits of planning expertise, supported by local citizens, would provide a defensible framework for an extension of the hard-to-define limits of the police power.

The model Act was hugely popular. The first edition, published in 1924, became a bestseller with sales of more than 55,000 copies. "Within a year of its original issue, nearly one quarter of the states in every corner of the nation had passed enabling acts which were modelled substantially on the Standard Act" (Toll 1969: 201).

And so, though planning languished, zoning boomed: by 1926, forty-three of the (then forty-eight) states had adopted zoning enabling legislation; some 420 local governments containing nearly a quarter of the population had adopted zoning ordinances, and hundreds more were in process of preparing them (Mandelker and Cunningham 1990: 166). By 1929, there were 754 local governments who had adopted zoning ordinances: these contained about three-fifths of the urban population of the country (Hubbard and Hubbard 1929: 3).

Euclid

Sixty years after the Court's approval of zoning, *Euclid* endures as substance and symbol, despite waves of demographic, economic, and political change.

(Wolf 1989)

Despite the growing popularity of comprehensive zoning, it was not until 1926 that the Supreme Court dealt with its constitutionality.[10] In that year, in the case of *Village of Euclid, Ohio, v Ambler Realty Co.*, the court argued:

The ordinance now under review, and all similar laws and regulations, must find their justification in some aspect of the police power, asserted for the public welfare. The line which separates the legitimate from the illegitimate assumption of power is not capable of precise delimitation. It varies with circumstances and conditions. A regulatory zoning ordinance, which would be clearly valid as applied to the great cities, might be clearly invalid as applied to rural communities. In solving doubts, the maxim *sic utere tuo ut alienum non laedas* [use your own property in such a manner as not to injure that of another], which lies at the foundation of so much of the common law of

nuisances, ordinarily will furnish a fairly helpful clew (*sic*). And the law of nuisances, likewise, may be consulted, not for the purpose of controlling, but for the helpful aid of its analogies in the process of ascertaining the scope of the power. Thus the question whether the power exists to forbid the erection of a building of a particular kind or for a particular use, is to be determined, not by an abstract consideration of the building or of the thing considered apart, but by considering it in connection with the circumstances and the locality. A nuisance may be merely a right thing in the wrong place, like a pig in the parlor instead of the barnyard.

(*Village of Euclid, Ohio, v Ambler Realty Co.* 1926)

The Supreme Court upheld the constitutionality of the Euclid zoning ordinance, though only barely (McCormack 1946: 712), and thus put its seal of approval on comprehensive zoning.[11] This represented a significant extension of the police power in that it enabled a municipality to prohibit uses which were not "nuisances" in the strict sense of the term. In particular, shops, industry, and *apartments* were excluded from single family zones. Justice Sutherland, delivering the opinion of the court, referred to reports of experts on zoning (who were not named), and noted that:

These reports, which bear every evidence of painstaking consideration, concur in the view that the segregation of residential, business and industrial buildings will make it easier to provide fire apparatus suitable for the character and intensity of the development in each section; that it will increase the safety and security of home life, greatly tend to prevent street accidents, especially to children, by reducing the traffic and resulting confusion in residential sections, decrease noise and other conditions which produce or intensify nervous disorders, preserve a more favorable environment in which to rear children etc. With particular reference to apartment houses, it is pointed out that the development of detached house sections is greatly retarded by the coming of apartment houses, which has sometimes resulted in destroying the entire section for private house purposes; that, in such sections very often the apartment house is a mere parasite, constructed in order to take advantage of the open spaces and attractive surroundings created by the residential character of the district. Moreover, the coming of one apartment house is followed by others, interfering by their height and bulk with the free circulation of air and monopolizing the rays of the sun which otherwise would fall upon the smaller homes, and bringing as their necessary accompaniments, the disturbing noises incident to increased traffic and business, and the occupation, by means of moving and parked automobiles, of large portions of the streets, thus detracting from their safety and depriving children of the privilege of quiet and open spaces for play, enjoyed by those in more favored localities – until, finally, the residential character of the neighborhood and its desirability as a place of detached residences are utterly destroyed. Under these circumstances, apartment houses, which in a different environment would be not only entirely unobjectionable but highly desirable, come very near to be nuisances.

(*Village of Euclid, Ohio, v Ambler Realty Co.* 1926)

This quotation gives a clear idea of the strength of the objection to apartments. At this time these were often (in fear if not always in fact) nineteenth-century tenement buildings built on single lots, overshadowing adjacent single family homes, and making no on-site provision for parking (Mandelker 1988: 56). The image was of slums occupied by immigrants. The feeling against tenements was widespread:

Once a block of homes is invaded by flats and apartments, few new single family dwellings ever go up afterwards. It is marked for change, and the land adjoining is forever after held on a speculative basis in the hope that it may all become commercially remunerative, generally without thought of the great majority of adjoining owners who have invested for a home and home neighborhood only.

(Cheney 1920a)

The import of the *Euclid* decision was that it held that the social implications of apartment living could be a "use" category for the purposes of land use planning. Thus another dimension was added to the exclusionary nature of land use regulation.

Conclusion

Since *Nectow*, innumerable zoning cases have demonstrated that where land and municipal planning are involved, there is little in the way of first principles except that each case is unique.

(Babcock 1973)

The issue before the Court in the *Euclid* case was a "facial" one: whether the ordinance was, on the face of things, constitutional; it was not asked to consider whether the ordinance "as applied" would be unconstitutional. This issue arose in a 1928 Massachusetts case (*Nectow v City of Cambridge*) where it was ruled that an otherwise constitutional zoning ordinance was unconstitutional as applied to Nectow's property. This property was zoned residential, thereby reducing its pre-ordinance market value for industrial use. However, an adjacent property was zoned industrial and the court found that the zoning of Nectow's property was "not indispensable to the general plan"; that "the health, safety, convenience, and general welfare of the inhabitants of the part of the city affected will not be promoted by that zoning"; and that, while the cost to Nectow was high, the public benefit was small.

Following *Nectow*, the U.S. Supreme Court did not take another planning case for almost half a century (that case was *Belle Terre*, discussed in Chapter 3).[12] At the state level, land use cases relied on *Euclid* for a "presumption of validity," which protected land use regulation from constitutional attack – until the 1960s when cases such as *National Land* were struck down (Mandelker 1989b: 384).

The legitimation of zoning did not end the arguments which surrounded its birth: on the contrary, they have continued to this day. In the next chapter, there is a broad discussion of some of the main features of zoning and the issues which it raises. It is quickly apparent that the avoidance of nuisances and the protection of property values guaranteed that zoning's discriminatory beginnings endured. Bruno Lasker's (1920: 279) argument is as relevant today as it was in 1920: "no height restriction, street width or unbuilt area will prevent prices from tottering in a good neighborhood unless it helps at the same time to keep out Negroes, Japanese, Armenians, or whatever race most jars on the natives."

Chapter 3

The objectives and nature of zoning

Introduction

> Despite the standard police power justifications given for zoning over many years, the fundamental significance of neighborhood zoning is that it creates a collective property right to the neighborhood environment that is effectively held and exercised by its residents.
>
> (R.H. Nelson 1977)

The spread of zoning was greatly facilitated by the publication, by Hoover's Department of Commerce, of a Standard State Zoning Enabling Act. This was prepared by a committee appointed by Hoover, who was convinced that he could make a significant contribution to the solution of the nation's problems of unemployment and housing by stimulating housing supply. Part of his policy was to encourage zoning as a means for protecting homeowners from destructive land uses, such as factories, which could invade their neighborhoods (Toll 1969: 201). The prestigious committee's Standard Act had a standing which was attractive to municipalities searching for solutions to the problems of their suburbanizing areas, and at the same time bore the stamp of authority and rationality which impresses courts. Not surprisingly, it proved to be immensely popular. The first edition was a bestseller, with over 55,000 sales. Within a year, nearly a quarter of all the states had passed legislation Enabling Acts modeled on the Standard Act (Toll 1969: 201).

The Standard Act listed six purposes of zoning: "to secure safety from fire, panic, and other dangers; to promote health and the general welfare; to provide adequate light and air; to prevent the overcrowding of land; to avoid undue concentration of population; to facilitate the adequate provision of transportation, water, sewerage, schools, parks, and other public requirements." This list of purposes, which is constantly paraded before the courts, omits the one which is by far the most important: the exclusion of unwanted people or uses, and thus the preservation of the status quo. These exclusionary objectives are seldom much below the surface, even when they are not explicit. Of course, all zoning is exclusionary; by definition zoning excludes *some* uses. The one exception (now rarely used) is where a zone is "unrestricted." Thus, in the 1916 New York ordinance, areas not zoned residential or business were unrestricted. The village of Euclid also had an unrestricted zone, as does any similar "cumulative" zoning system. *Euclid* provided, first, for an exclusively single-family house zone. The second use zone provided additionally for two-family houses; the third further included apartment houses; the fourth offices and shops; and so on until the final zone could accommodate all uses (and was therefore, in effect, an unrestricted zone). In each zone, development could take place that accorded not only with its specific categorization but also with all "higher" uses. Under this system, a single-family house could be built in any zone, but elsewhere only the uses specified for the particular zone and all "higher" zones were permitted.

This cumulative or pyramid zoning system (which has now largely given way to single use districting) had some oddities; for example, while it is clear why an industrial use might be banned from a residential area, it is not apparent why a residential development should be permissible in an industrial area. It certainly flouts the zoning philosophy of "a place for everything and everything in its place." Hagman suggests that, at the time, it might have been considered that market forces would do the necessary job. Moreover, there might have been constitutional difficulties since the courts were inclined to declare that early zoning cases were invalid on due process grounds (i.e. that they did not serve the general welfare), and "such restrictions would be satisfied only where strong and universally held values such as the sanctity of the home were involved" (Hagman 1975: 107).

On the other hand, in drafting an earlier Berkeley zoning ordinance, Charles H. Cheney

(1917) heeded the appeals of manufacturers who argued that residences should be kept out of industrial areas: "We find in most cities the most abject poverty and the worst tenements and bad housing conditions in the factory neighborhood. When we want heavy traffic pavement for heavy hauling ... home owners appear before the city council and holler so loud that the improvements are held up. So we have dejected housing and hampered industry." The Berkeley ordinance accordingly provided that no new housing could be built in factory districts; but this was unusual in these early years. As a consequence, zoning ordinances did nothing to stop the continued mixture of uses which they were ostensibly conceived to phase out.

Initially, zoning was concerned essentially with "districts." Though lip service was paid to comprehensive plans (which were supposed to form the rational basis for the operation of zoning) this did not amount to much in practice. The term "comprehensive" came to mean little more than a zoning provision which covered all or most of the districts in a local government area. However, as we shall see later, the overriding concern with exclusion led to zoning policies which were concerned with the whole of an area. In this way, the term "comprehensive" took on a new meaning: safeguarding the status quo of a neighborhood was simply writ large.

It is important to stress that zoning is an inherently rigid instrument. This remains so in spite of the extraordinary ingenuity (discussed in the following chapter) which has been displayed in adapting it to the real moving world; and in this rigidity lies its enormous popular appeal. The planning ideal of flexibility (so wholeheartedly acclaimed by the British planning system) is anathema to protectionist homeowners. Rigidity gives them certainty and security. There is perhaps no clearer statement than the following quotation from a 1953 New Jersey case:

In order to be valid, zoning restrictions and limitations must have a tendency to promote the general welfare by prohibiting in particular areas, uses which would be detrimental to the full enjoyment of the established use for the property in that area. The real object, however, of promoting the general welfare by zoning ordinances is to protect the welfare of the individual property owner. In other words, promoting the general welfare is a means of protecting private property.[1]

In Nelson's words, "supported by the courts, zoning rigidity has provided neighborhood residents with a highly valued assurance that the community would not at some future date allow undesired uses into the neighborhood against their wishes" (R.H. Nelson 1977: 19). It is now time to turn to the specific ways in which zoning has been used.

The Traditional Techniques of Zoning

Zoning: . . . the workhorse of the planning movement in this country.

(Haar 1977)

Zoning is the division of an area into zones within which uses are permitted as set out in the zoning ordinance. The ordinance also details the restrictions and conditions which apply in each zone.[2] Thus, the ordinance for the city of Newark, Delaware, has seventeen classes of district including residential, business, and industrial. There are seven classes of residential districts, which are distinguished by house type and density. For example, the RH classification provides for districts with one-family, detached houses with a minimum lot area of a half acre, a minimum lot width of 100 feet, a building setback of forty feet, a rear yard of fifty feet, and two side yards with an indi-

vidual width of at least fifteen feet, and a combined width of thirty-five feet. Two other one-family detached residential districts (RT and RS) have somewhat lower standards; similarly with one-family semi-detached residential districts. In the three detached districts, the taking of boarders is restricted to:

> The taking of nontransient boarders or roomers in a one-family dwelling by an owner-occupant family resident on the premises, provided there is no display or advertising on the premises in connection with such use and provided there are not more than three boarders or roomers in any one-family dwelling.

For a one-family dwelling in which the owner is nonresident the limit is reduced from three to two.

Other residential districts are garden apartments (RM) up to three stories in height, high-rise apartments (RA) of more than three stories with an elevator, and row or town houses (RR).

Certain uses are permissible by "special use permit." These include police and fire stations, golf courses, professional offices in a residential dwelling, "customary home occupations," day-care centers, and private nonprofit swimming clubs.

The zoning ordinance also provides for a Board of Adjustment to which appeals can be made against the decision of the building inspector in enforcing the ordinance, or for a variance from the provisions of the ordinance where a literal enforcement would result in unnecessary hardship. A variance is permissible only if it is not contrary to the public interest.

This is a very abbreviated description of the Newark zoning ordinance, but it is intended only to provide a background to the discussion which follows.

The Single-Family Zone: What is a Family?

A quiet place where yards are wide, people few, and motor vehicles restricted are legitimate guidelines in a land use project addressed to family needs. This goal is a permissible one. . . . The police power is not confined to elimination of filth, stench, and unhealthy places. It is ample to lay out zones where family values, youth values, and the blessings of quiet seclusion, and clean air make the area a sanctuary for people.

(*Belle Terre v Boraas* 1974)

Since the protection of the single family home is a major reason for, and a major objective of, zoning it is clearly necessary to define "family."[3] Without a definition it would be possible for a group of unrelated students to live "as a family" and introduce discordant elements into a single family zone! But definitions can raise as many problems as they solve; and so it is in this case. The first difficulty is whether it is in fact constitutional to "penetrate so deeply ... into the internal composition of a single housekeeping unit." The answer seems to be in the negative except in a very few states. A 1966 Illinois Supreme Court case, *City of Des Plaines v Trottner*, held that, in Illinois, it was not. The Illinois legislature reversed this decision by specifically authorizing zoning classifications based on family relationships. However, in the 1974 *Belle Terre* case, the U.S. Supreme Court upheld a definition which required a family to consist of persons related by blood, adoption or marriage, or a maximum of two unrelated people. The court agreed with the local authority that "the regimes of boarding houses, fraternity houses, and the like present urban problems. More people occupy a given space; more cars rather continuously pass by; more cars are parked; noise travels with crowds." It could have been objected that the same would result from a family with four teenage children, but the court was carried away by its respect for

judicial deference, its overwhelming concern for the archetypal suburban family – and the poetry of its own words, a further oft-quoted sample of which is given at the head of this section.[4]

The matter did not end there, however, since a later case (*Moore v City of East Cleveland* 1977) concerned an embarrassingly nonsensical outcome. The city of East Cleveland, Ohio, had a complex definition of a family which had the result of making Mrs. Moore's occupancy of her house illegal. The oddity was that all the occupants were related by blood, but the degree of relationship was insufficient to satisfy the ordinance: the family consisted of Mrs. Moore, her son, and two grandsons who were first cousins rather than brothers. Mrs. Moore received a notice of violation from the city stating that one of the grandsons was an "illegal occupant." Mrs. Moore refused to remove him, and the city filed a criminal charge. Upon conviction she was sentenced to five days in jail and a $25 fine. The city argued before the Supreme Court that its decision in *Belle Terre* required it to sustain the ordinance. The usual case was made about the need to prevent overcrowding, to minimize traffic and parking congestion, and to avoid an undue financial burden on East Cleveland's school system.

Surprisingly, at least to those who are unfamiliar with the element of unpredictability which is to be found in the workings of the Supreme Court, the justices had great difficulty with this case. Justices Stewart and Rehnquist, quoting *Belle Terre*, opined that East Cleveland's definition of "family" did not offend the Constitution. In any case, a line had to be drawn somewhere; but it could never be drawn in such a way as to guarantee that it "would not sooner or later become the target of a challenge like the appellant's." The majority, however, grappled with constitutional matters and concluded that the ordinance was an "intrusive regulation of the family," and that it was distinguishable from *Belle Terre* in that the latter case dealt with *unrelated* persons.

The issue is not, however, settled; and perhaps, like many other zoning matters, it may never be. State courts are free to interpret cases in the light of state constitutions. In fact, generally there seems to be a trend to liberalize the meaning of the term "family" to take into account the freer modes of conjugality that are now more common. The test appears to be whether there is "a legitimate aim of maintaining a family style of living" (Wright and Wright 1985: 179). But no such zoning generalization is likely to stand up; as the *Ladue* case quoted in Chapter 1 demonstrates.

A similar test has been applied to group homes for foster children, the mentally retarded, and like groups. The rationale here is that the essential purpose of a group home is to provide a family-like environment (in contrast to the custodial character of an institution). The case is clear where a foster home consists of a married couple, their two children, and ten foster children (seven of whom are siblings) as in *City of White Plains v Ferraioli*. The court distinguished this from *Belle Terre*, noting that "the children, natural and foster, live together as if they were brothers and sisters," and that the home was "structured as a single housekeeping unit and is, to all outward appearances, a relatively normal, stable, and permanent family unit."

The issue is more difficult when a home is staffed by professionals, and court decisions are conflicting. In *Culp v City of Seattle*, the court considered a proposal to establish a home for retarded children which would have a professional staff. It was held that there was a significant difference "between a dwelling occupied by a family which takes in and cares for children and one which is occupied by children supervised by a staff. The former is compatible with the traditional notion of a family; the latter is compatible with the traditional notion of an institution." Thus the proposed home could not be considered as being of a family type. In a similar case in New York (*Group House of Port Washington Inc. v Board of Zoning and Appeals*), however, a staff-supervised home for unadopted disturbed children was held to be "functionally equivalent" to a family.

In recent years, many states have passed legislation to overcome exclusionary zoning of group homes (Hopperton 1980). The nature of the legislation varies: some measures are restricted to certain types of home while others are much broader. Some designate group homes as a "special exception" under the zoning ordinance; others classify group homes as a separate use to which special standards apply.

In a 1985 Supreme Court case, *City of Cleburne v Cleburne Living Center*, a zoning regulation which excluded group homes for the mentally retarded from residential areas was held to be unconstitutional as violating the equal protection clause. It remains to be seen what effect this will have on future cases. Mandelker suggests that it "may encourage state courts to review the denial of conditional uses for group homes more stringently" (Mandelker 1988: 139). Cases that come before the courts are, by definition, cases involving conflict. It is not easy to obtain an overall picture of the situation regarding group homes. However, it is interesting to note the findings of a report by the U.S. General Accounting Office (1983). This documents the difficulties which the sponsors of group homes have experienced with planning bodies and with citizen groups; but it concludes that other problems were more severe: "inadequate funding, unsuitable locations and facilities, and certain other factors caused problems more frequently and hindered the development of group homes more often than zoning problems." Moreover, most group homes were established in residential areas without great difficulty, and more often than not communities accepted group homes. This is not to say, of course, that there is no problem: even a casual reading of a local paper will show otherwise; and the GAO study inevitably had to omit from its survey those homes that were closed or never opened because of community pressures. Nevertheless, the report does provide a corrective to the impression that is gained by focusing on court cases.[5]

The Single-Family House: Should there be a Minimum Size?

A recognized public health expert . . . testified that the living floor space in a dwelling had a direct relation to the mental and emotional health of its occupants and that he had developed scientific standards for different size families.
(*Lionshead Lake Inc. v Wayne Township* 1953)

The ratio of occupants to space obviously can affect public health, family stability and emotional well being.
(*Home Builders League of South Jersey Inc. v Township of Berlin* 1979)

Photographs of insanitary, tiny, crowded tenements leave one in no doubt that there are standards below which society will not, in all conscience, wish families to live. These standards vary over time and space. What is considered intolerable in 1990 is very different from what was so considered in 1790. Similarly, contemporary standards in the United States are very different from those in Bangladesh. Every society has to define for itself the standards at which it expects (and will assist) its people to live. There is nothing scientific about this: it is a matter for judgment and political decision. Yet the zoning system frequently brings these matters before the courts for adjudication; for example, is the minimum lot size or the minimum floor area prescribed by a zoning ordinance acceptable? Unfortunately, the question is more narrowly conceived than this since the courts commonly operate on "the

presumption of validity": that an act of a legislative body cannot be challenged unless it is blatantly unfair. This makes it difficult to challenge minimum area requirements because the onus of proof is on the plaintiff to show that the provision could not have had a valid purpose (Babcock and Bosselman 1973: 30). As a result, the argument before (and of) the court is usually couched in terms of a dispute between an individual developer and the local inhabitants: wider issues of exclusion and of regional housing needs tend to be pushed into the background, even if they surface at all.

In a 1953 New Jersey case (*Lionshead Lake Inc. v Wayne Township*), an ordinance provided that residential areas should have a minimum square footage of 768 for a one-story dwelling, 1,000 for a two-story dwelling having an attached garage, and 1,200 for a two-story dwelling not having an attached garage. Despite the so-called "expert" evidence referred to in the quotation at the beginning of this section, the trial court concluded that the requirements "were not reasonably related to the public health, were arbitrary and unreasonable, and not within the police powers" of the township.[6] (In so doing, it challenged the presumption of validity.) The New Jersey Supreme Court disagreed:

Has a municipality the right to impose minimum floor area requirements in the exercise of its zoning powers? Much of the proof adduced by the defendant township was devoted to showing that the mental and emotional health of its inhabitants depended on the proper size of their homes. We take notice without formal proof that there are minimums in housing below which one may not go without risk of impairing the health of those who dwell therein. One does not need extensive experience in matrimonial causes to become aware of the adverse effect of overcrowding on the well-being of our most important institution, the home. Moreover, people who move into the country rightly expect more land, more living room, indoors and out, and more freedom in their scale of living than is generally possible in the city. City standards of housing are not adaptable to suburban areas and especially to the upbringing of children. [Moreover, on the question of the general welfare,] the size of the dwellings in any community inevitably affects the character of the community and does much to determine whether or not it is a desirable place in which to live. It is the prevailing view in municipalities throughout the state that such minimum floor area standards are necessary to protect the character of the community. . . . In the light of the constitution and of the enabling statutes, the right of a municipality to impose minimum floor area requirements is beyond controversy.

(*Lionshead Lake Inc. v Wayne Township* 1953)

The court soon had cause to regret these words: its decision gave rise to extensive academic discussion which caused it to rethink its position in later cases.[7] Thus, in *Home Builders League of South Jersey Inc. v Township of Berlin* the New Jersey Supreme Court quoted liberally from the academic articles which criticized its *Lionshead Lake* decision. In the *Home Builders* case, there was a challenge to a municipal zoning ordinance which, like that of *Lionshead Lake*, imposed minimum floor area requirements for residential dwellings irrespective of the number of occupants living in the dwelling, and unrelated to any other factor such as frontage or lot size. After an extensive hearing, the trial court again challenged the presumption of validity, and found that the "nonoccupancy based" floor area minima were "unrelated to the public health, safety or welfare, and hence an arbitrary, capricious and unreasonable exercise of the municipal zoning power." On appeal, the Supreme Court of New Jersey gave prominence to the issue of "economic segregation." It noted that, in the quarter-century since *Lionshead Lake*, changes had taken place

which were reflected in legislative and judicial attitudes: "As we have stated previously, once it is demonstrated that the ordinance excludes people on an economic basis without on its face relating the minimum floor area to one or more appropriate variables, the burden of proof shifts to the municipality to show a proper purpose is being served." This was an issue which did not arise in *Lionshead Lake*, but did in *Home Builders*.

One further point raised in the *Home Builders* case is significant, namely the importance of the specific facts of a case: "with respect to every zoning ordinance . . . the question remains as to whether or not in the particular facts of the case and in the light of all of the surrounding circumstances the minimum floor area requirements are reasonable." This point arises frequently: so much so in fact that the student of zoning law is on occasion presented with the puzzle of how the circumstances of one case really differ from that of another. Moreover, it needs to be noted that *Lionshead Lake* and *Home Builders* are New Jersey cases and, as will become clear as our analysis proceeds, New Jersey is by no means representative of the fifty states. In fact, most states have upheld minimum size restrictions; and most do not regard economic discrimination as being invalid (Wright and Wright 1985: 185).

As a footnote to this discussion, mention can be made of an unusual *maximum* house size proposed in 1988 in the town of Greenwich, Connecticut (Ravo 1988). Troubled by the development of new houses averaging 7,000 to 9,000 square feet on relatively "small" lots of two or four acres, the local Planning and Zoning Commission proposed a maximum house size of 4,356 square feet. A zoning commissioner is reported as saying: "The houses built within the last few years are becoming larger, and the lots they've been built on are becoming smaller. More and more residents and residential associations are becoming afraid that the character of the area is changing. These kind of homes belong on ten acres, not on a little land 110 feet from the road."

Even the affluent areas have their problems!

Large Lot Zoning: Maintaining Community Character

> A zoning bylaw cannot be adopted for the purpose of setting up a barrier against the influx of thrifty and respectable citizens who desire to live there.
>
> (*Simon v Needham* 1942)

Large lot zoning has the ostensible purpose of safeguarding the public welfare, for example by ensuring that there is good access for fire engines, that roads do not become unbearably congested, or that there is adequate open space. These and similar worthy objectives appear frequently in zoning cases, as does an alternative formulation: to keep out undesirable (that is different) people, and to maintain the social and economic exclusiveness of an area. A leading case is *Simon v Needham*. Needham is a suburb of Boston, some twelve miles from the downtown. To control the amount of development in the area, the town passed an ordinance which provided for a minimum lot size of one acre over much of the area. Though declaring that insular interests must give way to the wider good, the court held that the zoning was valid and reasonable. It was swayed by the fact that "many other communities when faced with an apparently similar problem have determined that the public interest was best served by the adoption of a restriction in some instances identical and in others nearly identical with that imposed" by Needham.

Other apparently similar cases have been decided differently (Wright and Gitelman 1982: 843). A well-known Pennsylvania case is *National Land and Investment Co. v Kohn* in which Easttown Township had a four-acre minimum lot size. The arguments used to justify this were couched in terms of ensuring adequacy of sewage disposal and of roads, and of preserving the character of the area. The court cut through these contentions and stressed "the township's responsibility to those who do not yet live in the township but who are part, or may become part, of the population explosion of the suburbs" – which of course was precisely what Easttown wished to exclude from its area. The court declared that the township could not "stand in the way of the natural forces which send our growing population into hitherto undeveloped areas in search of a comfortable place to live." Cases in New Jersey have been decided likewise (the most dramatic of which is *Mount Laurel*, which is discussed at length in Chapter 5). On the other hand, in the Maryland case of *County Commissioners of Queen Anne's County v Miles* a five-acre minimum lot requirement was upheld. The reasoning was that the area was small (only 6.7 percent of the county) and that the zoning was done within the framework of a long-range plan to preserve an unusually beautiful country estate section of a river.

An interesting case where large lot zoning (six acres minimum) was upheld by the courts arose in 1972 in Sanbornton, a recreation area in the rural hills of New Hampshire (*Steel Hill Development Inc. v Town of Sanbornton*). The opening of an interstate highway brought this idyllic area within an easy ride of the Boston area. Not surprisingly, development pressures, particularly for vacation homes, grew and one company purchased land (amounting to half the area of the town) for the construction of some 500 homes. The land in question was zoned as "general residence and agricultural," requiring a minimum lot size of 35,000 square feet, or about three-quarters of an acre. There was a public outcry, and the planning board rezoned the land in such a way as to require lots of six acres over 70 percent of the area and three acres over the remainder. The company filed suit alleging, *inter alia*, that the minimum lot sizes bore no rational relationship to the health, safety, morals, or general welfare of the community and therefore were unconstitutional (the essential "substantive due process issue").

The district court found against the company on all counts. An appeal was made to the United States Court of Appeals for the First Circuit, which found that there was no precedent for determining this particular case. The town was not attempting to stem the "natural" pressures of growth; rather, a developer was seeking "to create a demand in Sanbornton on behalf of wealthy residents of Megalopolis who might be willing to invest heavily in time and money to gain their own haven in bucolic surroundings." Nor was the town simply attempting to keep people out and maintain open space and a rural community. Its concerns were different, and were related "to the construction and integration of hundreds of new homes which would have an irreversible effect on the area's ecological balance, destroy scenic values, decrease open space, significantly change the rural character of this small town, pose substantial financial burdens on the town for police, fire, sewer, and road service, and open the way for the tides of weekend 'visitors' who would own second homes."

On the other hand, though it was proper for the town to consider these factors, the court thought that it had acted in "a most crude way." Moreover, the court had:

> serious worries whether the basic motivation was simply not to keep outsiders, provided that they wished to come in quantity, out of the town. We cannot think that expansion of population, even a very substantial one, seasonal or permanent, is by itself a legitimate basis for objection.... Where there is a natural population growth it has to go somewhere, unwelcome as it may be, and in that case we do not think that it should be channeled by the happenstance of what town gets its veto in first.
>
> (*Steel Hill Development Inc. v Town of Sanbornton* 1972)

In this particular case, however, there was insufficient evidence of "pressure from land-deprived and land-seeking outsiders," and "at this time of uncertainty as to the right balance between ecological and population pressures" the town's ordinance could properly stand – but only as a stopgap measure. In short, the ordinance was upheld as an interim one pending adequate study and planning.

The *Sanbornton* case is an interesting illustration of the problems experienced by small rural towns faced with sudden development pressures for which they are unprepared (Moss 1977: 332). In a later chapter we will be examining a number of cases (such as *Ramapo* and *Petaluma*) where towns have devised regulatory mechanisms which operate to phase growth over a period of time.

But again it is necessary to stress that decisions from courts in the northeastern states are not necessarily nationally representative. On the contrary, as a leading legal digest expresses the matter: "the validity of large lot zoning is likely to vary depending on the size of the lot, the circumstances of the community or area involved, and the hostility or lack of it to large lot zoning in a particular jurisdiction" (Wright and Wright 1985: 183).

Apartments and Mobile Homes[8]

> Apartment living is a fact of life that communities like Nether Providence must learn to accept.
>
> (*Appeal of Girsh* 1970)

The reader who has come this far will not be surprised to find that apartments and mobile homes are the targets of particularly explicit exclusionary practices.[9] However, courts differ in their attitudes to these. Some have gone so far as to approve the restriction in an entire jurisdiction of all uses except single-family dwellings. By contrast, other cases, particularly in Pennsylvania and New Jersey, have ruled that municipalities must provide adequately for all types of dwellings in their area. A leading case is *Appeal of Girsh*, which invalidated a zoning ordinance which totally excluded apartments from the Philadelphia suburb of Nether Providence, Delaware County.[10] This case was very similar to *National Land*, and the Supreme Court of Pennsylvania ruled likewise: "Nether Providence Township may not permissibly choose to only take as many people as can live in single family housing, in effect freezing the population at near present levels. Obviously if every municipality took that view, population spread would be completely frustrated. Municipal services must be provided *somewhere*, and if Nether Providence is a logical place for development to take place, it should not be heard to say that it will not bear its rightful part of the burden."

So far as mobile homes are concerned, one might have expected that attitudes would have changed as the character of mobile homes has changed. A "Note" in the *Yale Law Journal* expressed such a view in 1962:

> Community fear of blight can be traced back to the low quality of both the early trailers and their parking facilities. Economic conditions of the thirties, followed by wartime housing shortages and rapid relocations of the labor force, pressed many thousands of unattractive trailers into permanent use. Often these units were without running water or sanitary facilities. There were no construction standards to insure even minimum protection against fire or collapse. They were parked in areas which were usually crowded, poorly equipped, and generally unsuited to residential use. As a result, conditions in these parks seldom exceeded minimum health and sanitation standards. The specter of such parks teeming with tiny trailers made community apprehension understandable. But substantial improvements in the quality of both mobile homes and park facilities may have undermined the bases for this antipathy today. The mobile home currently produced is an attractive, completely furnished, efficiently spacious dwelling for which national construction standards have been adopted and enforced by the manufacturers' associations.
>
> (*Yale Law Journal* 1962a: 702)

This was written in the same year (1962) as the notorious *Vickers v Township Committee of Gloucester Township* (which is more renowned for its dissent than for the majority view). This case concerned the township of Gloucester, New Jersey, which, after receiving an application from Vickers for a permit to develop a trailer (mobile home) park in an industrial zone, amended the zoning ordinance to bar trailer camps from the industrial zone. (They were already banned from all other zones.) The stated objective was to improve the appearance of potential industrial areas and to attract industry. The trial court upheld the exclusion, but the appellate division reversed, saying: "surely in this vast rural area, there must be some portion in which the operation of a trailer park would be compatible with the scheme of zoning the township has seen fit to select."

The case went to the New Jersey Supreme Court which found the exclusion of trailer parks to be valid. With an extreme respect for the presumption of validity, the court declared that since the amendment presented a fairly debatable issue "we cannot nullify the township's decision that its welfare would be advanced by the action it took." The eloquent dissent was written by Justice Hall, who strongly opposed the breadth of the court's rationale: "The import of the holding gives almost boundless freedom to developing municipalities to erect exclusionary walls on their boundaries." In his view, the majority had defaulted in their duty: "to evaluate and protect all interests, including those of individuals and minorities, regardless of personal likes or views of wisdom, and not merely to rubberstamp governmental action in a kind of judicial laissez-faire." On its approach, it was difficult to conceive of any local action which would not come within the "debatable" class. As we shall see, Justice Hall later went much further, and carried the majority of the court with him, in the 1983 *Mount Laurel* II case (which reversed the *Vickers* decision).

One obvious question which arises with mobile homes is a definitional one: is not a mobile home a single-family dwelling? Certainly, modern well-equipped mobile homes in an attractive park may be difficult to distinguish from the stereotypical single-family home which, in fact, nowadays can be largely factory-produced. The point becomes one of particular significance with "manufactured housing" intended for a permanent siting. This type of housing has been built since 1976 under a national code of health and safety requirements supervised by HUD (Lahey, Diskin, and Lahey 1989: 27). Indeed, in two Michigan cases decided in the early 1970's (*Bristow v City of Woodhaven*, and *Green v Lima Township*), the court held that mobile homes are a "preferred use" because they meet a need for low-cost housing. Prejudices are hard to overcome, particularly by so-called "facts." An observer might have thought that locating an immobile manufactured house on a permanent site would have translated a "mobile home" into a "singe family dwelling." Not so, for example, in the village of Cahokia, Illinois, where the zoning ordinance not only restricted manufactured housing to mobile housing parks, but also prohibited such housing from being permanently fixed in such a way as would prevent its removal (Rose and Duncan 1985: 1). The Illinois Supreme Court, in *Village of Cahokia v Wright*, upheld the ordinance on the grounds that a mobile home might be detrimental to the value of adjacent conventional single-family homes, stifle development in the area, or create potential hazards

to public health. A later Michigan case (*Robinson Township v Knoll* 1981) held that the exclusion of mobile homes from all areas not designated as mobile home parks had no reasonable basis under the police power and was based on reasoning which was "no longer valid in light of improvements in the size, quality and appearance of mobile homes."

Three 1987 cases demonstrate the ingenuity which can be devoted to devising methods for excluding mobile homes. In Minnesota, a minimum width for all dwellings was imposed (*City of Bemidji v Beighly*); in Maine, a three-acre minimum lot size was introduced (*Long v Town of Eliot*); and in Michigan, a requirement was imposed for a minimum of "core living space" for all dwellings of twenty by twenty feet (*Painter v Comstock Charter Township*). All three were upheld.

A study of mobile homes in Delaware noted that objections to this form of housing included the following: "(a) mobile home parks depreciate the value of surrounding residential properties; (b) mobile home parks generate local public service costs far in excess of their contributions to local public revenues; and (c) residents of mobile home parks are not desirable neighbors: they are transient people, often on welfare or unemployment, have large numbers of children, sit in front of their unkempt homes in lawn chairs drinking beer, and clutter their small yards with old pieces of automobiles and motorcycles" (Stapleford 1985: 1). A review of data from the 1980 public-use micro sample for New Castle County revealed that the typical mobile home household in the county was white, relatively small in size (the average was 2.33 persons compared to 3.10 for households living in single-family houses), and had relatively few children. About a half consisted of married couple families, and a large proportion of the rest were headed by widowed, divorced, and separated adults, many of whom had children. Relative to other types of owner-occupied housing, mobile homes in the county contained a disproportionate percentage of young and of elderly households. Over nine-tenths of mobile homes were owner-occupied, and they had mobility rates far lower than the renter dominated apartment sector. Incomes were lower than average, and it was clear that "mobile homes provide an important means for lower income households to become homeowners, including some households with incomes below the poverty level and many households receiving social security income. Mobile homes are also an important avenue to home ownership for households headed by persons with less than average formal education who work steadily in primarily blue collar occupations." A survey in one mobile home park gave rise to these conclusions:

> There are, in fact, very few mobile home park residents who are living as singles or nonfamily households, collecting unemployment compensation or public assistance, and breeding hoards of children who drain the social service dollars of local government. Beyond these household characteristics, however, the Millcreek survey demonstrates that there exists an extremely stable and close-knit community of households within a mobile home park, who choose to live together for a variety of very positive reasons, and who are committed to maintaining the quality of their properties and their community.
>
> (Stapleford 1985: 13)

The picture clearly does not accord with popular prejudices, but it does reveal sufficient differences from the single family home areas to fuel, rather than allay, fears.[11]

Prior to the Fair Housing Amendment Act of 1988, it was common for local governments to restrict mobile homes to adults and seniors only.[12] This Act makes it unlawful to discri-minate against families in the sale, rental, or financing of housing (with some exceptions in the case of housing communities for senior citizens). With changes in design and layout, mobile housing (now more commonly termed manufactured housing) has become more acceptable in recent years. Indeed it is often difficult to identify what is and what is not "manufactured." To quote a report from California:

> Manufactured housing has departed from its traditional use in mobilehome parks and on individual rural lots to become compatible with permanent installation in subdivisions and urban infill lots. New mobilehome parks can now be designed as condominium or cooperative developments. Even the term "mobilehome" appears antiquated, since this housing in reality is typically permanent: fewer than seven percent of manufactured housing units are ever moved again after first being installed.
>
> (California 1990b: 1)

The report from which this quotation is taken clearly illustrates the dramatic change which has taken place with this type of housing.[13]

Conclusion

It has been shown in this chapter that traditional zoning was essentially of a mechanical character. There are, however, zoning techniques (some well-established, some newer) which provide a much greater degree of flexibility for those zoning agencies that wish to make use of them. These are discussed in the following chapter.

Chapter 4

Zoning with a difference

Introduction

Ever since basic zoning and land use regulation was upheld by the United States Supreme Court in *Village of Euclid v Ambler Realty Co.*, local legislatures, administrative bodies, and courts have been groping in the dark for ways to make the rather rigid "Euclidian" formula more flexible and adaptable.

(Scott 1973)

Zoning provides certainty: that is its major characteristic: so much so in fact that the traditional zoning system was often referred to as "self-executing." To quote from the report of the Douglas Commission:

> The originators of zoning anticipated a fairly simple administrative process.... After the formulation of the ordinance text and map by a local zoning commission and its adoption by the local governing body, most administrations would require only the services of a building inspector who would determine whether proposed construction complied with the requirements. This official was not expected to exercise discretion or sophisticated judgment. Rather, he was to apply the requirements to the letter. In the case of new construction, he was to compare the builder's plans with the requirements governing the particular piece of land and either grant or deny a permit. Even today, this nondiscretionary permit process is at the heart of zoning administration.

(Douglas Report 1968: 202)

Matters have changed dramatically since the early days of zoning, but even the earliest zoning schemes provided some relief from the strict letter of the law. However carefully drafted, a zoning ordinance and map can never cover all the circumstances that might arise or all the eventualities that might come to pass. There thus has to be some way of providing for the unforseen. There are three main methods of doing this: by way of special exception, variance, and amendment. The first two are quasi-judicial acts, and are performed by the board of adjustment; an amendment on the other hand is a legislative act which is properly the responsibility of the local governing body. As might be expected, matters are not as neat as this in the real world.

Special Exceptions or Conditional Uses

> Is zoning by men replacing zoning by law?
> (Green 1955)

There are some uses which, though permissible (and perhaps necessary), require review to ensure that they do not have an undesirable impact on an area. Hospitals, schools, day-care centers, and clubs, for example, are needed in a community, but their specific location may give rise to traffic congestion and dangers, or to severe parking difficulties. Similarly with gas stations in commercial districts, and multifamily dwellings in a single-family district. Zoning ordinances typically make specific provision for such "special exceptions"[1] (to use the term employed by the Standard State Zoning Enabling Act). Terminology varies in a confusing way. The Newark ordinance quoted earlier refers to a "special use," while "conditional" is more generally used (Mandelker 1988: 252). The latter is favored here.

It should be noted that an exception is different from a nonconforming use (discussed later): the use is explicitly allowed but subject to the conditions detailed in the zoning ordinance. This is why the term conditional use is preferable.[2] A classic, though not particularly elegant, statement of the meaning of conditional zoning is to be found in the 1959 case of *Tullo v Township of Millburn*.

> The theory is that certain uses, considered by the local legislative body to be essential or desirable for the welfare of the community and its citizenry or substantial segments of it, are entirely appropriate and not essentially

incompatible with the basic uses in any zone (or in certain particular zones), but not at every or any location therein or without restrictions and conditions being imposed by reason of special problems, the use or its particular location in relation to neighboring properties presents from a zoning standpoint, such as traffic congestion, safety, health, noise, and the like.
> (*Tullo v Township of Millburn* 1959)

A conditional use is permissible if it meets the conditions set out in the zoning ordinance. Often these conditions are expressed in very general terms. For instance, the Newark, Delaware, ordinance requires that the conditional use will not:

(a) Affect adversely the health or safety of persons residing or working in the neighborhood of the proposed use;
(b) Be detrimental to the public welfare or injurious to property or improvements in the neighborhood; and
(c) Be in conflict with the purposes of the comprehensive development plan of the city.

By contrast, the New Jersey Enabling Act requires municipalities to spell out the conditions "according to definite specifications and standards which shall be clearly set forth with sufficient clarity and definiteness to enable the developer to know their limit and extent."[3]

It is difficult to summarize the attitudes of the courts to conditional uses since the judgments have been so diverse. Most courts have held that there is little discretion in determining whether a conditional use should be granted but others, following the path of judicial deference, have upheld conditional use denials even when the evidence showed no adverse impact on surrounding uses (Mandelker 1988: 255).

Variances

Anyone who attempts to organize and set
forth a clear picture of the American law on
variances either (a) has not read the case
law, or (b) has simply not understood it.
(N. Williams 1985–90)

While a conditional use is one which is permissible under the conditions of the zoning ordinance, a variance involves a relaxation of the provisions of the ordinance. The Standard State Zoning Enabling Act confers on the board of adjustment the power "to authorize upon appeal in specific cases such variance from the terms of the ordinance as will not be contrary to the public interest, where, owing to special conditions, a literal enforcement of the provisions of the ordinance will result in unnecessary hardship, and so that the spirit of the ordinance shall be observed and substantial justice done."

Variances are of two types: "area" (or "bulk") variances and "use" variances. The former involve a departure from the requirements of the ordinance in relation to such matters as lot width, lot area, setback and the like. By contrast, a use variance allows the establishment (or continuation) of a use which is prohibited by the ordinance. Allowing a house to be built closer to the lot line laid down in the variance would be an area variance; allowing a multifamily house in a single-family district would be a use variance. In many states, the distinction is of no consequence since the same conditions have to be met (as is the case with the provisions of the Standard State Zoning Enabling Act). In others (such as California, Indiana, and Virginia), the distinction is crucial since use variances are totally prohibited. In yet others (such as Illinois, New York, and Ohio), while a test of "unnecessary hardship" is used for a use variance, an (arguably) relaxed test of "practical difficulties" is used for an area variance.

The hardship theoretically has to be one which applies to a particular property, not to the personal circumstances of the owner. The rationale for this is that the matter for consideration is the relationship between the particular plot and the wider area. In the words of an Atlanta ordinance, "personal hardship shall not be considered as grounds for a variance, since the variance will continue to affect the character of the neighborhood after title to the property has passed" (Bair 1984: 33). In fact, many variances are given precisely because of personal hardship. One board had an explicit policy of allowing any use variance requested by a disabled veteran – including automotive repair and body work at homes in a residential area, and the sale of groceries in the front room of a residence (Bair 1984: 33). This may be unusual, but there is plenty of evidence to show that boards frequently do consider personal circumstances. One writer describes how home occupations were permitted by the Philadelphia board because of personal hardship: the apparent view was that "the harm to the particular neighborhood was far outweighed by the economic hardship to the applicant" (*University of Pennsylvania Law Review* 1955: 533).

The hardship should also be specific to the property for which a variance is sought. It is the task of a board to satisfy itself that the property has certain characteristics which are substantially different from those of other properties in the area. If the characteristics are general to the area (soil conditions for example) the proper course of action is to amend the ordinance. The classic statement of the hardship test appears in the 1939 New York case of *Otto v Steinhilber*:

Before the board may . . . grant a variance
upon the ground of unnecessary hardship,
the record must show that (1) the land in
question cannot yield a reasonable return if
used only for a purpose allowed in that zone;
(2) that the plight of the owner is due to
unique circumstances and not to the general
conditions of the neighborhood which may

reflect the unreasonableness of the zoning ordinance itself; and (3) that the use to be authorized by the variance will not alter the essential nature of the locality.

(*Otto v Steinhilber* 1939)

These tests have been widely (though certainly not universally) adopted. In particular, the requirement that there be an inability to make a reasonable return "has become standard requirement in variance law" (Mandelker 1988: 243). However, "reasonable" does not mean "maximum": in *R-N-R Associates v Zoning Board of Review*, the court affirmed the denial of a variance to allow a laundromat in a commercial district where the ordinance did not allow them:

> It is important, in our opinion, that zoning boards of review be informed that the right of a landowner to a variance turns upon a showing of unnecessary hardship, that is, a showing that an ordinance restriction deprives him of *all beneficial use of his land* [emphasis added].
>
> (*R-N-R Associates v Zoning Board of Review* 1965)

But it would be misleading to quote a series of cases, since the differences among (and perhaps within) the states on the issue are great. As one critic has put it, "if a system can be judged by the frequency of departures from it, zoning fails spectacularly" (*Harvard Law Review Note* 1969: 673).

It was the original intention that variances would be exceptional. It has not worked out that way. The variance is a popular tool of the boards of appeal which "see their function as a broker for the individual citizen against the inevitable comprehensiveness of the law" (Frost 1958: 277). One writer has suggested that the board of appeals operates as a kind of jury, dispensing rough justice in its hearings of variance applications, resulting in decisions which "are very apt to reflect the conscience of the community – a close approximation of what

most people in the community would think the proper course of action" (Green 1961: 165).

Various studies have convincingly shown that boards of adjustment commonly operate according to their own sense of what is right, with little regard for the law or even their local planning department. Most applications are in fact approved. Thus in a study in Lexington, Kentucky, the board granted 76 of 102 applications though 75 had been recommended by the planning staff for denial. Other studies have reported variance approvals ranging from 63 percent to 85 percent of applications (Anderson 1962, Dukeminier and Stapleton 1962, and Bryden 1967). In Alameda County, California, the board granted 208 applications which the planning department recommended be denied. The decisions were "drafted with a view toward satisfying the letter of the law [but] their indiscriminate use certainly does not satisfy its spirit" (Donovan 1962: 108).

Often, little time is taken in considering applications: there are just too many of them. A Philadelphia study revealed that, during a nine-week period, the board heard 569 cases, of which over 80 percent were for variances. Though a case might take over half an hour when a large number of protesters were present, in most instances the board took only three to five minutes to hear a case (*University of Pennsylvania Law Review* 1955: 527). Bryden comments that:

> It is difficult to see how such a summary disposition of an application could assure both the applicant and the community that the variance would be granted or refused in accordance with the standards laid down by the ordinance.... It would appear either that zoning ordinances are so enacted as to operate with unnecessary hardship against a large number of landowners in the municipality, which would indicate the need for more thorough and competent planning before enactment of the ordinance, or that many applications are not supported by

48 Zoning

evidence of unnecessary hardship, in which case the high percentage of board approvals would be unjustified.

(Bryden 1967: 293)

In the Alameda study, Donovan (1962: 107) found that only 15 of 284 applications granted by the board produced evidence of special circumstances adequate to sustain a legal review. Generalizing, Blucher (1956: 100) has expressed the opinion that 50 percent of all the rulings of zoning boards of appeal are probably illegal usurpations of power. Dunham finds an explanation for the lack of adherence to the law in the character of the membership of the boards.

> Variances are considered by scores of boards of adjustment which, in turn, may have scores of standards for considering variances. In addition, the members of those boards of adjustment are ordinary citizens who are probably appointed because they wanted to serve and not because of their expertise in zoning matters. Thus, boards of adjustment frequently are made up of people, usually business people, who are interested in granting or denying variances for their own personal reasons rather than in accordance to legal requirements.
>
> (Dunham 1988: 39.01)

The situation is particularly unsatisfactory when the board of adjustment is opposed (either explicitly or by implication) to the legislative body. At the extreme, the board can defeat the legislative body's intentions. A member of the Grand Rapids Board of Zoning Appeals has commented:

> Too often variances have been granted in the past because individual members of the board, without regard for the conditions set up in the zoning ordinance, felt that variances were best for the city, would remove an eyesore, or that the city commission was behind the times in changing a zone.
>
> (Souter 1961: 29)

In the same area, the board's granting of variances to permit business uses in a residential zone so changed the character of the neighborhood that rezoning was subsequently necessary to make the regulation conform to the actual uses to which property in the area was being put. The curious aspect of this case is that the city commission had considered rezoning the area for business uses, but decided against it because of its residential character with its "fine old houses" (Souter 1961: 29).

An important issue in the debate on the usefulness, or otherwise, of the variance (though now generally obscured) is of a constitutional nature. In the early days, there was a concern that the courts would declare an ordinance unconstitutional on the ground that it imposed severe hardship on an individual owner, even though the ordinance generally operated for the good of the community. The variance provides a safety valve: it applies only to that part of the ordinance which is subject to complaint on constitutional grounds. But, as Bryden points out, the *Nectow* case (summarized in Chapter 2) settled the problem: the Supreme Court there declared that an ordinance could be unconstitutional as applied to a single property, without affecting the constitutionality of the whole ordinance.

Since the evidence supports the view that "illegal issuance is a widespread phenomenon nationwide" (Hagman and Juergensmeyer 1986: 173), there is support for the argument that variances should be abolished. Bryden (1967: 314) argues that this would lead to better drafted zoning ordinances: without the safety valve of the variance, "zoning ordinances would have to be drafted with considerably more thought and care if they were to be acceptable to the public." This, together with the elimination of the groundless applications, would lead to a vast reduction in the number of cases which would be brought before the courts. Where the municipality is held to have acted unconstitutionally a rezoning would be required (a legal form of spot zoning); where a property is incapable of being put to use in a manner which is compatible with surrounding uses, the municipality should be empowered to acquire the property by eminent domain (Bryden 1967: 318).

Other critics have suggested that variances be subject to review by a higher authority such as a state review board or specialized courts having metropolitan jurisdiction, or that the power to grant variances be taken away from boards of appeal (Anderson 1962). But, however much lawyers attack the legal deficiencies of the variance, its popularity at the local level assures its continuance as a major feature of the zoning system. There is more to planning than law.

Spot Zoning

> Although spot zoning is a neutral land use control technique, it is capable of abusive application.
>
> (*Maine Law Review* 1972)

One of the difficulties which can arise with upzoning is the charge of "spot zoning": the unjustifiable singling out of a piece of property for preferential treatment. Spot zoning is not a statutory term: it is a judicial epithet signifying legal invalidity. In N. Williams' words (1985–90: 27.00), "In contrast with the statutory requirement, spot zoning has been a judge-made doctrine, obviously intended to give the courts a way to hold invalid certain zoning changes which for various reasons seemed arbitrary and inappropriate." It is important to note that it is not synonymous with piecemeal zoning (and for this reason it might be thought to be a misnomer). The following cases are illustrative.

In a Connecticut case, *Bartram v Zoning Commission of City of Bridgeport*, the court warned that an amendment to the zoning map "which gives to

a single lot or a small area privileges which are not extended to other land in the vicinity is in general against sound public policy and obnoxious to the law." Such spot zoning is frowned upon by the courts but, if a planning commission decides "on facts affording a sufficient basis and in the exercise of a proper discretion, that it would serve the best interests of the community as a whole to permit a use of a single lot or small area in a different way than was allowed in surrounding territory, it would not be guilty of spot zoning in any sense obnoxious to the law."

In this particular case, a landowner requested a rezoning of a small piece of land near (but not adjacent to) some new development. The existing zoning was for residential use, but the owner saw a need for some shops, and he proposed to erect a drugstore, and hardware and a grocery store, a bakeshop, and a beauty parlor. He requested an appropriate change of zoning, which the planning commission granted. On appeal, the trial court concluded that the requested change amounted to spot zoning, but the appeal court reversed. Its argument was that, on the facts of the case, there was by no means unanimous opposition from surrounding owners to the proposed development and that, even had there been, "it was the duty of the commission to look beyond the effect of the change upon them to the general welfare of the community." This the planning commission had done in deciding to support the rezoning: there was a need for additional stores in the area; and it was the policy of commission "to encourage decentralization of business in order to relieve traffic congestion and that, as part of that policy, it was considered desirable to permit neighborhood stores in outlying districts" (Wright and Gitelman 1982: 674).

In a similar case, *Kuehne v Town of East Hartford*, where a town council agreed a rezoning (but where there was a nearby site already zoned for business), the same court took the view that the council had had excessive regard to the benefit to the owner, and had "given no consideration to the larger question as to the effect the change would have upon the general plan of zoning in the community ... and cannot be sustained."

Finally, reference can be made to an oft-quoted Florida case, *Oka v Cole*, where a council agreed to rezone an area from single family to multiple family use. The neighbors protested strongly that their properties would decline in value (though an adjacent site was zoned multifamily). The case found its way to the Supreme Court of Florida, which ruled that it could find no authority for the proposition "that vested rights can accrue to neighboring owners, or that ordinances altering zoning restrictions are to be tested by any standard other than that applicable to zoning classification generally, i.e. that the restriction imposed shall not be arbitrary but reasonably related to the public health, safety or welfare." Though there were arguments both for and against the rezoning, the differences expressed were only those which might arise between reasonable men; they certainly did "not warrant judicial condemnation of the action as erosive or spot zoning" (Wright and Gitelman 1982: 683).

N. Williams (1985–90: 27.01) has argued that spot zoning is the antithesis of the comprehensive plan: "if a zoning change is in accordance with the comprehensive plan, then it is not spot zoning; and if it is spot zoning, then it is not in accordance with the comprehensive plan." This a neat formulation, though it does not meet all situations particularly, of course, where no comprehensive plan exists.

Downzoning

The law on downzoning clearly is unsettled, in part because the number of downzoning decisions is limited.

(Mandelker and Cunningham 1985)

While an upzoning may well raise the wrath of the neighborhood, an amendment to rezone to a use of lower intensity (a "downzoning")[4] is often the result of neighborhood pressure. Since a downzoning is likely to reduce the value of undeveloped land, an objection is likely on the part of the landowner.

A good illustration is a 1983 Iowa case, *Stone v City of Wilton*, where the city of Wilton downzoned some six acres of land on which the owner was intending to build a federally subsidized housing project. The downzoning took place after public opposition, but ostensibly on the ground that the city's electrical, water, sewer, and road systems were inadequate for a concentration of multifamily dwellings in the area. Not surprisingly, the owner claimed that the reasons given were mere pretext, and that the downzoning was racially motivated. The court, however, held that the city's decision had been taken for valid reasons, i.e. the inadequacy of the utility systems. Furthermore, there was no evidence that the city had a discriminatory purpose. Thus, applying the "fairly debatable" rule, the downzoning was upheld. The court added: "zoning is not static. A city's comprehensive plan is always subject to reasonable revisions designed to meet the ever-changing needs and conditions of a community. We conclude that the council rationally decided to rezone this section of the city to further the public welfare in accordance with a comprehensive plan."

Similarly, in a 1967 Connecticut case, *Chucta v Planning and Zoning Commission*, the court upheld a downzoning which increased the minimum lot size and frontage requirements in a rural area of a township on the ground that this "was desirable in view of existing schools and other public facilities, and to provide adequately for a safe water supply and proper disposition of sewage."

By contrast, another Connecticut case, *Kavanewsky v Zoning Board of Appeals*, was decided the opposite way. The court rejected a downzoning which affected the whole of one of the two districts into which the town was divided. It noted that the downzoning was "made in demand of the people to keep Warren a rural community with open spaces and keep undesirable businesses out." In California, in the case of *Ogo Associates v City of Torrance*, a downzoning intended to exclude low income housing was invalidated.

It is difficult to make any general sense of the decisions on downzoning, but one point can be made with a moderate degree of certainty: piecemeal downzonings are likely to be examined much more carefully by the courts, without the usual assumption of validity.

Conditional or Contract Zoning and Site Plan Review

Contracts have no place in a zoning plan .... Legislation is not and ought not to be for sale.

(Bassett 1940)

Euclidian zoning requires uniform conditions within districts. Uniformity, however can lead to undesirable rigidity. It may be to the benefit of both the owner and the community to depart from a uniform regulation. It is here that conditional (or contract) rezoning can be useful.[5]

Essentially, conditional rezoning is, as the term suggests, the rezoning of a property subject to conditions. Typically, the conditions are negotiated between the owner and the local government following a specific proposal by the owner. There is much learned discourse on the validity and the desirability of conditional rezoning. The argument in favor holds that conditions can render acceptable a use which otherwise would be unacceptable. The contrary argument is that the police power cannot be subject to bargaining, that conditional rezoning is illegal spot zoning, and that local governments have no power to enact conditional

zoning amendments. Courts differ widely in their attitudes, and overall the position is confused to say the least. Until recently, the courts have tended to be hostile, and they remain so in some states such as New Jersey; but the trend seems to be toward a more favorable stance where the conditions are reasonable and of community benefit (Rathkopf and Rathkopf 1988: 12.05). Thus in Georgia (*Cross v Hall County*) it has been held that "such conditions will be upheld when they were imposed pursuant to the police power for the protection or benefit of neighbours to ameliorate the effects of the zoning change."

Many of the conditions that have been imposed are now normally included in "site plan review." This is the preparation of a site plan for approval by the planning board. In built-up areas it is analogous to subdivision control on vacant land, and urban renewal in an area of redevelopment, though it usually relates to a single lot rather than an entire development. It can be a normal zoning requirement, or a requirement for particular types of development such as cluster zones and planned unit developments (discussed below), or in connection with conditional rezoning or floating zone amendments. Although most statutes make no specific provision for it, some states have passed legislation of which an illustration can be given from the New Jersey Municipal Land Use Law.[6] This provides that a site plan review shall include and shall be limited to ... standards and requirements relating to:

a) Preservation of existing natural resources on the site;
b) Safe and efficient vehicular and pedestrian circulation, parking and loading;
c) Screening, landscaping, and location of structures;
d) Exterior lighting needed for safety purposes in addition to any requirements for street lighting;
e) Conservation of energy and use of renewable energy resources; and

f) Recycling of designated recyclable materials.

The relationship between site plan review and other zoning controls is by no means crystal clear, and to date there has not been much case law to clarify (or confuse) the matter. In a 1986 New York case, *Moriarty v Planning Board of Village of Sloatsburg*, the court noted that though site plan review undoubtedly had many origins, "its most likely sources seem to have been the vast expansion of public interest in environmental and aesthetic considerations, the need to increase the attractiveness of commercial and industrial areas in order to invite economic investment, and the traditional impulse for controls that might preserve the character and value of neighboring residential areas." N. Williams (1985–90: 152.01) notes that recent case law "makes it clear that in actual practice site plan review, like subdivision control, has been concerned as much (or more) with shifting the burden of the cost of new facilities, needed to serve new development, as it has been with the appropriate physical layout of such development." This is discussed in a broader context in Chapter 6.

Cluster Zoning and Planned Unit Development

> Traditional zoning located structures on similarly sized lots in similar locations much as Washington's head might be similarly located on each dollar bill.
>
> (Wright and Wright 1985)

Traditional zoning is based on the assumption that residential development will take the form of single-family houses on individual lots. New patterns of development emerged in the postwar years which require much more flexibility. This is provided by cluster zoning (sometimes called open space zoning or density zoning) which involves the clustering of development on one part of a site, leaving the remainder for open space, recreation, amenity or preservation. The overall density of the site is unchanged but, of course, the density of the developed part is increased. This has a number of advantages: the cost of paving and of supplying utilities is reduced; attractive landscape features (or wetlands) can be protected; open space can be provided for recreation; and housing can be provided of a type suitable for "non-traditional" households who do not want the bother of maintaining a large lot.

The basic idea, of course, is not a new one. It goes back to Clarence Stein, Ebenezer Howard, Clarence Perry, and Frederick Law Olmstead. Perhaps its most famous prototype is Radburn, New Jersey: a 149-acre development with a strict separation of road systems, traffic-free residential culs-de-sac, and a continuous inner park (Schaffer 1982).

Though there has been some success with cluster developments, they have not been widely accepted. Babcock gives an entertaining account of one of the reasons for this:

> An articulate developer in the East told me he can chart the reaction of the plan commission of any municipality in Westchester County to his proposal for a cluster subdivision . . . . The initial response is one of enthusiasm for the novel plot plan. It does have a catchy design. The local commissioner is as intrigued as he would be by four-color copy in an ad in the New Yorker magazine. The emotional empathy rises. Then there is a pause. This proposal represents people, not tonic water. The emotional graph levels off. And down it zooms as some practical soul asks: "What kind of nut would move to Wedgewood and not want his own backyard?" Another citizen asks: "Well, if your costs are less, then, of course, you expect to reduce your prices?" With that the jig is up. It is apparent to all that the one sure result of this

departure from accepted subdivision design is the introduction of persons who do not prefer life in detached single family dwellings and presumably do not have social interests and attitudes in common with families now in the community. The double indictment is sufficient to counteract the most exquisite architectural rendering.

(Babcock 1966: 31)

A refinement of the cluster concept is the "planned unit development," affectionately known as a PUD.[7] This differs from cluster zoning in that it is more than a design and planning concept: it also provides a legal framework for the review and approval of development. It also can incorporate (or even be confined to) commercial and industrial development.[8] The study by Moore and Siskin (1985: 5) points to several factors which contributed to the growth of PUDS in the 1960s:

1 development of consumer preferences for housing variety and amenities that could only be provided in a project planned as an entity rather than lot by lot;
2 recognition that typical zoning and subdivision controls were often inadequate in controlling large-scale development and did nothing to encourage high quality and innovation in such development;
3 growth of the homebuilding business primarily as a result of extensive postwar development; firms had become increasingly sophisticated, and capital and organization were sufficient to accomplish large-scale PUDS that might take five, ten, or more years to complete;
4 abatement of the severe housing shortage of the immediate postwar years, which allowed consumers to be more selective, and developers to devote more time and resources to the careful planning required for a successful PUD;
5 enactment of the National Housing Act

of 1961 which provided Federal Housing Administration (FHA) mortgage insurance for condominiums, and the development of the 1962 FHA model condominium statute for states, which provided translatable solutions to some of the legal problems faced by developers who wanted to build cluster housing but did not have an opportunity to obtain PUD zoning;
6 the economics of PUDS, recognized by rapidly growing municipalities that frequently could not afford to purchase and maintain large amounts of open space or to build recreational facilities, and saw PUD as a method of obtaining desirable amenities for the community; and
7 an explosive rise in land costs in the late 1980s, which made cluster housing and other innovative arrangements more desirable.

Bosselman (1985: 25) has noted that, in some instances, "the PUD concept swallowed up the zoning ordinance that gave it birth and regurgitated it in the form of standards for evaluating the impact of proposed PUDS." These standards carried names such as "impact zoning"[9] or "performance zoning,"[10] and were designed to replace traditional regulations. Instead of adhering to preset regulations, PUD gives developers the freedom to design developments which satisfy market demands. Wakeford explains:

Instead of attempting to draw up and categorize a comprehensive list of all the potential uses in an area, the ordinance simply classifies according to whether the use can meet criteria relating to noise, vibration, smoke, odors, heat, glare and traffic generation. Those criteria are often quite complex and make use of standards prepared by state or federal agencies, or standards associations. The zoning ordinance might, for example, contain a table showing for different octave bands the amount of noise permissible at different times of day, and according to whether the lot abutted on a residential or industrial district.... The difficulty comes in monitoring the uses established and negotiating appropriate action should the standards be breached.

(Wakeford 1990: 74)

Of course, this approach necessitates negotiations between the developer and the municipality: this is the mechanism by which flexibility is achieved. This is a far cry from zoning.[11]

One other feature of PUDs needs to be mentioned: the role of homeowners' associations in managing commonly held property. Membership in such an association can be mandatory for the owners of dwellings in a PUD. Whether the result is a happy, democratic way of managing the local environment, or a financial, administrative and political nightmare depends on the particular circumstances of the development – and the accidents of time, place, and neighbors.

Floating Zones

The term "floating zone" seldom appears in ordinances but must be inferred from its special procedural requirements.

(Meshenberg 1976)

A "floating zone" is a use which is provided for in the zoning ordinance, but which is not located on a specific site on the zoning map. It is "unmapped" and "floats" until the local legislative body determines where it shall settle.[12] The device is particularly useful for large developments which cannot be readily foreseen, which are acceptable, but over which more than the usual controls are considered desirable. Examples are large commercial or mixed uses and planned unit developments.

Though similar to a conditional use, a floating zone is different in a number of significant ways:

First, the fact that floating zones can be established at various locations in a municipality permits greater flexibility than the conditional use procedure. Second, conditional uses are the product of administrative action while floating zones are enacted by the legislature. Therefore, a conditional use generally must be granted if a landowner meets the conditions set forth in the ordinance. Establishment of a floating zone, on the other hand, is discretionary and can be conditioned upon compliance with additional restrictions. Furthermore, a floating zone alters the zone boundaries of an area by carving a new zone out of an existing one.

(Rathkopf and Rathkopf 1988: 27.06, fn 2)

A floating zone is also of use when a municipality wishes to allow the provision of lower-cost housing at high density without pushing up the price of land. Designating particular areas for high density housing can lead to major increases in land prices, with the existing landowners capturing a "windfall" that the municipality intended to go to the benefit of the lower income housing – a type of internal subsidy (R.A. Williams 1985: 18). In a 1977 study carried out in Princeton, New Jersey, it was shown that land in a district zoned for two-acre large lot development was valued at $6,000 per acre, but could be expected to increase to nearly $28,000 per acre if the unit density was increased to 4.2 units per acre (New Jersey 1977). "Thus, pre-designation of a density bonus district, usually required under Euclidian zoning paradigms, may seriously impede municipal efforts to capture the increase in land value to use as leverage in inducing the developer to subsidize lower income housing" (R.A. Williams 1985: 18).

The courts have held that the use of floating zones, if provided for in a comprehensive zoning plan, does not constitute spot zoning.[13]

Nonconforming Uses

As the years go by and as present nonconforming uses are abandoned, the benefits of zoning will become even more apparent. Then, America shall not continue to see homes and tire repair shops, apartments and stores, factories and houses intermingled and clustered. Good housekeeping for municipalities as in homes, will find an orderliness where industry, business, and houses will each have their proper places; each benefiting therefrom, and the community being much better as a result thereof.

(Metzenbaum 1955)

Because of the simple fact that most nonconforming uses (such as a store or a filling station in a residential district) have a high earning capacity of a well-situated monopoly created and protected by law, nonconforming uses typically have had an extraordinary vitality.

(Donovan 1962)

The introduction of a zoning scheme presents obvious problems with regard to existing uses which thereby become nonconforming. It is impracticable to have these uses removed: indeed, any such threat would have been sufficient to kill off any idea of zoning. The general approach taken has been to hope that, in time, the nonconforming uses would pass away. This has typically proved not to be the case, and municipalities have strived to find ways to speed up the process. They have had little success.

The courts have been unsympathetic to municipalities which attempt to "zone retroactively," though some of the early landmark decisions (such as *Hadacheck*, which is discussed in Chapter 2) apparently provided the constitutional basis where the offending use became a nuisance. The most common method of applying a control over nonconforming uses (limited though it is) forbids "expansion" or "alteration." Sometimes, a restriction is imposed on rebuilding if a nonconforming use is "destroyed," for example by fire. Moreover, a use which is "abandoned" may be refused permission for resuscitation (after a certain number of years). Unfortunately, all these terms have been – and continue to be – subject to intense debate and judicial differences. Perhaps the clearest case in which a nonconforming use may be eliminated is where a billboard is amortized over a number of years. Amortization is sometimes seen as the most painless way of ridding an area of an undesirable use, and the courts have been sympathetic, particularly where the nonconformer is given a reasonable amount of time. However, the political problems remain, as is dramatically illustrated by the success of the billboard lobby in preventing the use of amortization in connection with the federal highway advertising program. Not only was amortization prevented, but the Act actually requires the payment of compensation for the removal of billboards (a matter which is discussed further in Chapter 8). There could be no clearer example of the force of politics in land use planning.

Zoning Amendments

If there is anything to be learned from the history of zoning to date, it is that development tends to occur more through a series of modifications in the preestablished rules than through an automatic satisfaction of them.

(American Law Institute 1968)

A zoning amendment (or "rezoning," or "map amendment") is similar to a use variance in that it permits a use which is not allowed by the provisions of the zoning ordinance. However, while a use variance grants the owner an exemption (and leaves the ordinance intact), an amendment changes the ordinance itself. An amendment should be of greater consequence than a variance but, as already noted, practice does not always conform to theory, or even legality.

Amendments can be made to the zoning ordinance or to the map. The former deals with the written provisions of the ordinance; the latter with its detailed designation on a map of the area. The most common is a map amendment which allows a more intensive use of a particular area. Such an "upzoning" is usually to a more profitable use, and it is typically made in response to a request by the landowner. It is also regularly opposed by nearby residents: a more profitable use for an owner (for example, an increase in the permitted density of development; multifamily instead of single family houses; or a commercial instead of a residential use) can arouse fears of unwelcome neighbors and a fall in property values – commonly expressed in terms of "a change in the character of the area." Such rezonings are characteristically on a piecemeal basis:

> The desire of nearby property owners to maintain neighborhood lands in lower intensity use and the desire of local government officials to get reelected virtually guarantees, as a matter of politics, that a local government body will rarely upzone land on a comprehensive basis. Rather, the tendency will be to upzone land only on the specific request of the property owner seeking rezoning and, then, only if the upzoning is not met by too much neighborhood resistance or is thought necessary to defeat a potential claim of confiscatory zoning by a landowner.
> (Rathkopf and Rathkopf 1988: 27.02)

There are, however, circumstances in which a local government might "upzone" on its own initiative, as for example, "where a town is seeking to increase its tax base by attracting industry or development, or in response to dramatically changed circumstances which rendered the prior designation inappropriate, for example, in the case of a newly constructed highway" (Rathkopf and Rathkopf 1988: 27.02).

The Standard State Zoning Enabling Act does not provide any mechanism for a zoning amendment: the amendment typically has to go through the same process as an original zoning ordinance. This is rather curious; as Kelly notes:

> That a map amendment, or rezoning, is a legislative act raises troublesome issues. The rezoning applicant and interested neighbors view the rezoning application as similar to subdivision or liquor license applications, which are reviewed in accordance with established standards, and approved or disapproved on an objective basis. But the rezoning "application" is a request to change the law, and in changing the law the governing body can do what it likes, subject only to broad constitutional and state legislative limitations. Because the granting or denial of rezonings is generally performed without reference to significant standards, it is often unpredictable and can be unfair. Zoning law is a remarkable anomaly in this respect: in no other area of American administrative law are requests for amendments to the law so frequent that there are specific application forms and fee schedules for requesting the amendments.
> (Kelly 1988: 258)

Though most courts hold that a zoning map amendment is a legislative act, some take the opposite view, which implies that it is judicially reviewable. The Oregon courts are particularly well known on this account. The leading case is *Fasano* (noted in the Introduction), where the court found that the determination of whether the use of a specific piece of land should be changed was an exercise of *judicial* authority. The case is discussed in all the standard legal texts.[14] Callies, Neuffer, and Caliboso hold that the argument in support of rezoning as a quasi-judicial act is a strong one:

> First, rezoning amounts to the implementation of policy, usually in connection with a particular parcel of land with one owner. Thus, it more closely resembles the granting of a permit than the enacting of a general law. Second, and closely related to the first point, rezonings apply to a small group of people. Only rarely does much of a legally-cognizable interest exist, except perhaps that of some neighbors, beyond that of the property owner. Third, the courts that have characterized rezonings as legislative appear to be engaging in judicial overreaching in order to subject zoning to the initiative process. Typically, such courts are located in states with a long history of ballot box initiative and referendum measures, and in which their state constitutions reserved to the people the power of initiative and referendum. Even pro-initiative and referendum commentators have firmly drawn the line on rezonings, preferring the *Fasano* test which makes such rezonings referendum-proof.
> (Callies, Neuffer, and Caliboso 1991: 83)

This is a complex matter, on which there is much disagreement. Its importance stems, of course, from the fact that the courts have a major role to play in certain zoning matters. Definitions thus are important.[15] But the substantive issues concerning the rezoning of the land in question are ignored. So it is with much of the law of zoning.

Special District Zoning

> More and more, the response to a particular localized problem has been a particular "fix" without regard to uniformity, comprehensiveness, or any of the other theoretical principles of sound planning and plan implementation.
> (Babcock and Larsen 1990)

The term "special districts" may be confusing since it has more than one meaning. Traditionally, special districts are governmental units established to perform specific functions which, for one reason or another, cannot be performed by the existing general purpose local governments. Examples from the nineteenth century are the toll road and canal corporations. Today there are special districts for education, social services, sewage, water supply, and natural resources. (Perhaps the most famous is the Port Authority of New York and New Jersey.) Special districts which are so designated for zoning purposes, however, are very different. These are areas to which an amendment of the zoning ordinance applies: they thereby become subject to "special" zoning controls.

In an interesting and illuminating monograph on special districts, Babcock and Larsen (1990) examine their contemporary use in a variety of contexts including, in New York, the Theater District (designed to preserve the area as such by forcibly bribing developers to build new theaters); the Special Fifth Avenue District (designed to stop the influx of banks and airline offices, and to encourage profitable residential uses above the stores); the Special Garment Center District (designed to prevent the conversion of manufacturing space to office uses and to safeguard the garment industry – which covers 8 million square feet); and the ill-conceived Special Little Italy District (designed to preserve the Italian character of the community, in disregard for the Chinese residents). San Francisco has sixteen special districts, several of which are Neighborhood Commercial Special Districts. This designation is applied to relatively small commercial corridors in residential areas, with the objective of "preserving upper-floor residential units in commercial buildings, and keeping fast-food restaurants from taking over the street."

Chicago has a generic special district: the Planned Manufacturing District, designed to prevent the loss of industrial and manufacturing land to residential and commercial uses. This can be applied wherever it is needed, i.e. wherever the local electors pressurize their alderman for one.

In all cases, the intention is to shield the area from market forces. There is nothing "special" in this: much of American zoning is essentially of this protectionist and exclusionary nature. The curiosity of "special" districts is that most of them have little that is special about them. They are used as a means of dealing with local political problems. The residents complain about unwelcome changes, or the threat of changes, and the zoning authority responds by giving them a "special" status. What appears to be "special" is the large degree of citizen involvement, not only in the designation of the area but also in enforcement:

> The residents get a psychological lift from residing in an area that has a tag to it. . . . They know the special regulations; some of them know the twists and bends of the provisions of their districts as well as the lawyers do and probably better than most of the administrators of the ordinance. They become, as Norman Marcus put it, "zoning freaks". Their zoning is the one part of the hopelessly complex myriad of municipal laws and policies that city residents believe they can understand. . . . They can immediately spot a sign that violates the regulations of their special district or quickly detect a commercial establishment that operates in a way that is in violation of the labyrinthine district regulations.
>
> Thus it appears that the professionals are losing a zoning conflict to the amateurs, a not unheard of event in the zoning arena.
> (Babcock and Larsen 1990: 97)

A particularly revealing case is the Special Yorkville–86th Street Special District in New York, more popularly known as the "Anti-Gimbels District." This was created as a result of the reactions of residents to a new 400,000 square-foot Gimbels store. The provisions of the zoning ordinance amendment included the imposition of a maximum frontage of twenty-five feet for any new establishment. A majority of the New York Planning Commission believed that this and other restrictions "would halt the decline of the area enough so that the locals could still enjoy a good wiener schnitzel or a cool stein of beer in their favorite watering place for some time to come" (Babcock and Larsen 1990: 98). But the ideas were flawed. The only uses which could squeeze into a twenty-five foot frontage while at the same time paying escalating rents) were fast-food outlets – and they proliferated.

The attempt to avert change failed: in fact all that the commission had done was to accelerate the advent of fast-food shops, and hasten the demise of ethnic businesses. Despite widespread agreement that the scheme was a failure, it took fifteen years to overcome local resistance to the abolition of the special district. Babcock and Larsen's book is replete with such telling examples of misplaced support for citizen initiatives.

It seems clear that special districts are being used in areas where they have no justification; and, once established, their popularity with the citizenry makes abolition extremely difficult. Moreover, they involve a fragmentation of a city's planning policy. They are not determined on the basis of a comprehensive plan (though there is no reason, other than the rarity of comprehensive plans, why they should not be). They arise simply as a result of vociferous citizen pressures.

They have another important characteristic: they run counter to market forces. Though this is by no means necessarily undesirable, it is a matter that requires analysis stretching further than the assessment of the degree of local anger.

But all this is equally true of American zoning generally. It is planless, rudderless, fragmented, uncoordinated – the very opposite of planning. Babcock and Larsen propose that special districts be permitted only where they meet specific requirements which should be written in the zoning ordinance. This might relieve "the chaos surrounding the creation and use of special zoning districts" (Babcock and Larsen 1990: 145). But it would have little impact on the more general problems. The authors recognize this when they suggest that "probably the only real solution is for the state government to reassert its latent power over planning and zoning, and insist that every city have a plan and a zoning ordinance both consistent and in compliance with state policies" (Babcock and Larsen 1990: 145). This is an issue which is discussed at length later.

Conclusion

> The ability of zoning battles to turn ordinarily reasonable people into wide-eyed fanatics is a never-ending source of amazement.
>
> (Crawford 1969)

It is abundantly clear that zoning is not the rigid, simple system of land use regulation that it is supposed to be. In fact it is not rigid: it displays remarkable flexibility. It is not simple: it is increasingly complex. And it is not, in any true sense of the term a system: it is a bag of tricks used in a locally operated land use regulatory process. Above all, it is not used as an instrument of public policy to serve the public interest. Zoning began as a device for exclusion; and it has continued its exclusionary pathways ever since. An early instance was the zoning system prepared for Atlanta by the nationally prominent planning consultant, Robert Whitten:

> His scheme, which met with little opposition in Atlanta, was said to have been the first to

enact segregation along straight social lines by dividing the residential districts into three types: white, colored, undetermined. Without proposing that an anthropologist be appointed zoning administrator, the ordinance made it illegal for whites and colored to move to sites in each other's districts. To prevent excessive rigidity, however, the law also provided that a white employer might maintain colored servants in residential quarters on his building lot. And to give the whole thing a touch of even-handed fantasy, it guaranteed the colored resident in his district the same rights with respect to his white servants. Whitten defended the arrangement on two grounds. Prior to zoning, he said, racial segregation had existed, but it was achieved in a way which both aggravated race hatred and brought economic loss because of the uncertain future of much undeveloped land. Now there would be room for all, and on a predictable basis. The threat of race riots had been removed. That was the policy argument. The empirical one was that "race zoning . . . is simply a common sense method of dealing with facts as they are."

> (Whitten 1922: 418)

This was by no means a unique case, particularly in the southern states (Rabin 1989: 106). Louisville, for example, had an ordinance "to prevent conflict and ill-feeling between the white and colored races in the city of Louisville, and preserve the public peace and promote the general welfare by making reasonable provisions requiring as far as practicable, the use of separate blocks for residences, places of abode, and places of assembly by white and colored people respectively." In *Buchanan v Warley* (1917) the Supreme Court declared this to be unconstitutional, rejecting several arguments including the contention that the acquisition by colored people of houses in areas occupied by whites would depreciate property values. Though it did not deny that this might happen, the court noted that "property might be acquired by undesirable white neighbors, or put to disagreeable though lawful uses with like results." Interestingly, the court drew a distinction between this case and the landmark 1896 *Plessy v Ferguson* case where it was decided that "separate but equal" accommodation for white and colored railroad passengers did not run counter to the constitution: in that case there was a "basis of equality for both races."

Such racial opposition has a long and lamentable history. Indeed, the tradition was well established with the prezoning anti-Chinese cases in California, where municipalities, having failed to exclude Chinese openly, devised the simple alternative of excluding laundries from large areas: these were not only an obvious nuisance and fire risk, they also happened to be Chinese. In 1885 the city of Modesto ordained that: "it shall be unlawful for any person to establish, maintain, or carry on the business of a public laundry or wash house where articles are washed or cleansed for hire, within the city of Modesto, except that part of the city which lies west of the railroad track and south of G Street." As Delafons (1969: 20) comments, the "wrong side" of the tracks was thus given statutory definition in the earliest land use controls.

The discussion so far has shown that zoning techniques have become increasingly sophisticated (or at least complex), but their basic exclusionary character remains as strong as ever.[16] The three chapters in Part II will demonstrate how the exclusionary ethic has taken on new challenges such as pressures for affordable housing, and also explore new avenues of exclusionary tactics such as charging developers for the costs which they impose on the existing community.

Part II
Planning control, charges, and agreements

Chapter 5

Exclusionary zoning and affordable housing

Introduction

> Exclusionary zoning: local land use
> controls that have the effect of excluding
> most low income and many moderate
> income households from suburban
> communities and, indirectly, of excluding
> most members of minority groups.
> (Mandelker and Cunningham 1985)

Though all zoning is by definition exclusionary, some is more exclusionary than others. Some blatant examples are given in Chapter 3, and there are references to more in several other chapters. In this chapter, the focus is on measures to combat exclusionary zoning and to facilitate, through the zoning process, the building of affordable housing.[1] The first part of the chapter is devoted to governmental attempts to legislate against exclusionary policies. Though a separate "policy area" from land use regulation, the federal legislation designed to prevent, or at least to discourage, discrimination operates alongside efforts to make land use controls nondiscriminatory. This has assumed greater importance in recent years. However, it is a large and complex field of law which cannot be adequately dealt with here, and only a brief reference is possible.

Particular attention is paid to the renowned *Mount Laurel* cases, in which the powers of the judiciary to enforce the provision of affordable housing were stretched to the limit. This was a New Jersey case and therefore does not have the "universality" of a U.S. Supreme Court opinion, but it is of particular interest in demonstrating the extent to which the courts can (and cannot) impose their views on local and state governments. Finally there is a discussion of the inclusionary and "linkage" policies followed by a number of municipalities. These are mechanisms available to local governments to encourage, or even compel, the provision of affordable housing by private developers. This raises some difficult questions about the "sale" of development permissions. As with so many similar issues, there are no easy answers, and continued debate can be guaranteed.

The Pervasiveness of Exclusionary Zoning

> The purpose to be accomplished is to regulate the mode of living of persons who may hereafter inhabit [the village of Euclid]. In the last resort, the result to be accomplished is to classify the population and segregate them according to their income or situation in life.
>
> (Judge Westenhaver, *Euclid* 1924)[2]

From the date of the Supreme Court case *Euclid* (1926) to the 1960s, the exclusionary practices of municipalities received little critical attention from the courts or from federal or state governments. The presumption of validity (which essentially gives municipalities the benefit of any doubt) held sway. At the same time, policies in relation to both land use and housing were predominantly local in character. There was little concern for regional needs. Municipalities were "tight little islands" (Sager 1969). Those needs of the exploding cities which could be met by low-density housing were welcomed by suburban municipalities, but low-income housing was nobody's responsibility – in practice even if not in theory.[3] There were, of course, numerous critics,[4] but they had little political impact.

During the 1960s, however, various attempts were made to promote the provision of low-cost suburban housing. For example, the New York State Urban Development Corporation was established in 1968 with powers to override local exclusionary policies.[5] In the following year Boston passed its "anti-snob" law which provided a process of appeal for developers against obstructive municipalities.[6] Both Acts failed in their purpose because of implacable opposition from suburban municipalities (Babcock and Bosselman 1973). Furthermore, a number of states began introducing planning requirements for the zoning of land for low-income housing. California was one of the first

(in 1970); it required municipalities to provide "regulatory concessions and incentives" to enable moderate-cost housing to be built. Later legislation introduced zoning bonus incentives for the building of low-cost housing (Schwartz and Johnston 1983: 5).

At the federal level, there were several initiatives aimed at securing comprehensive planning – "comprehensive" in terms of both the issues and the area covered. Section 701 of the 1954 Housing Act (which had authorized assistance to small communities in the development of plans) was broadened to encourage the establishment of areawide planning agencies such as councils of governments and county planning boards. The 1966 Demonstration Cities and Metropolitan Development Act required regional planning agencies to "comment" on applications for certain federal grants: "such comments shall include information concerning the extent to which the project is consistent with comprehensive planning." These review requirements were extended by the 1968 Intergovernmental Cooperation Act to cover a wider range of federal programs. The Bureau of the Budget issued Circular A-95, which called for the designation of regional clearing houses and more extensive reviews of the relationship of local applications for federal grant aid to statewide or areawide comprehensive plans. A "housing element" was added to the requirements of section 701 plans, and the A-95 process was extended to cover civil rights. The housing element was as follows:

> Planning carried out with assistance under this section shall include a housing element as part of the preparation of comprehensive land use plans and this consideration of housing needs and land use requirements for housing in each comprehensive plan shall take into account all the available evidence of the assumptions and statistical bases upon which the projections of zoning, community facilities, and population growth are based; so that the housing needs of both the region

and the local communities studied in the planning will be adequately covered in terms of existing and prospective immigrant population growth.

(quoted in Rubinowitz 1974: 162)

Despite this spurt of federal activity, most of the good intentions were in vain and, as often happens in the American political system, it was left to the courts to cope with this sensitive issue. (In Franklin's words [1983b: 1], "we often transform political issues into legal issues so that they can be decided by judges.")

In the landmark *Euclid* case, though the district court held that the village's zoning ordinance was unconstitutional, the Supreme Court overruled and declared that the ordinance was a valid exercise of police power in the interest of the general welfare. For the next four decades most state courts followed *Euclid* on the basis that municipal zoning satisfied the general welfare. But the conception of the general welfare was a *local* one: no consideration was given to the regional implications of local decisions (Burchell *et al.* 1983: 3). As stressed in earlier chapters, attitudes differ between states, and changes in attitude differ even more. A few states, such as New Jersey and Pennsylvania, have been in the forefront of change.

In 1952 the New Jersey State Supreme Court affirmed a five-acre minimum lot size and minimum floor area requirements; in 1962 local prohibition of mobile homes was also upheld.[7] These decisions characterize the attitudes which were then general across the states, but signs of change were evident. There is no clear pattern: none could be expected in the absence of any federal Supreme Court decisions.

An early sign of impending change was the dissent of Justice Hall in the 1962 New Jersey case of *Vickers*. Though the majority opinion held that a municipality had the constitutional right to exclude mobile homes, Justice Hall expressed a dissent in strong and compelling terms:

Legitimate use of the zoning power by such municipalities does not encompass the right to erect barricades on their boundaries through exclusion or too-tight restriction of uses where the real purpose is to prevent feared disruption with a so-called chosen way of life. Nor does it encompass provisions designed to let in as new residents only certain kinds of people, or those who can afford to live in favored kinds of housing, or to keep down tax bills of present property owners. When one of the above is the true situation, deeper considerations intrinsic in a free society gain the ascendancy, and courts must not be hesitant to strike down purely selfish and undemocratic zoning enactments.

(*Vickers v Township Committee of Gloucester Township* 1962)

This view increasingly became the majority opinion. Thus in the 1965 case of *National Land*, the Pennsylvania Supreme Court found that a zoning ordinance establishing a four-acre minimum size lot was unconstitutional. In the 1970s there were growing numbers of such decisions.[8] In *Oakwood at Madison Inc. v Township of Madison*[9] the New Jersey Superior Court overturned Madison's zoning plan which almost completely excluded multifamily housing from the district and also stipulated minimum lot sizes of one and two acres. This was declared to be invalid on the grounds that it ran counter to *regional* housing needs:

In pursuing the valid zoning purposes of a balanced community a municipality must not ignore housing needs, that is its fair share proportion of the obligation to meet the housing needs of its own population and of the region.... Large areas of vacant and developable land should not be zoned as Madison Township has into such minimum lot sizes and with such other restrictions that regional as well as local housing needs are shunted aside.

(*Oakwood at Madison Inc. v Township of Madison* 1971)

This concern for "fair share" housing was becoming widespread at this time.[10]

Meanwhile, efforts were being made on the professional front to devise fair share housing programs – by planners such as Dale Bertsch in Dayton and Trudy McFall in the Twin Cities.

> Bertsch's "Dayton Plan," conceived when he was director of the Miami Valley Regional Planning Commission, included an allocation formula by which each locality in the region would take its allotted share of low and moderate cost housing. The idea was to offset the concentration of such housing in Dayton, the central city of the region.
>
>   Within a short time, the plan was known from coast to coast, and HUD officials in both the Ford and Carter administrations responded to Bertsch's entreaties to make it a national showcase. The approach was crystalized in the requirement for Areawide Housing Opportunity Plans (AHOPs) by HUD Assistant Secretary Robert Embry and, later, in the Regional Housing Mobility Plan by Trudy McFall, who directed HUD's planning office from 1977 to 1981. At HUD, she infused the housing and planning programs she directed with the inclusionary doctrine enunciated in *Mount Laurel*, making it possible for lawyers to cite the intent of federal programs in litigation against exclusionary land use controls.
>
>   (Erber 1983: 8).

The AHOPs program (supported by only minimal funding) never had more than a marginal effect: "The combination of the nearly inexorable workings of economic trends, once set in motion, coupled with the underlying values and attitudes of large parts of American society, have to date largely neutralized this entire aspect of American housing policy" (Mallach 1984: 5).

The issue of fair shares arose to prominence in the *Mount Laurel* case: a dramatic illustration of how little had been accomplished by judicial decisions over the previous decade.

Mount Laurel I

> Nobody thought that *Mount Laurel* was going to be an important case.
>   (N. Williams 1984a)

> The *Mount Laurel* case ... threatens to become infamous. After all this time, ten years after the trial court's initial order invalidating its zoning ordinance, Mount Laurel remains afflicted with a blatantly exclusionary ordinance.
>   (*Mount Laurel II* 1983)

With the exception of *Euclid*, there is perhaps no better known zoning case than *Mount Laurel*. It is, of course, far more than a single case: it is more appropriately described as a saga which, starting in the 1960s, continues to unfold.[11] A large number of cases fell under the *Mount Laurel* umbrella, but the beginnings of the story start well before the first *Mount Laurel* decision of 1975 (Erber 1983). Indeed, given the rapid and widespread suburbanization of New Jersey since the end of World War II, coupled with the state's extreme balkanization of municipal government, wedded to the principle of home rule, it is not surprising that it became one of the first battlegrounds in exclusion. Another factor was also relevant: the tradition of a highly qualified and independent judiciary (Franklin 1983a: 10).

Mount Laurel is a most pleasant area, and it has attracted large numbers of people from nearby Philadelphia and Camden. Its population doubled between 1950 and 1960 (from 2,800 to 5,200), doubled again between 1960 and 1970, and grew by another half by 1985 (to 17,600). In common with many of the 567 municipalities in New Jersey,[12] Mount Laurel

imposed minimum lot size and other restrictions. There was no doubt as to the intention, as well as the effect, of these: it was to keep out low-income families and other undesirables. Concerted effort by activist groups of academics, lawyers, and others (including Paul Davidoff's Suburban Action Institute, the National Committee Against Discrimination in Housing, and the stalwart lawyer, Norman Williams) had already been taken against other municipalities, including Madison, which proved – in legal, if not in practical terms – a success (*Oakwood at Madison* 1971). Mount Laurel, with a blatantly exclusionary zoning policy, was an obviously important target, and in 1975 a group of poor Black and Hispanic residents of the township, joined by various public interest groups and led by the local NAACP, brought a case against the township[13]. The case eventually found its way to the New Jersey Supreme Court – with the defendants attempting to discredit the plaintiffs and challenge their right of standing, by alleging that they lived in "dilapidated or substandard housing." Unfortunately for Mount Laurel, the New Jersey courts have a liberal attitude with regard to standing in exclusionary zoning cases – unlike the U.S. Supreme Court and most state courts, which have generally held that, in order to have standing to challenge the validity of a regulation, the party must be able to show that a personal legal right has been invaded (Mandelker and Cunningham 1990: 336).

The *Mount Laurel I* case was a landmark. For the first time, a court proclaimed the doctrine that a municipality's land use regulations had to provide a "realistic opportunity for the construction of its fair share of the present and prospective regional need for low and moderate income housing." This new doctrine was based on the New Jersey constitution, which provides that "all persons are by nature free and independent, and have certain natural and unalienable rights, among which are those of enjoying and defending life and liberty, *of acquiring, possessing, and protecting property*, and

of pursuing and obtaining safety and happiness" (emphasis added). (Since the court relied exclusively on the New Jersey constitution, there was no likelihood of a review by the U.S. Supreme Court; and, in fact, the Supreme Court refused to hear an appeal from the township.)

Much litigation followed the court's opinion and there were varying interpretations of what it really meant. Indeed, far more time and effort was spent on this than on actually doing anything substantively. Additionally, Governor William Cahill sought to encourage municipalities voluntarily to increase opportunities for lower-income housing within their areas.[14] Nothing came of this, or of the efforts of developers. The court had not provided any effective remedy, and developers were largely powerless against municipal stalling tactics. Mount Laurel's response to the court was, to put it mildly, niggardly: it rezoned twenty acres on three widely scattered plots owned by three separate individuals, two of whom were not even residential developers. Its zoning ordinance entirely prohibited the construction of apartments, townhouses, and mobile homes.

Mount Laurel II

Judicial function includes whatever legislators do not have the guts to deal with.
(N. Williams 1984a)

Under the state's constitution, the need for adequate housing for all categories of people is so important and of such broad public interest that the general welfare which municipalities like Mount Laurel must consider extends well beyond their boundaries.

(*Mount Laurel II* 1983)

It became inevitable that the issue would have to return to the courts. In fact, six cases involving challenges to separate suburban exclusionary zoning ordinances were consolidated and heard by the Supreme Court of New Jersey at the end of 1980. Mount Laurel was "the most symbolically significant" of the cases (R.A. Williams 1985: 7), and they are collectively known as *Mount Laurel II*.

The situation – and the court's approach – was neatly summed up by the court in these words:

> Papered over with studies, rationalized by hired experts, the ordinance at its core is true to nothing but Mount Laurel's determination to exclude the poor. Mount Laurel is not alone; we believe that there is widespread non-compliance with the constitutional mandate of our original opinion in this case. To the best of our ability, we shall not allow it to continue. This Court is more firmly committed to the original *Mount Laurel* doctrine than ever, and we are determined, within appropriate judicial bounds, to make it work. The obligation is to provide a realistic opportunity for housing, not litigation. We have learned from experience, however, that unless a strong judicial hand is used, *Mount Laurel* will not result in housing, but in paper, process, witnesses, trials, and appeals. We intend by this decision to strengthen it, clarify it, and make it easier for public officials, including judges, to apply it.
> (*Mount Laurel II* 1983)

In short, the court saw the necessity for introducing effective means for enforcing the *Mount Laurel* doctrine, and it boldly and unanimously reinforced its earlier decision establishing an affirmative inclusionary zoning obligation requiring municipalities in New Jersey to adopt land use regulations that would permit a realistic possibility for a fair share of housing opportunities for lower income households, based on the regional need for such housing (Franklin 1983b: 1). The opinion is a quite extraordinary one in its scope and the role which it took on

for the judiciary. It is also characterized by its great length – 120 pages – and the long delay in its issuance – twenty-six months (N. Williams 1984a: 838). The object of the opinion was stated in these terms:

> First, we intend to encourage voluntary compliance with the constitutional obligation by defining it more clearly. We believe that the use of the State Development Guide Plan and the confinement of all *Mount Laurel* litigation to a small group of judges ... will tend to serve that purpose. Second, we hope to simplify litigation in this area. While we are not overly optimistic, we think that the remedial use of the State Development Guide Plan may achieve that purpose, given the significance accorded to it in this opinion. Third, the decisions are intended to increase substantially the effectiveness of the judicial remedy. In most cases, upon determination that the municipality has not fulfilled its constitutional obligation, the trial court will retain jurisdiction, order an immediate revision of the ordinance (including, if necessary, supervision of the revision through a court appointed master), and require the use of effective affirmative planning and zoning devices. The long delays of interminable appellate review will be discouraged, if not completely ended, and the opportunity for low and moderate income housing found in the new ordinance will be as realistic as judicial remedies can make it. We hope to achieve all of these purposes while preserving the fundamental legitimate control of municipalities over their own zoning, and, indeed, their destiny.
> (*Mount Laurel II* 1983)

Mount Laurel II was a remarkable case. It was almost tantamount to a usurpation of municipal powers (it was certainly attacked as such by both local and state government). However, the court was at pains to stress that, though it would have preferred not to take extreme measures, it had no alternative:

> A brief reminder of the judicial role in this sensitive area is appropriate, since powerful reasons suggest, and we agree, that the matter is better left to the Legislature. We act first and foremost because the Constitution of our State requires protection of the interests involved and because the Legislature has not protected them. We recognize the social and economic controversy (and its political consequences) that has resulted in relatively little legislative action in this field. We understand the enormous difficulty of achieving a political consensus that might lead to significant legislation enforcing the constitutional mandate better than we can, legislation that might completely remove this Court from those controversies. But enforcement of constitutional rights cannot await a supporting political consensus. So while we have always preferred legislative to judicial action in this field, we shall continue – until the Legislature acts – to uphold the constitutional obligation that underlies the *Mount Laurel* doctrine. That is our duty. We may not build houses, but we do enforce the Constitution.
>
> (*Mount Laurel II* 1983)

In such magisterial terms did the court make it clear both that it intended to enforce the *Mount Laurel* doctrine and that it was now up to the Legislature to tackle the difficult political problem from which it had so far shied.

Not surprisingly, the reaction against the court's opinion was vociferous, and there was a great deal of activity aimed at both the implementation and the obstruction of the court's wishes. One particularly symbolic act was undertaken by Governor Kean who, within days of the *Mount Laurel II* decision, dismantled the Department of the Division of State and Regional Planning and abolished the State Development Guide Plan for which it was

responsible – and which the court used in fashioning its decision (Stockman 1986: 581).

However, a factor of great importance was that *Mount Laurel II* came at a time when interest rates were falling and there was an upsurge in the housing market. "Because market-rate housing development was again profitable, builders were more than willing to abide by the requirements of *Mount Laurel II* and dedicate one lower-income housing unit for every four market-rate units they produced. Moreover, the prospects of such profits led developers to sue municipalities which had denited them the right to build at market rates, and to offer to build the four-to-one ratios" (McDougall 1987: 630).

Large numbers of law suits were brought under *Mount Laurel II* to force recalcitrant municipalities to accept the "builder's remedy." Twenty-two lawsuits, involving 14,000 lower-cost housing units, had reached virtually final settlement by January 1986, when *Mount Laurel III* (discussed below) was argued.[15]

The growing effectiveness of *Mount Laurel II* enforcement by the courts increased the political pressures by the municipalities on the State of New Jersey to take control of the matter of the provision of lower income housing – which the court had strongly urged it to do. After months of argument and compromise, including the adoption of the amendments to Governor Kean's conditional veto (which significantly weakened the legislation), the Fair Housing Act finally became law.

The New Jersey Fair Housing Act

> The Legislature declares that the statutory scheme set forth in this Act . . . satisfies the constitutional obligation enunciated by the Supreme Court.
>
> (New Jersey Fair Housing Act 1985)

I would rather go to jail than allow my
community to be overrun with *Mount Laurel*
housing.

(New Jersey Senator Peter Garibaldi 1987)[16]

New Jersey's Fair Housing Act was designed to
retrieve the role which the court had usurped in
relation to the provision of lower income
housing. It was intended, above all, to "disarm"
the judiciary (Kent-Smith 1987: 943). This
achievement – against strong opposition in the
Legislature and from the Governor – was made
possible only by crafting the provisions of the
Act in such a way that it was more acceptable
(or, to be more precise, less unacceptable) than
the system being operated by the court. Above
all, the "judicial monster in the form of a *Mount
Laurel II* lawsuit" was replaced by a system of
voluntary compliance. This, of course, implied
a considerable weakening in the system of
control over exclusionary zoning.

The Act established a Council on Affordable
Housing, charged with carrying out the *Mount
Laurel II* obligation. It also provides for a mora-
torium on builder's remedies, and a transferring
of most existing (and future) suits to the
Council. Transfer is dependent upon a municipality including in its zoning ordinance a
"housing element" which contained a fair share
plan. If this is acceptable to the Council it is
approved (certified), and the municipality is
then shielded from successful builders' suits –
unless a builder can provide "clear and
convincing evidence" that the zoning ordinance is exclusionary.

As a part of its fair share plan, a municipality
can transfer up to half of its fair share obligations to another municipality. In early 1991,
twenty-one suburban communities were negotiating such transfers to sixteen older urban
areas such as Newark. Over 3,000 dwellings,
costing some $60 million, were involved.[17]

Though accurate figures are difficult to
collate, the following notice, published in
March 1991, provides the best available indication of progress under the Fair Housing Act:

Since passage of New Jersey's 1985 Fair
Housing Act . . . between 8,000 and 12,000
lower income housing units have been
constructed in the state, according to
Douglas Opalski of the New Jersey Council
on Affordable Housing. Nearly half of the
state's municipalities are in the process of
complying with *Mount Laurel* – which
ultimately could add another 41,000 units.

(Gallagher 1991: 15)

The Fair Housing Act tips the balance of
advantage greatly in favor of a municipality. If
it can satisfy the Council that it is making an
appropriate fair share allocation (under provisions much less demanding than under the
builder's remedy system), the only risk it faces is
that an aggrieved builder will go to the time
and expense of appealing to the court. Few are
likely to do this for reasons which are clear from
an examination of a number of cases which
have been labelled *Mount Laurel III*.

Mount Laurel III

Most objections raised against the Act
assume that it will not work, or construe its
provisions so that it cannot work, and
attribute both to the legislation and to the
Council a mission, nowhere expressed in the
Act, of sabotaging the *Mount Laurel*
doctrine. On the contrary, we must assume
that the Council will pursue the vindication
of the *Mount Laurel* obligation with
determination and skill. If it does, that
vindication should be far preferable to
vindication by the courts, and may be far
more effective.

(*Mount Laurel III* 1986)

Recalcitrant municipalities that have
opposed their fair share through bad faith

opposition and noncompliance have been amply rewarded.

(Kent-Smith 1987)

All the cases involved the issue of the transfer of litigation to the Council's jurisdiction, and the court ruled that all were properly transferable. Thus the builder's remedy is, in practice, not available, even as a last resort. Kent-Smith (1987: 948) has termed this "the court's retreat from responsibility in enforcing this constitutional obligation." At the time of writing, the *Mount Laurel* doctrine is in limbo. It remains to be seen whether opposing municipalities, or the Legislature, will eventually kill it, or whether the court will be spurred into a new round of activism.[18]

The omens are certainly not good. This is not the place to document them, but one further illustration of the resistance that can be expected is given by the way in which the Council on Affordable Housing quickly devised tactics familiar to the battle-scarred but winning veterans of *Mount Laurel*. The illustration concerns the boundaries of the regions within which "fair shares" of affordable housing are supposed to be provided:

> The Council exuberantly followed the legislature's mandate that a housing region consist of no more than four cities, and it has drawn housing region lines so that the greatest sources of need (in the older, urban counties) are walled off from the greatest sources of supply (in the rapidly growing central part of the state).

Such bureaucratic games are part of political maneuvering which is characteristic of this uneven battle. The system demonstrates the skill with which the opponents on affordable housing can nullify any attempts to make them do what they absolutely refuse to do.

The *Mount Laurel* saga, though dramatic, was restricted in its impact (small though that was) to the northeastern states. More universal, though no more encouraging, was the 1977 case in *Arlington Heights*, a suburb of Chicago.

Arlington Heights and the Fair Housing Act

> Official action will not be held unconstitutional simply because it results in a racially disproportionate impact.... Proof of racially discriminatory intent or purpose is required.

(*Arlington Heights* 1977)

The limits which courts will impose upon themselves are particularly clear in the *Arlington Heights* case. This concerned a proposal by a religious order, which owned eighty acres of land near the center of the village and proposed to develop this for a federally subsidized multifamily, racially integrated housing project. A rezoning was necessary, which, following fierce public opposition (only twenty-seven of the village's 64,000 residents were black), was denied by the local plan commission.

The case came before the courts and eventually to the Supreme Court of the United States. There are a number of complexities (as usual) which are ignored here: the essential point is that the Court held that there simply was not sufficient proof that discrimination was a motivating factor in the village's decision. As the quotation at the head of this section illustrates, it is not sufficient to show discriminatory results: it is necessary to show intent.

The case legally was concerned with the equal protection clause of the Constitution, and the decision was technically correct. But on any broader approach, it gave support to blatant discrimination. That this is so is demonstrated by a later development of the same case which was dealt with under the Fair Housing Act. This legislation differs significantly from the general constitutional requirements in that it requires proof only of a racially discriminatory *effect*. It was clear to the parties concerned which way the wind was blowing, and a settlement was reached out of court. The development went ahead (Callies and Freilich 1986: 583).

Clearly the specific provisions of the Fair Housing Act are a stronger tool to combat housing discrimination than the Constitution. Regrettably, they have not so far proved to be adequate.[19]

Inclusionary Zoning

The amendment in establishing maximum rental and sale prices for fifteen percent of the units in the development, exceeds the authority granted by the enabling Act to the local governing body because it is socio-economic zoning and attempts to control the compensation for the use of land and the improvements thereon.

(*Fairfax County v DeGroff Enterprises* 1973)

Another policy approach to housing discrimination is by way of "inclusionary zoning." The term "inclusionary" was initially used in relation to any housing and zoning programs which fostered the provision of housing for minorities and lower-income households (Franklin, Falk, and Levin 1974). The term derives from its much more common opposite, exclusionary zoning. Mallach (1984: 2), in a book which has become a classic in this field, uses the term in a narrower sense: "under an inclusionary housing program, the provision of housing for lower-income households becomes part of the overall residential development of the community, constructed as a direct outcome of the construction of more expensive housing, and in some cases, even as an outcome of nonresidential development." This "integrative" aspect is in striking contrast to the typical separated (even ghettoized) low-income developments. However, as is so usual in American planning, there is no agreement on a common set of meanings. For example, many so-called inclusionary programs allow for payments in lieu: which can produce segregated rather than integrated housing (Mallach 1984: 166). Thus

the reader will find that different techniques and policies are called by the same or similar names.

Nevertheless, the essential feature of inclusionary zoning is that it seeks the provision of lower cost housing whether by offering a developer a higher density in return ("incentive zoning" or "bonusing"), or by a mandatory requirement (which may or may not provide an incentive). Its purpose may be to increase the provision of lower-cost housing anywhere in the municipality, to "open up the suburbs" to lower-income households (Downs 1973) or in a particular area of predominantly higher-cost housing development (the integrative approach). Alternatively, it may seek to ensure the provision of housing in an area which is predominantly commercial, thus "bringing life back to the city after office hours" or increasing the supply of lower cost housing in an area which is being "gentrified."

Typically, however, the objective is simply to provide "affordable" housing. This has become more prominent as the cost of housing has increased more rapidly than incomes, thus presenting increasing numbers of middle-class households with an affordability problem which had previously been essentially confined to the poor (Frieden 1979; Johnson 1982). Thus the lower-income middle-class have come into competition with the poor for limited benefits. There are no prizes for correctly guessing who usually wins!

Inclusionary housing, like any U.S. housing policy, is liable at some point to stumble over the law. Objectors will complain that the policy is unconstitutional, or that the administration is unfair. There seems to be no limit to the barriers which opponents can erect against a body attempting to meet the needs of minorities. There is always a chance of winning: the quotation at the beginning of this section is from the opinion of a Virginia court, which (so far) no other court has followed.[20] In striking contrast is the *Mount Laurel II* decision of the New Jersey Supreme Court:

It is nonsense to single out inclusionary zoning . . . and label it "socio-economic" if that is meant to imply that other aspects of zoning are not. . . . Practically any significant kind of zoning now used has a substantial socio-economic motivation. It would be ironic if inclusionary zoning to encourage the construction of lower income housing were ruled beyond the power of a municipality because it is "socio-economic" when its need has arisen from the socio-economic zoning of the past that excluded it.

(*Mount Laurel II* 1983)

This is one aspect of the *Mount Laurel* cases that has become generally acceptable.[21]

The issue of nexus is theoretically as relevant and important in inclusionary zoning as it is in the levying of development charges. Thus charges are in principle imposed to relieve the municipality of some of the burdens caused by development: sewer and water systems, for example. Similarly, an inclusionary housing requirement is intended to relieve the municipality of some of the burden of providing for lower-income housing needs which have been increased by a particular development. Office and commercial development, so the argument goes, create an increase in the need for such housing. As an embellishment to this, it can be suggested that an upper or middle-income housing development should, in part at least, "support" an appropriate amount of lower-income housing. This is particularly so if the intent is to achieve a degree of social integration in residential districts.

Clearly there is here a nexus of a kind, but it is somewhat tenuous; and the imposition can readily be seen as a tax. The fact that developers may be prepared to pay this to save time (and therefore money) and to keep on good relationships with the municipality does not change the matter. In practice, the more successful schemes have included incentives as well as impositions. There is further discussion on this nexus issue in the following chapter,

when the matter of "charges" is dealt with more broadly.

California in the Lead

California now has the most stringent affordable housing and anti-exclusion statutes of the fifty states.

(Schwartz and Johnston 1983)

Californian municipalities have been in the lead in this field as with many other innovative planning techniques. State legislation requires that municipalities prepare a general plan. The mandatory coverage of the plan includes a housing "element" which "shall identify adequate sites for housing, including rental housing, factory-built housing, and mobile homes, and shall make adequate provision for the existing and projected needs of all segments of the community." Moreover, each locality has to assume its fair share of regional housing needs. In attaining this, use is made of the state density bonus program (California 1990a). Where a developer agrees to provide 25 percent of dwellings in a housing development for low- or moderate-income households (those with 80–120 percent of median county income), or 10 percent for low-income households (80 percent or less of median county income), or 50 percent of the units to be restricted to elderly residents (age 62 or older) the municipality must grant him either a density bonus of at least 25 percent or another incentive of equal financial value.[22]

Naturally, the municipalities of California have approached this with varying degrees of enthusiasm. Of particular note is Orange County, which offers incentives such as density bonuses, streamlined processing of development applications, a reduction in some development standards (especially for parking), and below-market interest financing (Schwartz and

Johnston 1983: 8; Bozung 1982). One of the most interesting aspects of the Orange County scheme is its transfer credit program:

> This program allows builders of five or more units who choose not to build affordable housing to purchase credits from other builders who have built and delivered more than their share of deed-restricted, price-controlled, low and moderate income units. These credits, each of which carries with it the right to build three market units in a development of over four units without the obligation to build low or moderate income units, have sold at prices ranging up to $15,000.
>
> (Hill 1984: 149)

Interestingly, a study by Stanford University School of Business Administration shows that it is possible for a developer to obtain a higher return from development which includes a bonus density provision than she can from a normal project. The key factor was that the bonus units were cheaper to construct (Fox and Davis 1976: 1028 and 1071; Schwartz and Johnston 1981: 62). Corroboratory results were obtained in a statewide survey in California (Johnston *et al.* 1989). Such results challenge the classic critique of Ellickson (1981).

Conclusion

> Deeply institutionalized racism perpetuates residential segregation.
>
> (Taeuber 1988)

The provision of housing for lower-income groups has never been a popular policy among the American electorate. Indeed, the first foray into this area by the federal government was declared to be unconstitutional.[23] As a result, the role of the federal government shifted from directly supplying housing to indirectly promoting production by subsidizing other agencies. Regrettably, for reasons which cannot be analyzed here, this provision has frequently fallen short of success, and fell into disfavor. The federal government's role shifted to the promotion of "fairness" in the private market and various subsidy schemes to make housing more affordable. All this made a logical link between land use and housing policies, but neither was well articulated and certainly not cordially received at the local level.

This chapter, like others in the volume, provides testimony to the weakness of government in the face of strong socioeconomic forces. Exclusionary zoning is prevalent because it is widely desired by those who have acquired, or are in process of acquiring, their share of the American Dream. Like all dreams, it is easily shattered: hence the vociferous opposition to any development which carries such a threat.

Mount Laurel is unusual in that it assumed epic proportions, but its elements are repeated time and time again across the nation, typically with the same result – the exclusion of minorities. *Mount Laurel* does, however, demonstrate that, no matter to what extremes the courts are prepared to go, they are no match for the wiles of municipalities whose power base extends from their tiny townships to the state capital. As in other fields, the courts are "of limited relevance," to use Rosenberg's term (1991: 157). They are constrained by the limited nature of constitutional rights, the lack of judicial independence, and the judiciary's lack of powers of implementation.[24] To quote Rosenberg's final paragraph:

> American courts are not all-powerful institutions. They were designed with severe limitations and placed in a political system of divided powers. To ask them to produce significant social reform is to forget their history and ignore their constraints. It is to cloud our vision with a naive and romantic belief in the triumph of rights over politics.

And while romance and even naivete have their charms, they are not best exhibited in courtrooms.[25]

(Rosenberg 1991)

The Fair Housing Act offers little more encouragement, though it is a step ahead of the *Arlington Heights* philosophy. Professor Schwemm (1990: viii), an authority in this area, notes that "housing discrimination has proven far more intractable a national problem than the sponsors of the Fair Housing Act of 1968 anticipated."[26] The amending 1988 legislation provides enforcement "teeth" which the earlier Act was lacking, but it remains to be seen whether they have any bite.

Inclusionary zoning offers a different approach, but it is (at least so far) slight in its impact, even when it is implemented with Californian enthusiasm. But it does have the advantage that it adopts the coinage of American zoning: money. Thus those concerned in dealing with it, either to promote it or to avoid as much as possible as they can of it, are talking the same language. The following two chapters explore other avenues which adopt the same ethic.

Chapter 6

Development charges

Paying for the Costs of Development

> If ever there was an issue that automatically pitted builders and developers against other elements of the community, it is the topic of development exactions.
>
> (Frank and Rhodes 1987)

The costs of development include not only the construction costs of buildings (houses, shops, offices, etc.) but also the costs of the services and facilities which are needed to serve these. Sewage disposal, water supply, and other utilities are the most immediately obvious, but the full list ranges much more widely – highways, schools, day-care centers, hospitals and other social services, public transit, the provision of housing (or transportation) for low-income workers needed to service the development, and so on.

Who is to pay for these? and how? – the existing property owners through their property taxes, the developers through exactions, the new residents through special assessments? The possibilities are theoretically almost endless and, not surprisingly, the whole subject bristles with difficulty and controversy.

The story of development charges (or exactions or imposts – there is no standard terminology)[1] in recent decades is one of an ever-expanding net, bringing more and more services within its grasp.[2] The simplest, and oldest, is the development charge levied to pay for the provision of basic utilities on the site. These charges arose in connection with subdivision control, and were legitimated in the Standard City Planning Enabling Act of 1928 which explicitly included a requirement for the provision of infrastructure internal to the development.[3] Such services were normally limited to streets, sidewalks, street lighting, and local water and sewage lines. Services external to the development were paid for by the appropriate suppliers.

This system worked satisfactorily until the housing boom of the post-World War II period which placed a great strain on the budgets of the municipalities and school districts (and on the tolerance of property tax payers). Existing property owners were unhappy (sometimes vociferously so) at having to pay increased taxes for the benefit of newcomers and, increasingly, municipalities required developers to make contributions ("dedications") of land for such purposes as schools and playgrounds (or, particularly in small developments, cash payments in lieu).

The next step was to extend these contributions to other services which are necessary to serve the development. Typically, these are "off-site," such as sewerage and water supply systems, and arterial roads. These "impact fees"[4] have become increasingly popular for two reasons. First, the reluctance of existing property owners to pay for the servicing of new development grew substantially as federal aid to localities was reduced. At the extreme (as with California's Proposition 13), taxpayer "revolts" brought matters to a head. More broadly, there has been the expansion of popular concern for the environment which has eroded the traditional American belief in the benefits of never-ending growth: this culminated in an articulate and sometimes blinkered no-growth ethic.

In some areas, municipalities have for long required developers to provide or finance infrastructure which benefits not just a particular development but a wider area, or even the public at large. This has been particularly so in California, where the state courts have taken an unusually relaxed view on the matter. As we shall see, this relaxed view is now significantly affected by the 1987 Supreme Court case of *Nollan v California Coastal Commission*.[5]

Impact Fees

Many communities have sought to shift the burden of paying for growth from the community at large to new development through impact fees. Those attempts are often challenged in court, usually by developers. Although some state courts

sympathize with the plight of communities and have found ways to justify impact fees, no one can predict the outcome of court involvement. Sometimes the courts uphold the enactment of fees; other times they do not.

(Lillydahl *et al.* 1988)

An impact fee is a sophisticated mechanism for shifting from a municipality a part of the cost of the capital investment necessitated by new development. (The question as to whom the cost is shifted is discussed later in this chapter.) In tune with the spirit of the age, impact fees are much more complicated than the earlier charges. On the one hand, they can be far more wide-ranging, extending to any municipal capital expenditure required to meet the needs of the inhabitants of the new development. On the other hand, they are subject to the restraints of a new calculus which attempts to calibrate the marginal impact of the new development upon a municipality. This is a field which has been extensively dealt with in the courts[6] and in a newly developed area of planning expertise.[7]

The crux of the matter is the determination of the "rational nexus" – or, more simply (and therefore less appealing to lawyers) the "connection" – between the charge levied upon a developer and the burden placed on the municipality by the development. Thus a new development might necessitate the building of a major arterial highway in an adjacent area, but it would be wrong to relate the whole cost of the road to the new development if, as is likely, the highway were required not merely for the new development, but for the area as a whole. There is a useful analogy in the last straw that broke the camel's back. The new development is the last straw: but the main burden on the camel is the straw that is already there.[8]

The rational nexus rule is a cost accounting method which was first elaborated by Heyman and Gilhool (1964), and refined in academic papers such as Ellickson's *Suburban Growth*

Controls (1977). It has become increasingly popular with the courts. Basically, it uses cost accounting methods for calculating what share a new development has in creating the need for a facility. That proportionate share then becomes the basis for a charge. The calculation usually involves determining:

1   The cost of existing facilities;
2   The means by which existing facilities have been financed;
3   The extent to which new development has already contributed, through tax assessments, to the cost of providing existing excess capacity;
4   The extent to which new development will, in the future, contribute to the cost of constructing currently existing facilities used by everyone in the community or by people who do not occupy the new development (by paying taxes in the future to pay off bonds used to build those facilities in the past);
5   The extent to which new development should receive credit for providing common facilities that communities have provided in the past without charge to other developments in the service area;
6   Extraordinary costs incurred in serving the new development;
7   The time–price differential in fair comparisons of amounts paid at different times.

(Nicholas and Nelson 1988: 173)

Some of the difficulties involved with this are illustrated by the *Nollan* case.

The *Nollan* Case

Nollan is merely a judicial request for planners and politicians to repent and atone for the sins of our brethren in California who have gone too far with too little to support their ever-growing appetite for land use regulation.

(Kramer 1987)

The case of *Nollan v California Coastal Commission* is one of a number of recent cases which signify a reawakening of the interest of the Supreme Court in land use cases.[9] It was the first exactions case heard by the Supreme Court, and its decision has been followed by an avalanche of writings reflecting a range of differing opinions (not all of which can be said to have clarified matters).[10]

The case concerned an application by the Nollans to the California Coastal Commission for a permit to demolish their dilapidated beachfront bungalow and replace it with a new and larger house. The commission, which has a policy of increasing access to and along the beach, gave permission conditional on the Nollans providing public access between the sea and their seawall. The matter went through the state courts and finally to the U.S. Supreme Court, where the majority held that the commission's requirement was an unconstitutional taking of property. The essence of the argument was that there was no nexus between the condition and the permit.

> The commission's imposition of the access-easement condition cannot be treated as an exercise of land-use power *since the condition does not serve public purposes related to the permit requirement* [emphasis added]. Of those put forward to justify it – protecting the public's ability to see the beach, assisting the public in overcoming a perceived "psychological" barrier to using the beach, and preventing beach congestion – none is plausible. Moreover, the commission's justification for the access requirement unrelated to land-use regulation – that it is part of a comprehensive program to provide beach access arising from prior coastal permit decisions – is simply an expression of the belief that the public interest will be served

by a continuous strip of publicly accessible beach. Although the State is free to advance its "comprehensive program" by exercising its eminent domain power and paying for access agreements, it cannot compel coastal residents alone to contribute to the realization of that goal.

(*Nollan* 1987)

The decision is not a model of clarity, but fortunately it is not necessary here to enter into any detailed discussion. It is sufficient to note that it is widely accepted that the decision will persuade municipalities to have closer regard than heretofore to the nexus between a development decision and any condition or charges which are imposed:

> To avert the evil decree of invalidation and compensation, local governments will have to plan better, utilize a more professional approach to land use regulation and subdivision exactions, and leave a paper record that substantiates the required nexus between the condition or exaction and the police power objective.

(Kramer 1987: 2)

A clearer, though less well known or reported, case is that of *Parkes v Watson*. Here, a developer in Klamath Falls, Oregon, requested a rezoning of his land to allow the construction of 214 garden apartments. His site included a geothermal well which he hoped to use to obtain steam to heat the apartments. The city, eager to use the well for the assembly of a geothermal utility district to provide heat and power to the general public, asked the developer to dedicate the land on which the well was located for a street widening. Naturally, the developer refused, and the case went to court. It was ruled that the requirement for the well donation had no rational relationship to the city's interest in street widening (Bosselman 1985: 28).

The Incidence of Charges

Economic theory indicates that the burdens of suburban antigrowth programs usually fall primarily on landowners, not housing consumers. When consumers *are* injured, the worst affected (in dollar terms) are not the households that have been excluded (as has usually been thought) but those that actually buy at prices inflated by the antigrowth measures. The chief beneficiaries of a suburb's development charges and development restrictions are the individuals who own its existing houses and apartments.

(Ellickson 1977)

It is frequently assumed that charges imposed on developers are passed on to housing consumers.[11] A classic case in support of this is the huge (100 percent) increase in house prices in the Washington metropolitan area which took place in the early 1970s following the simultaneous takeover of the counties of Fairfax, Montgomery, and Prince Georges by antidevelopment politicians (Ellickson 1977: 434). However, a closer examination of this illustration shows its falsity. There are two important points. Firstly, the *simultaneous* action of these three large counties created a regional land shortage. More typically, a single small municipality (or even a group of municipalities) around, say, Philadelphia or Chicago, could not exert such a market influence: housebuilders would simply move to more accommodating areas. (As Schnare and Struyk [1976: 164] have demonstrated, suburbs are often essentially identical.)

Secondly, it is suggested, home buyers in the Washington area were able to afford higher prices because of the generosity of federal pay scales (Ellickson 1977: 434).

The main issue is an elementary one: the incidence of a charge will be determined by market forces (Delaney 1987). These vary over time and space. The demand for housing in a particularly attractive area may be highly price-inelastic, and thus charges could readily be passed on to buyers. In an area with a plentiful supply of land of similar amenity, the tendency will be for charges to be passed "backwards" to landowners: developers will pay less for land than they would have done in the absence of charges. At a time of rapid house price inflation, homebuyers are tempted to pay high prices in the expectation that they will rise even further: here the developer should have no difficulty in simply passing on a charge to the buyers. On the other hand, high mortgage interest rates may make buyers resistant to prices which are increased on account of charges.

In short, there is no single answer to the question, who pays?[12]

Existing v New Homeowners

Local government . . . speaks for the people who get there first.

(Reilly 1973)

One often hears the argument from residents of a growing community: "Why should we pay for the expansion of public facilities? Let the developers or the newcomers pay for them. We have already paid our fair share."

(Snyder and Stegman 1986)

Why indeed? But again on reflection, the obvious is not so. It is equally easy – and persuasive – to argue that since growth benefits the community as a whole (not just those involved in the new growth), no extra charges should be imposed on newcomers. In any case, is it not unfair for established households to change the rules for newcomers? Existing residents did not have to pay charges when *they* moved in: why should those who follow them be penalized?

These questions of "intertemporal fairness"

(Beatley 1988: 340) or "intergenerational equity" (Snyder and Stegman 1986: 39) are not easy to deal with, and there is no single answer which will fit the situation of widely differing municipalities. Much depends on the historical development of the particular community and the methods employed over time to finance infrastructure. For example, if annual growth is constant, and capital requirements incremental, the debt (both for replacement of facilities for existing residents and for expansion of facilities for newcomers) is spread over an ever-increasing population.

> With time, residents who were established in the community and those who arrived during that year will pay a decreasing share of the cost of facilities that were built for their use. Furthermore, the share of the cost they will bear declines even more as growth rates and financing periods increase.
>
> On the other hand, growth requires new capital outlays for future residents, and established residents must help pay the debt for these capital expansions if they are to be publicly financed. Higher growth rates mean higher rates of facility expansion and, therefore, higher costs to established residents. This impact works as a counteracting balance against the dilution effect described above. The net effect of these forces is not easily determined. It depends upon the magnitude of the growth rate, the interest on borrowed funds, and the length of the financing period.
>
> (Snyder and Stegman 1986: 42)

However, some of the investment is for future residents who will not pay their "share" until further growth takes place. The net effect is problematic. Moreover, continued growth encounters "thresholds" (Kozlowski 1986) which require "lumpy" investments (sewage treatment plants for example – which cannot be expanded incrementally). The capital cost of these has to be carried for a lengthy period before the full complement of users (and therefore taxpayers) has arrived – by which time, of course, investment in the next "lump" is necessary.

Clearly, it is no easy matter to unravel all these (and similar) matters. The analyst's difficulties are compounded by the impact of inflation, which has a habit of destroying any notion of equity. However, the conclusions reached by Snyder and Stegman are interesting:

> In the case of incremental investments where the economic life of facilities exceeds financing periods, inequities between established residents and new development occur only when growth rates exceed the real interest rate. Few cities are experiencing sustained levels of growth of that magnitude. Only in rapidly growing cities can one justify exactions from developers for investments of this type. And even in those instances, exactions need not be large relative to per capita outlays to balance the burden on the two generations of users.
>
> On the other hand, lumpy investments with excess capacity to meet future needs do place an inequitable burden on established residents. Development fees to pay for these kinds of investments are justified for all rates of growth. Appropriate magnitudes for fees in these instances depend upon a number of factors, including the rate of growth, inflation, time intervals for expansion, the length of financing periods, and the physical lives of facilities.
>
> (Snyder and Stegman 1986: 49)

Clearly there remains plenty of scope for argument at both the academic and the individual municipal levels.

Of course, the issue of equity is affected by the incidence of any development charge. Though, for purposes of exposition, these two issues have been separated, they are in fact interrelated. And there are other related matters. One is that of the "ability to pay." Whatever theoretical concept of equity may be satisfied in an analysis of intertemporal fairness,

it will not of itself resolve this. If the charges (or property taxes) are so high that numbers of potential homebuyers are squeezed out of a local housing market, regard should be had to the alternatives which are (or could be made) available. To use other words, the effects of the development on the housing market should be "mitigated" – as in the San Francisco scheme discussed in the following section. In the last chapter, programs of inclusionary housing were discussed. We now examine other "inclusionary" programs which are related to downtown development.

Linkages

> Inclusionary zoning moves downtown.
> (Merriam, Brower, and Tegeler 1985)

The inconstant way in which planning terms are used is illustrated by the use of "inclusionary zoning" to refer to housing (and other facilities) which developers of major downtown projects are required to make before permission to develop is given. This "inclusionary housing downtown" is also, and better, termed "linkage": it is "included" in a scheme only in the sense that it is linked to it. However, there is some doubt as to how real this link is. Certainly, it is linked in the minds of the municipal officials who maintain that downtown projects draw new employees to the area, thus increasing the demand for housing. More fully:

> Large scale downtown projects – notwithstanding the benefits they provide, such as new jobs and additional tax revenues – have negative effects that their developers should ameliorate. In tight urban housing markets, new development attracts office and white collar employees, who create a demand for central city housing. Since they can outbid lower income city residents,

gentrification ensues, leading to
displacement of poorer residents.
(Keating 1986: 134)

As might be expected, developers see the
matter in a different light. Among their many
arguments is that downtown development *follows* and accommodates demand: it does not
create it. As one critic has nicely put it: "addi-
tions to the supply of office space do not create
office employment any more than cribs make
babies" (C. Gruen 1985: 34). Moreover, there
is a danger that linkage fees will kill the golden
goose of downtown development, particularly if
they are set at a level which will finance significant amounts of housing.

There is merit in both sets of arguments and,
as usual, which is valid will depend upon the
particular circumstances of the time and place.
In the case of San Francisco, there was a downtown boom: indeed some vocal elements in the
city wanted to see a slowdown of downtown
development; at the least there was an ambival-
ence to redevelopment which is not common in
American cities. Boston and Santa Monica
share San Francisco's stance, but many sunbelt
cities with strong downtown markets do not:
for instance, Atlanta, Dallas, Denver, Houston,
Miami, and New Orleans (Keating 1986: 134).

San Francisco's linkage policy was introduced in 1981 as the Office-Housing Production Program (OHPP).[13] The term "mitigation"
is also employed – signifying that the levy is
intended to mitigate the harmful effects of the
development. Developers erecting over 25,000
square feet of office space are required to
construct housing or pay an in-lieu fee into a
housing trust fund.[14] A certain percentage of
the housing must be affordable and available to
low- or moderate-income households for a
period of twenty years. If the developer elects to
pay the in-lieu fee, the rate (in 1990) is $5.80
per net additional gross square foot of office
space (Hausrath 1988: 215).[15] Between 1983
and 1989 forty-one developers committed $76
million to housing projects, of which 2,900

units had been built by the end of 1989; 84
percent of these were for low-income households (King 1990: 5).Boston's linkage program
was modeled on San Francisco's. It imposes a
fee of $6 per square foot (originally $5) on
office developments over 50,000 square feet.
Between 1983, when it was introduced, and
1989 it had resulted in forty-one developers
committing $76 million to housing projects,
and the building of 2,900 dwellings, most of
which are aimed at low-income households
(King 1990: 5).

The viability of these linkage programs –
indeed, any linkage program – depends on the
strength of the local economy. Many local
governments have been reluctant to use them
because they have been afraid of scaring away
private investment (Goetz 1989: 75). Similarly
in Boston, the program "has been successful in
large part because its implementation occurred
concurrent with a period of economic growth,
in which there was an increasing demand for
office space. Now [1990] that the period of
growth has ended, there is speculation that the
linkage program will act as a disincentive to
developers" (Taub 1990: 682).

Also important is the pressure to adopt a
program (typically from community groups or a
low income housing lobby). A study of San
Francisco noted the importance of "a strong
office market coupled with an aggressive,
usually community-based advocacy for linkage"
(Goetz 1989: 75). By contrast, in Seattle a
linkage proposal was strongly opposed by development interests; the city "opted instead for an
incentive zoning policy under which developers
can make housing contributions and, in return,
obtain density bonuses" (Keating 1989: 219).
Keating has analyzed the objections to linkage
schemes, and concludes that it is unlikely that
more than a few cities "will be in the economic,
political, and legal position necessary to achieve
their adoption." Nevertheless:

[Linkage schemes] represent a significant
advance for urban planners concerned with

downtown planning. In an era of public–private partnerships, linkage policies represent a legitimation of the idea that local government is entitled to and should demand that private commercial developers contribute to a better planned and more equitable revitalization of our central cities.

(Keating 1986: 141)

It seems probable that linkage schemes will operate in a broader framework of public–private agreements.

It is interesting to note that the trend toward greater private "participation" in the financing of infrastructure is not restricted to the United States. In a comparative study of the United States and Britain, it was noted that in both countries "the external costs of private land development have, over the past fifteen years, been increasingly borne by private land developers rather than public agencies" (Callies and Grant 1991: 221). There is evidence of a similar trend in Canada.[16] The root cause is the inability of local governments to shoulder the increasing demands being made on them. They are therefore searching for new sources of revenue. Imposing levies on new development is a politically painless way of obtaining extra funds. Some of the initiatives to which this is giving rise are examined further in the following chapter.

Chapter 7

Planning by agreement

Development Agreements

The most sophisticated form of exaction is the development agreement.

(Crew 1990)

Theoretically, traditional zoning involved a largely passive role for local governments. They did not seek to influence the course or pattern of development; neither did they attempt to strike any kind of financial bargain with a developer. Their requirements were set out in an ordinance and a map, and simply "applied" to each case as it arose. The system was essentially a "rule-application" process and, indeed, departure from the rules carried the danger of litigation. In practice, of course, a certain amount of "agreement" or even "bargaining" has often taken place; nevertheless, the system was characterized by certainty and a lack of scope for discretion.

Several factors conspired to change this approach and to encourage planning by negotiation and bargaining: the fiscal problems of local governments; changes in the structure of

the economy such as the huge growth of relatively footloose tertiary activities; demographic changes which have led to demands for new types of residential development; increasing concern for environmental quality and protection of natural resources; and a rise in public concern about "growth" and the quality of life. Such factors have transformed the land use regulation scene. Though there is still, perhaps over a large part of the country, (or, to be more precise, over those parts of the country which have zoning), a prevalence of traditional "rule-application" zoning, it is now much more common to see developers and local governments debating and reaching agreement on development proposals.

Of course, in one sense, all planning is by agreement, however reluctant that agreement might be. The question is how much scope there should be for debate, and what constraints should operate. There are no simple answers here, as numerous court cases demonstrate.

Some techniques of "planning by agreement" were discussed in the previous chapter. Here, the focus is on development agreements and incentive zoning.[1] Development agreements exist in a variety of forms, and can be used for a wide range of purposes: to facilitate agricultural land preservation, to compensate for lost tax revenues (as with Proposition 13), to foster community development, and to increase the supply of low and moderate cost housing (Wegner 1987: 994; Kirlin and Kirlin 1985: 31). However, before discussing these types of ventures, one particular class of "development agreement" needs to be discussed. This is an agreement for which specific legislative provision has been made by a number of states, including California, Hawaii, Nevada, and Florida. In each of these states, the legislation was adopted in response to state supreme court decisions which applied harsh judgments against landowners' claims for "vested rights" (the right to proceed with a development despite a change in regulations). The difficulty centers on determining the point at which a developer has proceeded sufficiently far with a project to acquire common law "vested" rights.[2]

There is a real difficulty here which tends to grow over time. Conditions change, and a local government has the power to amend its zoning ordinance to meet these changes. There is no absolute right to an existing zoning classification, and neither is there any ground for the expectation that a zoning classification will remain unchanged. On the other hand, developers need to operate on some assumptions about the stability of zoning regulations. But with major multiphased developments (or even lengthy delays in development due, for instance, to market conditions) this can be problematic. The difficulties are increased as the number of approvals required for a project multiply. More fundamentally, long-term assurance to a builder may involve an unlawful "contracting away" of a local government's police power. (No government can legislate in such a way as to bind its successors.)

California was the first state to enact provisions relating to development agreements,[3] in response both to the *Avco* case and also the

fiscal pressures following the passage of Proposition 13. As stated in the legislation:

(a) The lack of certainty in the approval of development projects can result in a waste of resources, escalate the cost of housing and other development to the consumer, and discourage investment in and commitment to comprehensive planning which would make maximum efficient utilization of resources at the least economic cost to the public.

(b) Assurance to the applicant for a development project that upon approval of the project, the applicant may proceed with the project in accordance with existing policies, rules and regulations, and subject to conditions of approval, will strengthen the public planning process, encourage private participation in comprehensive planning, and reduce the economic costs of development.

(California Code)

The legislation, which took effect on January 1, 1980, was initially viewed skeptically by local government officials as a "developers' bill," but it is now seen much more positively as a means of getting developers to "agree" to assist in the financing of public infrastructure (Holliman 1981: 44). In essence, the Act provides a degree of certainty for the developer,[4] and an explicit expectation that developers shall make a contribution to the wider costs of their development. In the words of the legislation, the contents of a development agreement are as follows:

A development agreement shall specify the duration of the agreement, the permitted uses of the property, the density or intensity of use, the maximum height and size of proposed buildings, and provisions for reservation or dedication of land for public purposes. The development agreement may include conditions, terms, restrictions, and requirements for subsequent discretionary actions, provided that such conditions, terms, restrictions, and requirements for subsequent discretionary actions shall not prevent development of the land for the uses and to the density or intensity of development set forth in the agreement. The agreement may provide that construction shall be commenced within a specified time and that the project or any phase thereof be completed within a specified time. The agreement may also include terms and conditions relating to applicant financing of necessary public facilities and subsequent reimbursement over time.[5]

Development agreements are often seen, from the local government viewpoint, simply as a convenient mechanism (similar to incentives and bonuses) which facilitate the private provision of infrastructure finance. On this approach, there can be anxiety that a local government may, in effect, "sell off" its police power. Questions about fairness also arise. In Wegner's (1987: 960) words, "unfair or inefficient outcomes may result from imbalances in power or skill that either distort the dealings of particular parties, or result in failures to consider the interests of affected nonparticipants."

On the other hand, there are some real advantages, quite apart from matters of finance. A development agreement allows consideration to be given to the particular features of a site and proposals for its development. No general set of rules can meet all circumstances, still less innovative ideas on development. Moreover, development agreements can be an important means for facilitating public participation in the development review process. Experience in Santa Monica demonstrates this:

> Since large scale projects necessarily are subject to change from future market forces, developers often wish to maintain certain design details for a later date. In collapsing the discretionary review process, development agreements tend to force the developer, city staff, and neighborhood residents to examine all aspects of a project and its long term impacts as part of one coherent decision making framework. This benefit of the development agreement process may prove even more valuable than the potential legal benefits to the contracting parties.
>
> (Silvern 1985: 9)

In a 1983 panel discussion on the subject (Silvern 1985: 6) a Sacramento attorney, William W. Abbott, gave the following reasons why local governments should employ development agreements:

a) to implement existing planning policies as expressed in a general plan, specific plan, or other existing policy document;

b) to help create new planning policies if they do not exist or if they appear dated;

c) to obtain public control over the phasing of public improvements;

d) to obtain a higher level of public improvements than would normally be required for a given project; and

e) to resolve unusual problems of substandard site conditions involving existing buildings or infrastructure.

The following two examples of development agreements from Santa Monica are taken from a paper by Silvern which details several others.[6]

*2701 OCEAN PARK BOULEVARD –
124,00 sq ft, mixed use (retail-office, residential)*

Deviations from planning standards: Three story building in Neighborhood Commercial Zone; Residential use in Neighborhood Commercial Zone

Public benefits: Eighteen affordable rental units; Ground floor commercial tenants must be neighborhood commercial; Day care center; Traffic control measures; Energy conservation measures; Specified infrastructure improvements; Prevailing construction wages

Time limits: Start within 18 months of building permit; Complete within 24 months of starting date; May be constructed in two phases (commercial-office, and residential); Contract term, 40 years

Subsequent approvals needed: Architectural, signage, and landscaping; Subdivision maps; Building permit

Level of environmental review: Negative declaration

Status: No construction initiated – successor owner currently seeking amendments to modified design

BAYVIEW PLAZA HOLIDAY INN – 134 room hotel addition and remodeling

Deviation from planning standards: Ten story hotel addition in high density residential zone

Public benefits: Landscape improvements to existing street median; Street improvements; Minimum exterior improvements to existing building; Targeted hiring program; Job training program; Shuttle bus system or in lieu payment; Fee to Civic Center Fund; Public art and craft displays; Room vouchers for emergency lodging; Free parking; Prevailing construction wages; Relocate two existing buildings and construct six new rental units on site

Time limits: Start within 18 months of building permit; Complete within 24 months of starting date; Contract term 39 years; Certain obligations terminate 25 years from Certificate of Occupancy, others in 40 years from Certificate of Occupancy

Subsequent approvals needed: Architectural, signage, landscaping; Subdivision maps; Coastal Commission approval; Building permit

Level of environmental review: Environmental impact report; Various mitigation measures required – related to noise, design, parking, circulation, public safety, and energy conservation

Status: Project completed

Clearly the measures on which agreements can be made are very wide indeed!

Incentive Zoning

It seemed a splendid idea. Developers wanted to put up buildings as big as they could. Why not harness their avarice? Planners saw a way. First, they would downzone. They would lower the limit on the amount of bulk a developer could put up. Then they would upzone, with strings. The builders could build over the limit *if* they provided a public plaza, or an arcade, or a comparable amenity.

(Whyte 1988)

And so New York embarked upon its policy of incentive zoning. At first, the scheme was across-the-board: bonuses were given as of right to developers who met the requirements set out in an ordinance. There was, therefore, no negotiation: thus for every square foot of plaza space provided, the developer could claim an extra ten square feet of office space. The scheme was a great success – in terms of the number of plazas provided. Indeed, it was really too successful: between 1961 and 1973, over a million square feet of new open space was created in this manner – more than in all of the other cities of the country put together.[7] A study by Kayden concluded:

Incentive zoning greatly exacerbated the overbuilding boom in New York City that gained momentum through the late 1960s, creating the oversupply and high vacancy rates of the 1970s. These in turn caused lower real estate tax assessments and occupancy tax revenues, costing the city over $8,000,000 in fiscal year 1973–1974, a cost indicative of a longer term trend begun in 1965. At the same time, many developers discovered that by employing some of the looser provisions of the incentive system, they would capture generous rewards approaching an undesirable and unnecessary level of windfall profits. Even the amenities provided have not escaped some criticism. In sum, the overall incentive zoning experience in New York City can best be characterized as mixed.

(Kayden 1978: 1)

Kayden's detailed cost–benefit calculations are even more damning (Kayden 1978: 65). Between 1961 and 1973, over 7.6 million square feet of "plaza bonus office space" was constructed, at a cost of $3.8 million. The capitalized value of the extra office space amounted to more than $182 million. Thus, for every dollar spent on plazas, developers obtained a return of nearly $48 in extra office space. The plazas conferred some public benefit, of course, even though they were obtained at such high cost. Unfortunately, the benefits are not easy to quantify: "although the economic value of the additional floor space to the developer and the cost of building the amenity can be calculated, the value of an amenity to the public cannot be measured in dollars" (Getzels and Jaffe 1988: 2). Whyte's study (1988: 234), however, showed that "a lot of the places were awful: sterile, empty spaces not used for much of anything except walking across." Whyte was not the only critic. Jonathan Barnett commented:

> The new principle of a zoning bonus for plazas proved far more popular with developers than had been anticipated; but the use of the plaza bonus, by itself, has created some serious design problems of its own. While plazas have introduced valuable

open space into the city, their proliferation has accentuated some of the defects of the underlying zoning, notably the tendency of the regulations to separate each new building from its surroundings. Beneath the language specifying setbacks, plazas, and open space rations are certain assumptions about what the resulting buildings should look like. Unfortunately, these implied architectural standards are based upon the "revolutionary" concepts of architecture expounded by Le Corbusier and others during the 1920s. Their vision of the city of the future as a park filled with orderly rows of towers of the same design and height does not seem to be adaptable to zoning and the implementation of different-sized buildings on a lot-by-lot basis.

> Zoning regulations that encourage plazas have had the effect of belatedly imposing a fragmentary version of the 1920s modernism on cities, creating towers that stand in their individual pools of plaza space, surrounded by the party walls of earlier structures that were planned to face the street. Shopping frontages are interrupted and open spaces appear at random, unrelated to topography, sunlight, or the design of the plaza across the way.

(Barnett 1982: 72)

The Negotiation Syndrome

> Developers love negotiation. Not only is it the way they make money; there is something about the process that appeals to the temperament of people who go into real estate in the first place.

(Barnett 1982)

As noted earlier, New York's incentive zoning scheme was originally of an "as-of-right" character, and thus there was no negotiation involved. Dissatisfaction with the outcome led to a wide review which, among others, included the eloquent William H. Whyte whose 1988 book *City: Rediscovering the Center* is used extensively in this discussion.

There were two ways of improving the situation. The first was to elaborate the guidelines. This was done, despite fears that it would be too inflexible and would unduly constrain architectural design.[8] The new (1975) guidelines spelled out the rules of the game in considerable detail: "the maximum height of the plaza, the amount of seating, the minimum number of trees, and so on" (Whyte 1988: 234). According to Whyte, the new guidelines had a salutary effect:

> The handwringing over their tightness proved unfounded. Once they were on the books the developers went along, quite equably. Over the ensuing ten years there was no complaint by a developer or architect over the basic guidelines. Builders put in benches and chairs, they planted trees, and some went beyond the minimum requirements of providing flowers and food kiosks and the like. The new zoning also encouraged the retrofit of existing plazas, and a number of hitherto dead ones were brought to life. The zoning guidelines were adopted by other cities, sometimes in such detail that they repeated the precise dimensions that had been derived from our New York prototypes. The formula for figuring the amount of sitting space to be required – one linear foot for every thirty square feet of plaza space – was a back-of-the-envelope compromise between linear-feet people and square-feet people. It has been enshrined in countless ordinances across the country and it works just fine.
> (Whyte 1988: 234)

The alternative to the mechanical as-of-right scheme is a special permit process which, as revised, was designated by the ungainly term Uniform Land Use Review Process – unaffec-tionately known by its acronym ULURP. Essentially this is a negotiated agreement. In place of fixed guidelines, there is considerable flexibility. To quote Whyte again:

> The special permit process . . . allows planners to tailor design requirements to particular situations – often with a better fit to the intent of the law than to the letter of it. In the negotiating sessions, furthermore, improvements in the design can be suggested – and adopted – that would not have been under the as-of-right approach.
> The planners have cards to play. The clock is ticking away on some high-cost borrowed money and the developer is anxious to expedite matters. The planners are in a position to suggest amenities and they usually push the developer quite hard in this respect – getting the developer to put in an extra escalator, for example, or public toilets on the ground floor.
> (Whyte 1988: 237)

By 1980, it was clear that zoning in New York City was in real difficulty. Anticipation of bonuses fed back into higher land prices (though developers sought some measure of protection by signing contingency agreements with landowners, with the higher price to be paid only if the anticipated bonus was granted), and buildings became larger and larger. Promised (negotiated) amenities were sometimes not provided. Citizen groups became increasingly loud in their complaints. Finally, in 1982, midtown zoning was subjected to a sweeping revision. Densities were reduced, and bonuses were largely dropped except for plazas and urban parks. Amenities which had formerly been obtained by way of bonusing now became mandatory. The cumulative effect of these provisions, it was anticipated, "would go a long way toward eliminating negotiated zoning. They would permit development to proceed on a more predictable and as-of-right basis" (New York 1982b: 11). Between May 1982 and May 1988, of the ninety-six buildings approved in

Midtown New York, seventy-eight were approved as-of-right (New York 1988b).

The Dangers of Bonusing

> Most land use regulatory decisions, not just incentive zoning transactions, pose trade-offs between neighborhood and citywide concerns.
>
> (Kayden 1990)

The lesson is clear: once introduced, incentive zoning is difficult to control.[9] In the absence of any overall official plan policy framework, the process "can engender considerable uncertainty respecting the city's intentions and can give the impression that the underlying basis of the plan is being subverted" (Toronto City 1988: 9). It can also be difficult to ensure that all land owners are being treated equitably and consistently. Moreover, the absence of basic ground rules results in a process which "can be extremely time consuming (and costly) and require extensive professional involvement as each application is negotiated."

Bucknall has written a major critique of bonusing schemes in which he argues:

> If our city planning theories about appropriate development, servicing and transportation have any validity, the extra density created in one place must either be denied somewhere else or paid for over decades in the expansion of services, utilities and transportation corridors. The costs are not as direct or quantifiable, but the taxpayers will bear them nonetheless. The transactions remain deals cut between the city [of Toronto], representing all of its citizens, and a single land owner or developer with very particular and narrow interests. Whereas voters in the past would be asked to approve such schemes where a

few thousands, or tens of thousands, of dollars were involved, the modern schemes are beyond the scrutiny of anyone but the municipal council.

> [What is created is] a circumstance in which one municipal goal (housing, for example) is traded off against another municipal goal (consistent planning) with no necessary relationship between them. If a municipal statement with regard to maximum densities is defensible by planning rationale why should the municipal need for a public swimming pool alter that rationale? Is there not a danger that the planning theory itself will come to be treated as arbitrary and unprincipled – simply one more chip to throw into the urban development poker game?
>
> (Bucknall 1988)

Bermingham (1988: 8) submits that the most fundamental problem with individually negotiated bonusing is that it leads to a situation in which "it becomes almost politically impossible for a municipality to approach a density increase without demanding the contribution of some amenity." Indeed, he argues that, since it would be difficult for a municipality to grant an increase in density to one owner on more favorable grounds than preceding owners, each grant of a bonus is likely to involve a demand for a higher contribution. "In other words, from the municipality's perspective, each deal becomes the starting point for the next deal."[10]

A different argument contends that "physical planning standards undermined by [incentive zoning] are not the only interests important to communities. Other values, including those represented by social amenities, contribute to the quality of life, and a city might reasonably resolve that it will tolerate taller buildings and greater congestion in return for more low income housing or daycare facilities" (Kayden 1990: 101). A now classic case of the unhappy times on which New York City bonusing fell is that of the Columbus Circle project, which eventually came before a New York trial court at the instigation of the Municipal Art Society of New York.[11] The agreement reached in this case provided for the acquisition by the developer of the city-owned site, a 20 percent increase in density, and the payment to the city of $455 million, plus another $40 million for improvements to the nearby Columbus Center subway station. The city would also have realized about $100 million in taxes each year from the 2.7 million square-foot development. Such riches were tempting indeed, and the city did not resist the temptation. But it fell foul of legal hurdles. The crucial point at issue was the fact that a substantial part of the payment to the city was to be for citywide purposes. Most damaging was the appearance of some $266 million of the proceeds in the city's 1988 budget, in advance of final approval of the sale. The court invalidated the sale on the grounds that the incentive provided by the city constituted improper "zoning for sale."[12] However, though much of the debate was focused on these financial matters, the underlying issue was the huge size of the proposed development, and the shadow it would cast over Central Park. (On October 18, 1987, the Municipal Art Society gathered more than 800 people with umbrellas to form a line from Columbus Circle to Fifth Avenue. On a given signal at 1:30 p.m. all the umbrellas were opened – thus demonstrating the shadow which the building would cause (Babcock 1990: 27)). The city eventually redesigned the project on a smaller scale.[13]

It would be quite wrong to regard this notorious case as the death knell of incentive zoning. Far from it, as can be seen by reference to Lassar's 1989 study, neatly entitled *Carrots and Sticks: New Zoning Downtown*. This documents in detail the wide range of incentive schemes which are being operated in many American cities. And, as far as the Columbus Circle case is concerned, New York transgressed because it "upset the delicate balance between competing public and private interests.... It made economic return the deciding factor, with scant attention to other public goals and land use considerations" (Lassar 1989: 38). In other words, incentive zoning is acceptable as long as it is kept within bounds and does not become a technique for raising additional municipal funds.

Purposes for Bonusing

All communities preparing zoning bonus provisions must deal with four elements. They must establish the purposes of the system; select amenities to achieve these purposes; determine the size and type of the bonus that will be granted; and develop a method of administering the system.
(Getzels and Jaffe 1988)

The quotation is expressed in imperative terms, but all too often bonuses are seen to be self-evidently beneficial. That this is not always so is apparent from the previous discussion. Here we discuss some of the useful purposes which bonusing can achieve.[14]

Ideally, a bonus should be an incentive for a developer to provide an amenity or facility which is of public benefit, and which the developer would not provide voluntarily. There is an immediate difficulty with this: even if there is an agreement about which benefits are desirable, how can it be determined that they will be provided only if an incentive is offered? It has already been noted that New York, in 1982, abandoned many bonuses for mandatory requirements (Whyte 1988: 250). San Francisco did likewise with matters of downtown design (Getzels and Jaffe 1988: 2). However, these two cities are hardly representative: many cities are extremely anxious to attract downtown development and are therefore far more inclined to provide incentives rather than disincentive conditions. Moreover, economic circumstances change, as they have done, for example, since Lassar made her wide-ranging review of downtown policies.

A favorite objective of bonusing is the promotion of lively street-level retailing in downtown areas – in contrast to the dead blank walls which so seriously diminish the attractiveness of a street. A good statement of purpose is provided in the Seattle ordinance:

> The intent of the retail shopping bonus is to generate a high level of pedestrian activity on major downtown pedestrian routes and on bonused public open spaces. While retail shopping uses ensure that major pedestrian streets are active and vital, a limit to the amount eligible is set in each zone in order to maintain the dominance of the retail core as the center of downtown shopping activity.
>
> (Getzels and Jaffe 1988: 3)

Seattle's downtown code provides brief "statements of intent" for each bonusable amenity. Shopping corridors, for example, are "intended to provide weather-protected through-block pedestrian connections and retail frontage where retail activity and pedestrian traffic are most concentrated downtown. Shopping corridors create additional 'streets' in the most intensive area of shopping activity, and are intended to complement streetfront retail activity." Lassar (1989) comments that bonus activities "run the gamut" and can be clustered around several general categories:

a) building amenities: urban spaces, ground-floor retail, retail arcades, artwork, sculptured rooftops, atriums, and day care;

b) pedestrian amenities: sidewalk canopies and other overhead weather protection devices, landscaping, multiple building entrances;

c) pedestrian movement: sidewalk widening and through-block connections;

d) housing and human services: employment and job training;

e) low-income health clinics; low-income, affordable, and market-rate housing;

f) transportation improvements: transient parking, below-grade parking, and transit station access and upgrading;

g) cultural amenities: cinemas, performing arts centers, art galleries, and live theaters; and

h) preservation: historic structures, theaters, and low-rent housing stock.

Lassar's book covers all these in useful detail. Here only one further issue (of particular topicality) is discussed, the provision of day-care facilities.

Day-Care Facilities

> This section is intended to ensure that adequate measures are undertaken and maintained to minimize the child care impacts created by additional office employment in the downtown, in a manner consistent with the objectives and policies of the Master Plan, by facilitating the development, expansion, and maintenance of affordable, quality child care programs and auxiliary services, the latter including, but not limited to, resource and referral services.
>
> (San Francisco Code)

The extension of planning concerns to embrace child-care facilities is of relatively recent origin.[15] The major reason, of course, is demographic. The number of working mothers has increased enormously – and particularly middle-class mothers who tend to be more demanding of services. In 1985 nearly two-thirds of all women with children under the age of eighteen worked outside the home. Even more striking has been the increase in labor force participation among women with very young children (under the age of three), to 51 percent in 1985. Moreover, most of these mothers worked full-time: about 82 percent of employed single mothers, and 68 percent of employed married mothers (Cohen 1989: 39).

The provision of child care is now seen by employers as a means of attracting (and retaining) needed staff, and by some local governments as a supplementary local economic development policy. Recognizing the need, however, does not necessarily indicate a willingness to make adequate provision.[16] To facilitate this, a number of measures has been taken, from incentive zoning to the appointment of child-care coordinators to link public, private, and voluntary efforts.

Day-care facilities raise various planning issues, depending upon their size and location. Small facilities in suburban areas (up to, say, six children) present the question as to whether they should be regarded as "home occupations," and therefore subject to zoning controls. In some jurisdictions they are; in others they are not. Public opposition (based on fears of noise, parking problems, and a fall in property values)[17] has bolstered outdated zoning provisions which impose inappropriate limitations on small scale facilities. In a 1984 national survey of 212 urban and rural communities, Cibulskis and Ritzdorf found that:

> The major zoning issue related to day care is the lack of municipal acknowledgement of the differences between large day care centers and small day care homes. This is true even though all fifty states require day care centers to be licensed, and differentiate between small and large facilities for that purpose. Only two-thirds of the communities acknowledged the existence of day care in their ordinances. Of the communities that mentioned day care, sixty-three percent did not distinguish between small homes and large centers. Often, the result of this lack of differentiation is the treatment of all day care facilities as if they were large, commercial

> operations. Indeed, forty-one of the communities required a special use permit to operate a small (six or fewer children) family care home in a residential zone.
> (Cibulskis and Ritzdorf 1989: 8)

There is here a clear case for action at the state level which has been undertaken in various ways. Some states have preempted the municipal zoning of homes. Thus, in California and Wisconsin, the use of a single-family residence as a small family day-care center is regarded as being a residential use of the property. Some states have legislated against the prohibition of large facilities on single-family lots (California) or in residential zones (Connecticut and Michigan).[18]

Provision of child-care facilities in downtown developments (or, often cash in lieu) may be subject to a bonus, and may be either optional or mandatory. Provisions vary: in Hartford, Connecticut, the bonus is an additional 6 square feet of floor space for every square foot of day-care space. San Francisco requires developers either to provide 2,000 square feet or the equivalent of 1 percent of a project's square footage, whichever is the larger. (There is an optional one-time fee which is paid into the city's Affordable Child Care Fund; this fee is so much lower than the cost of providing the facilities that developers tend to opt for it.)

Other municipalities have experimented with child-care facilities. Seattle, for instance, gives developers a density bonus (Colwell 1989: 13), but many other schemes are mandatory and provide no incentives.

There is considerable argument as to whether day care should be bonused or mandatory, or both. Though it seems that exactions may be more productive than the use of incentives, many cities may fear that these would deter development. It may be that comprehensive plans for day care will prove to be more effective than a mere extension of zoning techniques.

The Future of
Linkage Provisions

> Until the federal government recommits
> itself to assisting America's cities and to
> housing low income and moderate income
> citizens, housing-linkage will remain an
> important example of progressive urban
> policy that is attainable within the existing
> space for urban reform.
>
> (Dreier and Ehrlich 1991)

It is unclear how widespread exactions of these
kinds will become. Much depends, of course,
on the development climate, in both economic
and political terms. The success of existing
schemes is as yet uncertain. Different
commentators judge differently. For example,
on the San Francisco inclusionary housing
scheme two writers (in the same year) judged
progress in very different terms. Whereas one
wrote that the scheme "clearly . . . had worked"
(Sedway 1985: 168), the other judged it to be
"at best, modest" (N.J. Gruen 1985: 48).

Linkage schemes are, in general, limited.
There is much less activity in the real world
than the planning literature might suggest. (It is
the exceptional that makes good news.)

However, where these schemes carry an incentive or bonus, there is a danger that a municipality's desire to obtain contributions from the
developer might overwhelm the requirements
of good planning in the area. From this odd
point of view it is an advantage that American
municipalities have so little in the way of plans:
their absence means that they cannot be sabo-
taged. But where there is an effective plan,
bonusing can destroy it. Seattle provides a good
example of what can emerge as a result of an
assembly of bonuses. The Washington Mutual
Tower gained twenty-eight of its fifty-five
stories on account of the amenities offered by
the developer. As of right, the developer was
allowed twenty-seven stories. In addition to
this, he obtained thirteen stories for a $2.5
million housing donation, one story for a transit
tunnel entrance donation, two stories for a
public plaza, two stories to compensate for
mechanical space, a half-story for a public
atrium, a half-story for a garden terrace open to
the public, one story for a day-care facility, two
stories for space lost to a sculptured top to the
building, two stories for the provision of retail
space, and two and a half stories for a public
escalator to help pedestrians climb Seattle's
hills (Colwell 1989: 12). This was a remarkable
example of private munificence!

Part III
The quality of development

Chapter 8

Aesthetics

Introduction

> Parrot-like repetition of slogans such as "it is all a matter of personal taste" or "beauty lies in the eye of the beholder" is not a resolution of the aesthetic issue but an avoidance of the problem.
>
> (Royal Fine Art Commission 1990[1])

Despite the City Beautiful movement, aesthetic considerations have always been problematic in American land use planning. They involve questions of preference and taste on which opinions differ, as the following examples illustrate:

> The American Institute of Architects' choice of the best builder's house of 1950 was refused a mortgage by the Federal Housing Administration. Again, the Veterans Administration imposed a $1000 design penalty on an architect-designed house in Tulsa, Oklahoma, that *House and Home* had displayed on its 1954 cover. The Pruitt-Igoe public housing, which starred in a TV vehicle when HUD Secretary George Romney had it blown up, had won an architectural award in its day.
>
> (Haar and Wolf 1989: 533)

By contrast, designs once despised can become popular icons: the Eiffel Tower was once described in terms of "the grotesque mercantile imaginings of a constructor of machines." Now it is "the beloved signature of the Parisian skyline and an officially designated monument to boot" (Costonis 1989: 64).

The difficulties of aesthetics are great at both the practical and the philosophical levels[2] yet, in simple terms, Americans like their neighborhoods to be pleasant and attractive, free of noxious intrusions (and even apartments!). Fear of falling property values and unwelcome social groups play their role here too, but there remains a real, and increasing, concern for environmental quality. Certainly, there has been a major shift in public attitudes (even if less apparent on the ground) since the time when it was thought that, to coin a north English phrase, "where there's muck there's money."[3]

The changes can be seen, for example, in the increased use of controls over billboards; in the adoption of landscape ordinances, parking lot regulations, appearance codes, and design guidelines; and in the establishment of advisory or administrative design review boards. This chapter discusses a number of these planning mechanisms. In line with the historical developments, the first to be considered is the control over billboards.

Billboards

I think that I shall never see
A billboard lovely as a tree
Indeed, unless the billboards fall,
I'll never see a tree at all.
(Ogden Nash)*

Ogden Nash may never have seen
A billboard he held dear
But neither did he see
A tree grossing 20 grand a year.
(David Flint, Turner Advertising Company)

In one sense, all zoning involves aesthetic considerations even if they are as mundane as height and bulk; but other factors are also present, such as infrastructure, congestion, fire prevention, and so forth.[4] Aesthetics first arose explicitly with billboards – and initially the overwhelming judicial view was that controls imposed for such reasons would not pass constitutional muster. The 1905 New Jersey case of *City of Passaic v Paterson Bill Posting, Advertising and Sign Painting Company* is illustrative:

> Aesthetic considerations are a matter of luxury and indulgence rather than of necessity, and it is necessity alone which justifies the exercise of the police power to take private property without compensation.
> (*City of Passaic v Paterson Bill Posting, Advertising and Sign Painting Company* 1905)

Similarly, a Denver ordinance of 1898 was held to be unconstitutional because it had specific requirements solely for billboards, including a ten-foot setback from the street line. The wording of the decision became quite lyrical:

> The cut of the dress, the color of the garment worn, the style of the hat, the architecture of the building or its color, may be distasteful to the refined senses of some, but government can neither control nor regulate in such affairs. . . . Ours is a constitutional government based upon the individuality and intelligence of the citizen, and does not seek, nor has it the power, to control him, except in those matters where the rights of others are impaired.
> (*Curran Bill Posting and Distributing Company v City of Denver* 1910)

Nevertheless, a minority of courts did hold that aesthetics was a legitimate consideration in the exercise of the police power, and by the 1930s it was generally accepted that aesthetic factors could be taken into account. This involved a legal fiction, namely that while aesthetic regulations were not acceptable in themselves, they could be justified on the grounds of associated evils. A 1932 New York decision (*Perlmutter v Greene*) stated the view nicely: "Beauty may not be queen but she is not an outcast beyond the pale of protection or respect. She may at least shelter herself under the wing of safety, morality or decency." A classic statement of this view occurs in a Missouri case of 1913, in which it was stated that billboards:

> endanger the public health, promote immorality, constitute hiding places and retreats for criminals and all classes of miscreants. They are also inartistic and unsightly. In cases of fire they can often cause their spread and constitute barriers against their extinction; and in cases of high wind, their temporary nature, frail structure and broad surface, render them liable to be blown down and to fall upon and injure

those who may happen to be in their vicinity. The evidence shows and common observation teaches us that the ground in the rear thereof is being constantly used as privies and dumping ground for all kinds of waste and deleterious matters, and thereby creating public nuisances and jeopardizing public health; the evidence also shows that behind these obstructions the lowest form of prostitution and other acts of immorality are frequently carried on, almost under public gaze; they offer shelter and concealment for the criminal while lying in wait for his victim; and last, but not least, they obstruct the light, sunshine, and air, which are so conducive to health and comfort.

(*St. Louis Gunning Advertising Co. v St. Louis* 1913)

The majority of courts today hold that the police power can be used for aesthetic purposes, whether these have the ulterior purpose of promoting some other public good such as tourism or economic development, or for "pure" aesthetic objectives.[5] An important factor in this change was the 1954 U.S. Supreme Court case of *Berman v Parker*. In his decision, Justice Douglas delivered the following *dictum* (that is, it was a gratuitous comment, not crucial to the case in question):

> The concept of the public welfare is broad and inclusive. . . . The values it represents are spiritual as well as physical, aesthetic as well as monetary. It is within the power of the legislature to determine that the community should be beautiful as well as healthy, spacious as well as clean, well balanced as well as carefully patrolled.
>
> (*Berman v Parker* 1954)

A later case of some notoriety concerned a Mrs. Stover who, for several years, hung clotheslines of rags in the front yard of her house in Rye, New York, as a protest against the high taxes imposed by the city. Each year an additional line was festooned with a remarkable range of materials: tattered clothing, old uniforms, underwear, rags, and scarecrows. Neither the neighbors nor the city were amused, and after six years the city passed an ordinance prohibiting the erection and maintenance of clotheslines on a front or side yard abutting a street; exceptions could be granted where there were real practical difficulties in drying clothes elsewhere on the premises. Mrs. Stover applied for an exemption but was refused, but she retained her clotheslines. The case (*People v Stover* 1963) went to court, and it was ruled that the city was justified in preventing Mrs. Stover from her unusual form of protest: a form which was "unnecessarily offensive to the visual sensibilities of the average person."

Most courts now take the view that aesthetics alone is a legitimate public purpose and can be controlled by land use regulation.[6] It still remains, of course, for a municipality to ensure that the controls are properly applied.

As the *Berman* and *Stover* cases illustrate, some important court decisions on aesthetics are only indirectly concerned with signs. In the following pages, cases dealing specifically with the issues raised by signs (and billboards in particular) will be discussed.

Signs can be of various kinds: directional, political, on-site business, freestanding advertisements (billboards) and so on. The crucial distinction, however, is between "informational" signs and billboards. On-premise signs (which, of course, can be as obnoxious as the worst billboard) are generally accepted in principle, though restrictions are common on their size and number. Billboards, on the other hand, arouse a great deal of controversy – fueled by two active lobbies: one promoted by the wealthy and powerful billboard industry, and the other by Scenic America (formerly the Coalition for Scenic Beauty), dedicated to "curb an industry that . . . has run amok."

No holds are barred in the open warfare on billboards. The opposition is very strong, even by American standards. In his classic treatise, Norman Williams (1985–90: 118.02) refers to the billboard lobby as "quite intransigent in demands and quite ruthless in tactics." He comments that "it has been common gossip among leading planners that the billboard industry maintains (or used to maintain) a blacklist. It is certainly true that on occasion segments of the industry have intervened to try to keep a planner known to be 'uncooperative' out of an important job." Former New York Senator Thomas C. Desmond is quoted as saying that the billboard lobby "shrewdly puts many legislators in its debt by giving them free sign space during election time, and it is savage against the legislator who dares oppose it" by favoring anti-billboard laws (Blake 1964: 11).

The billboard industry endeavors to enhance its public image by donating billboards to good causes such as First Lady Barbara Bush's campaign to promote family literacy, and the boosting of morale in the San Francisco Bay area following the October 1989 earthquake. These public benefits are regularly reported in *Outlook: The Newsletter of the Outdoor Advertising Association of America*.

Rural Signs

Billboards are the art gallery of the public.
(B.L. Robbins, President, General Outdoor Advertising Company)

With rural signs, the focus of the debate is on the location of billboards in open rural areas alongside major roads and, to a lesser extent, in commercial areas. (There is relatively little controversy about the undesirability of billboards in residential areas though, as we shall see, there is a distinction to be drawn between on-site business signs and freestanding advertisements.) Billboards along highways and in rural areas have been objected to on aesthetic, safety, and other more ingenious grounds. Among the latter is the argument that regula-

tion of billboards takes away only that value which is created by the building of the road from which the billboard can be seen. Thus the erection of a billboard takes for private gain the value of an opportunity created by public expenditure (N. Williams 1985–90: 121.03). In New York, the state erected a screen on public land to hide a dangerously sited billboard. In *Perlmutter v Greene* (1932), the court upheld this action, claiming that no owner had a vested right for his billboard to be seen from the road.

A few states, such as Vermont, Hawaii, Maine, and Alaska have completely banned rural billboards. In some states, existing billboards can be amortized without compensation, but this policy has been affected by federal legislation concerning highways. Two years after the commencement of the building of the federal interstate highway system, the Federal-Aid Highway Act of 1958 (the "Bonus Act") provided for a voluntary program under which states could enter into an agreement with the federal government on the control of outdoor advertising within 660 feet of the edge of interstate highways. The incentive was a bonus federal grant of one-half of 1 percent of the construction cost of the highway project. The legislation provided for the prohibition of most off-premises signs, and some controls over on-premises signs. Later amendments exempted from control certain parts of the system:

1   areas that had been zoned or were in use for industrial or commercial purposes in September 1959, and
2   older rights of way which were incorporated into the interstate system.

Only half the states took advantage of this scheme. Three states used the power of eminent domain to eliminate nonconforming signs; seven used a combination of eminent domain and police power controls; and the remainder used police power controls alone. Six of the latter were challenged in court, but in only one case was the action declared unconstitutional: this was the highly conservative Georgia court (Floyd 1979b: 116).

A more elaborate system was introduced by the Highway Beautification Act of 1965 (sometimes known as the Lady Bird Johnson Act),[7] which, in President Johnson's words, would bring about a new approach to highway planning:

> In a nation of continental size,
> transportation is essential to the growth and prosperity of the national economy, but that economy, and the roads that serve it, are not ends in themselves. They are meant to serve the real needs of the people of this country. And those needs include the opportunity to touch nature and see beauty, as well as rising income and swifter travel. Therefore, we must make sure that the massive resources we now devote to roads also serve to improve and broaden the quality of American life.
>
> (quoted in Wright and Gitelman 1982: 1024)

The reality bore little relation to the rhetoric. The lofty intentions of the Act were assailed by the billboard lobby and, instead of a system of effective control over roadside advertising signs – and also junkyards (Moore 1988) – a vast number of signs were in fact removed from control (Floyd 1979b).

The Act was intended to make billboard control mandatory in all the states, and to extend the controls to major roads in addition to the interstates. The provisions of the Act were made mandatory (with a withdrawal of 10 percent of federal highway funds from states that did not comply), but the provisions themselves were emasculated by the efforts of the billboard lobby. Though new off-site signs are limited to commercial and industrial areas, the actual control in these areas (which include *unzoned* commercial and industrial areas) is minimal. The controls are agreed between the federal government and the individual states, but there are no *national* standards: the criteria

for control are based on state law and "customary use."[8] On-premises signs are totally exempted from control: hence the extremely high signs that are exhibited by gas stations close to the interstates.

The biggest victory for the billboard lobby, however, was the introduction of mandatory compensation for the removal of nonconforming signs. This precluded the elimination of billboards by amortization – a favorite technique among anti-billboard communities. The provision was extended in 1978 to require compensation for the removal of billboards under *any* legislation (not solely under the federal Act). This constitutes a boon to owners of obsolete and abandoned signs who can offload them on to the states and receive compensation!

A major problem here, as in the whole of this area, is that federal funds have been very small; as a practical result of this, many states have used all their funds for acquisition of signs voluntarily surrendered by their owners. A report by the U.S. Department of Transportation on the operation of the Highway Beautification Program in Florida and Alabama notes that:

> These voluntary sales resulted in many spot purchases from areas where other signs remained. Federal Highway Administration officials generally believed that the only signs acquired under the program were those that were no longer economically beneficial to their owners. The remaining nonconforming signs, presumably of value to the owners, are still visible to the travelling public, and little or no benefit can be seen from the spot purchases.
>
> (U.S. Department of Transportation 1984: 8)

The restriction of billboards to commercial and industrial areas is a much more limited provision than appears at first sight. Many municipalities (eager for the property tax on billboards – meager though it is)[9] have zoned large areas along interstate and other major highways as commercial. Moreover, an area can be regarded as commercial or industrial even if it is unzoned: all that is necessary is some adjacent activity that could be regarded as falling into one of these two land use categories. Floyd has described the ingenuity of some advertising companies:

> In Georgia one property owner erected a small shed in a rural area and put up a sign designating it as a warehouse. A large billboard was erected next to this "warehouse" and the outdoor advertising firm then applied for a permit based on the area's being an unzoned industrial area. In South Carolina, a large national advertising company helped set up a small radio repair shop in a residence that happened to be located near Interstate 95, and then used this "business" as justification to erect several large billboards.
>
> (Floyd 1979b: 119)

There are many similar stories. The problems are exacerbated by the widespread practice (whether permitted or not) of vegetation cutting undertaken to extend the economic life of signs,[10] misunderstandings (whether intentional or not) between the states and the federal government, and weaknesses in the enforcement of violations. Underlying these specific points, however, is the general lack of political support for the program. Despite the removal of a large number of nonconforming billboards, the legislation is a failure, and is more a testimony to the resourcefulness and power of the billboard industry than to effective controls.

Urban Signs

> Anything that tends to destroy property values of the inhabitants of the village adversely affects the prosperity, and therefore the general welfare of the entire village.
>
> (Supreme Court of Wisconsin 1955)[11]

Sign controls in urban areas present trickier problems than those in rural areas where protection of the character of the landscape is usually more clearly evident. But this is not always so: as was noted in *John Donelly & Sons v Outdoor Advertising Board*, "urban residents are not immune to ugliness." In residential areas, the problem arises only infrequently: aesthetic issues more often relate to the "harmony" or otherwise between new and existing developments. In commercial areas, the felt need to protect the view of a famous building, or mountain range, or vista can involve extensive controls, as can offensive satellite dishes. Some of these matters give rise to concerns about the infringement of the freedom of speech clause of the First Amendment.

On this, a distinction is frequently made between commercial and noncommercial free speech: commercial speech tends to receive less protection. A classic case in this field is *Metromedia*. Unfortunately, the case is a very complicated one and raised almost as many questions as it settled.[12] Nevertheless, a majority of the court agreed on several important issues:

1   Promotion of aesthetic objectives alone is sufficient basis for use of the police power to control signs.
2   A prohibition of all signs, both commercial and noncommercial, is permissible if confined to a relatively limited area of special interest such as a historic district.
3   It is lawful to make a distinction between on-premises and off-premises signs in an ordinance.
4   A prohibition of all off-premises signs is constitutional.
5   A sign control ordinance is not void even though it may put sign companies out of business.

6   An ordinance that is part of a
comprehensive city beautification effort
is more likely to withstand judicial
scrutiny.

(Duerksen 1986: 29)

The current situation (though by no means
entirely clear) can be summarized simply: most
federal and state courts now reject free speech
objections to sign ordinances; signs create
visual problems that justify aesthetic controls.
On-site signs advertising the business carried
on at the site tend to be exempt from prohibi-
tions though they may be banned from certain
areas for aesthetic reasons. Signs which are not
subject to a blanket prohibition can, nevertheless, be subject to controls over their placement
and size.

Architectural Design Review

Artistic expression does not easily lend itself
to police power restrictions.

(Kolis 1979)

Good design is an elusive quality which cannot
easily be defined. In the words of Hedman and
Jaszewski:

Short of requiring the builder to copy
specific prototypes, it is impossible to
legislate good design. No set of rules can
anticipate all the situations and conflicts that
will eventually surface, and there is a
tendency that rules designed to prevent
something bad will also prevent something
good from happening. At best, we stack the
odds against the worst and hope for the best.
However cleverly the controls have been
structured, designers have demonstrated an
uncanny ability to technically meet every
requirement and still evade the spirit of the
underlying design objectives.

(Hedman and Jaszewski 1984: 136)

This is of particular concern to British municipalities which decide on the design merits of
thousands of development proposals every year.
But, though design controls are much less
common in the United States, the same problem
arises: how can "good design" be defined and
obtained? Modest efforts have been made with
landscaping and screening parking areas; but
some municipalities have now adopted architectural design review ordinances.

These highlight the difficulties of articulating
standards which are comprehensible and clear.
If an owner cannot understand what is, or is
not, permitted under an ordinance, there is –
at least by American standards – a basic unfair-
ness. It provides too broad a discretion to the
municipality, permitting arbitrary action. On
the other hand, aesthetic matters cannot be set
out in the detail possible in, for instance, a
building code.[13] In the words of the New
Mexico court (in *City of Santa Fe v Gamble-
Skogmo Inc.*), "Literally setting forth every
detail would impair the underlying public
purpose." The problem is exacerbated by a lack
of clarity as to what "the underlying public
purpose" actually is. One survey concluded
that:

While most communities with design control
measures seem to know why they want such
a device, very few, if any, communities
demonstrate clear understanding of how the
concept can be translated into operational
means, how effective they are in attaining
objectives, and what may be the
consequence of implementation in the long
term.

(Habe 1989: 199)

One of the difficulties (as in many areas of
public policy) is that there is typically more
than one objective. Habe's survey of sixty-six
American cities showed that, in addition to
aesthetic considerations, each city had at least
two other objectives unrelated to aesthetic
concerns. These included general "economic"
and "public welfare," protection against urban

problems such as crime, slums, and traffic congestion, "psychological well-being, ecological concern, historic/cultural concern, facilitating the functional aspect of community life, accommodating user need, and manoeuvering migration" (Habe 1989). The vagueness of many of these objectives is noteworthy, and common in this field.

A particularly frequent objective is the preservation of community character. This can, in practice, mean anything from the perpetuation of an architectural style to the exclusion of different social groups. The latter is no longer expressed explicitly, but it remains as common as it was in Whitten's day.[14] Perhaps the most popular design control is the "no excessive difference" rule. This is typically expressed in terms such as "new buildings must reflect the existing character of the area," or "be sensitive to existing architecture."

> According to one community, harmony was defined as "pleasant repetition of design elements to provide visual linkage, direction, orientation and connection of areas." Often the concept is interpreted as similarity: "cornice lines, openings and materials of new structure to be similar to those of adjacent buildings" (Concord, California); or "retaining and freestanding walls should be finished with brick, stone or concrete compatible with adjacent buildings" (Rochester, New York).
>
> (Habe 1989: 202)

Habe (1989) comments that "such overemphasis on similarity of design encourages the trend toward specificity of standards, including setting specific architectural styles, rather than encouraging innovative solutions from designers." Rapoport has instanced a quite contrary design approach for an English village:

> Different styles, materials, roof pitches, buildings at different angles, "interesting" and "intimate" grouping. . . . natural vegetation such as gorse and heath grasses,

mixed age and income of people and lack of uniformity generally.

> (Rapoport 1982: 157)

This is an exaggerated version of the American "no excessive similarity" rule, originating with the explosive growth of tract housing (such as the Levittowns) in the 1950s (Poole 1987: 293).

"No excessive difference" seems to be generally acceptable, but "no excessive similarity" is more problematic (Poole 1987: 330). However, it is inappropriate to be dogmatic on this issue since remarkably few cases involving architectural review have come before the courts. Indeed, relative absence of litigation is a feature of aesthetic controls. The reasons for this, though speculative, are interesting. A major factor is that developers prefer to have community support for (or at least to avoid community opposition to) their proposals. They are therefore generally willing to negotiate: after all, the issue at stake is "only" one of design, not one of significant cost. And who wants to build, or live in, a dwelling to which neighbors are hostile? If a developer (or a developer's client) wants a dwelling that is unusual, the obvious path of least resistance is to choose a site occupied by, or being developed for, similar deviants. The lower the density, the easier it is to be different in peace.

The negotiation of good design is a striking feature of a number of control schemes. For instance, the Lake Forest, Illinois, ordinance provides for review by a five-member board before a building permit will be issued.[15] "The board has not denied a permit in its twenty years of existence, choosing instead to negotiate with designers and developers over points of disagreement. . . . The board's approach has been to seek improvement rather than censorship of design. The board is yet to be challenged through a lawsuit" (Poole 1987: 306).

Architectural design controls involve particular difficulties in the large cities affected by successive property booms. San Francisco can be quoted as an illustration. After a lengthy period of public controversy (Jacobs 1980), the city enacted a series of design related ordinances. Among other things these required that:

1 the upper portion of any tall building be tapered and treated in a manner to create a visually distinctive roof or other termination of the building facade, thereby avoiding boxy high rise buildings and a "benching" effect of the skyline;

2 new or expanded structures abutting certain streets avoid penetration of a sun-access plane so that shadows are not cast at certain times of the day on sidewalks and city parks and plazas;

3 buildings be designed so the development will not cause excessive ground level wind currents in areas of substantial pedestrian use or public seating;

4 the city consider the historical and aesthetic characteristics of the area along with the impact on tourism when issuing a building permit;

5 building heights downtown be reduced from 700 to 550 feet (from about 56 to 44 stories).

(Duerksen 1986: 14)

Whether the "fancy tops" controls have proved effective in improving the skyline of San Francisco is debatable: they have certainly produced some very curious buildings (perhaps a nice case of beauty being in the eye of the beholder?) Seattle has similar, though less detailed restrictions in the downtown area: "the requirements limit building heights, establish setbacks to maintain light and air, and ensure designs that reduce wind-tunneling and retain views of Elliott Bay" (Duerksen 1986: 14)

Boston has produced design guidelines for neighborhood housing. This is part of an ambitious project "to transform all of the city's vacant buildable lots into attractive and afford-

able housing" (Boston 1988). The guidelines emphasize existing neighborhood character and also cover such matters as the site, "the organization of the residences" (by which is meant "public and private territory and views, security and surveillance, and construction materials and maintenance"), and the residence itself.

Portland, Oregon, has received much publicity (deservedly so) for its urban planning and design. Of particular interest is the incorporation of design into the urban planning process. This came about in three stages:

During the 1960s, design issues were raised piecemeal in response to specific projects and problems. During the 1970s, design goals were incorporated into general planning policies. In the 1980s, design considerations have become an accepted part of the regulatory planning system.

(Abbott 1991: 1)

Though the city of Portland has its own particular character (which Abbott describes as its "orientation to a moralistic political style which accepts the possibility of disinterested civic decisions"), some other cities are moving toward a similar use of external standards and comparisons, and toward the integration of design review with other planning goals for the area (Duerksen 1986: 16).

It would, however, be wrong to give the impression that there is a widespread movement in this direction. Much of the United States has no design control (or certainly none that is apparent). There is considerable controversy on the reasonableness and effectiveness (and, despite evidence to the contrary, constitutionality) of design controls, except in areas with highly distinguishing features such as historic districts. One compromise is to have informal guidelines. In Denver's Lower Downtown, for instance, a group of citizens, developers, and business people worked on a scheme to protect this unique area's mix of old and new buildings. "Some of the key design elements that people have identified include absence of blank walls at street level; moderate building heights; continuity of exterior building materials; presence of historic buildings; pedestrian orientation; and facades that preserve a lot-by-lot appearance" (Duerksen 1986: 16). Unfortunately, the Denver attempt at "negotiated design" failed, mainly because of a dramatic change in economic conditions (Fulton 1989b: 10). Informal guidelines are no substitute for legal sanctions (even when, as in the case of Lake Forest, they are held in abeyance). A telling case in point is the city of Philadelphia, where an unofficial height limit was set at the top of William Penn's hat on the City Hall. This limit operated from 1894 to 1984, when it was exceeded by two buildings.

Poole (1987: 340) maintains that design controls directed at preventing the construction of excessively different buildings violate the First Amendment. Kolis (1979: 304) argues that "the general public welfare will be better served by recognizing the First Amendment rights of architects and their clients so that they may achieve great architecture." Habe (1989: 215) complains that in attempting to ensure legality (and also to achieve maximum efficiency) design controls tend to emphasize details (which are easier to define) and adopt the use of generalized conditions from a standard list. S.F. Williams (1977: 33) suggests

establishing criteria similar to those for obscenity (for example that the proposed design is "blatantly offensive" to community standards). Poole (1987: 340) has also argued that "architectural designs sufficiently distasteful to cause measurable harm to a neighborhood occur so rarely (if ever) that regulations to prevent them amount to making mountains out of molehills." Municipalities should "get out of the role of imposing majoritarian notions of tastefulness on the community at large. Tastefulness by a committee assures nothing more or less than mediocrity."

The final word can rest with a view from the science of economics: Hough and Kratz (1983) assert, on the basis of an hedonic price equation for office space in downtown Chicago, that "good" new architecture passes the market test: "tenants are willing to pay a premium to be in *new* architecturally significant office buildings, but apparently see no benefits associated with *old* office buildings that express aesthetic excellence." In short, the market can be left to look after new buildings; for historic buildings "those who value them must devise feasible non-market mechanisms so that their preferences for these buildings are revealed and their dollars are contributed."[16] Lake Forest is hardly likely to be impressed!

*The author and publishers would like to thank Curtis Brown Ltd for permission to reprint 'Song of the Open Road' by Ogden Nash.

Chapter 9

Historic preservation

Introduction

> Americans have no urban history. They live
> in one of the world's most urbanized
> countries as if it were a wilderness in both
> time and space. Beyond some civic and
> ethnic myths and few family neighborhood
> memories, Americans are not conscious that
> they have a past.
>
> (Warner 1972)

Planning involves the resolution of conflicting
claims on the use of land. This is particularly
clear in the case of historic preservation since
the nature of the conflict is so readily apparent.
Typically, one party (often more than one)
wants to preserve a historic structure for public
enjoyment now and in the future. The other
party (often one only) wants to use the site for a
new use which produces a higher profit. The
traditionalists use the language of culture and
history; the redevelopers speak in terms of
market trends and economic returns.

In the last century, the controversy was
normally between public and private interests.
This changed as it became evident that history
could be molded to produce profits and (what
amounts to the same thing) a good public
image. For example, there was capital to be
made out of a company's environmental
concerns if these were manifest in the preservation of an historic building for modern use.
Further profits were to be realizable from tourist
attractions. And, above all, changes in tax
provisions transformed the attitudes of
landowners and developers to preservation.

The new enthusiasm for historic preservation was not to everyone's liking. As in other
fields (national parks for instance – which in
the United States are in danger of becoming
theme parks or zoos) too many people seeking
to enjoy "a piece of history" can overwhelm it
and destroy the very experience which is sought
(Egan 1991). Moreover, both preservationist
and developer interests have become much
more sophisticated than in earlier times. The
step from preserving a physical structure to
preserving a community is not a large one (as
New York experience with landmark preservation clearly shows). Community groups and
preservation societies can be bought out by

generous contributions to their good work from developers. Preservationists sit on the boards of development companies; and their development interests in turn are to be found on the managing boards of voluntary bodies.

Thus the old lines of demarcation have become blurred. For the planner, the situation has become confused, and frequently an apparently simple clash of development and protectionist interests turns out to be something much more complex. In this chapter, a number of these issues are discussed, but the main focus is on the development from a simple approach to the historic preservation of landmarks toward a "planning perspective" on cultural matters. This perspective has now merged with a concern for urban design. On this, as Abbott has pointed out, "design considerations have become an accepted part of the regulatory planning system" (Abbott 1991: 1). Added to this there has been such an extraordinary expansion of the field of interest of what used to be simply called "historic preservation" that the very term is now of vintage stock.

The Early Days of Heritage Preservation

We have never had in the United States any governmental institution comparable to the French *Service des Monuments Historiques*, with its wide powers over the national historic and artistic heritage. Because of this lack of a comprehensive custodial agency, the task of managing the American heritage had originally fallen, *faute de mieux*, to the citizenry at large, especially to that small and heroic band of amateurs (to use the term in its original connotation of one who does the work for love and not for pay) without whom there would have been no historic preservation movement at all.

(Fitch 1990)

Historic preservation in the United States grew from the grass roots in an unorganized way. Its early development is the story of a large number of (predominantly private) endeavors to save individual structures or sites (Hosmer 1965: ch.1). Many of these failed, like the attempt to save the so-called "Old Indian House" in Deerfield, Massachusetts, which was the last home in the town which escaped the famous massacre of 1704 (demolished in 1848 because it had "no intrinsic value"). Similarly, the John Hancock House in Boston was destroyed in 1863. Others had a near miss, like Independence Hall in Philadelphia, which the city purchased for $70,000 in 1816 (Hosmer 1965: 30). One of the most notable successes was Ann Pamela Cunningham's crusade to save Mount Vernon. This particular success encouraged many more, but the preservation groups who tried to imitate Miss Cunningham's work:

all found that she had achieved something that was not likely to be repeated for years to come. Now, as then, far too few preservationists, overwhelmed by the importance of their particular projects, realize how many *other* buildings are supposed to be "second only to Mount Vernon."

(Hosmer 1965: 620)

The essentially indigenous character of the historic preservation movement in the United States was not changed by the occasional action of the federal government. This was restricted mainly to the acquisition of a small number of landmarks and individual park sites (such as Shiloh National Military Park in 1894, and Morristown Historical Park in 1933).

The national parks, of course, were already in the public domain and thus sites within these parks which needed public protection did not require acquisition.[1] Public lands, in fact, were the scene of another development in historic preservation. This was the preservation of "antiquities." The Antiquities Act of 1906 provided for the designation as National Monuments of areas *in the public domain* which contained "historic landmarks, historic and prehistoric structures, and objects of historic or scientific interest."[2] This was broadened in 1935 with the introduction of the Historic Sites, Buildings and Antiquities Act; this was aimed at fostering "a national policy to preserve for the public use historic sites, buildings and objects of national significance for the inspiration and benefit of the people of the United States." It called upon federal agencies to take account of preservation needs in their programs and plans and, for the first time, promoted the surveying and identification of historic sites throughout the country. This program became the base for the National Register of Historic Places some thirty years later.

These early endeavors in preservation were essentially concerned with history and cultural values, as distinct from architectural quality (though the line was sometimes blurred, as with Monticello, which had both historical and aesthetic features). At this time, buildings, structures and sites were of appeal because of their associative and inspirational values. There was, however, an increasing concern for architectural values toward the end of the nineteenth century, neatly expressed in William Sumner Appleton's statement of purpose of the Society for the Preservation of New England Antiquities, which he organized in 1910:

> to save for future generations structures of the seventeenth and eighteenth centuries, and the early years of the nineteenth, *which are architecturally beautiful or unique*, or have special historical significance. Such buildings once destroyed can never be replaced.
> (Hosmer 1965: 12)

The added italics emphasizes the primacy here accorded to architectural values. Of course, the historical and associative elements remain important today: in fact it is often difficult to disentangle them.

The interwar years were a lean time for historic preservation, although there were notable exceptions, such as the creation of preservation commissions in Charleston, South Carolina (1931), New Orleans (the Vieux Carre Commission),[3] and San Antonio (1939). World War II and the early postwar period was even leaner: indeed urban renewal and highway projects destroyed many buildings which a few years later might have been preserved. It was this very destruction which (together with the reaction to the sterility of the International Style in new architecture) acted as a catalyst to an unprecedented burst of activity in the mid-1950s. The culmination of this was the publication in 1966 by the U.S. Conference of Mayors and the National Trust for Historic Preservation of a powerful eloquent manifesto *With Heritage So Rich*.

With Heritage So Rich and Subsequent Legislation

> We do not use bombs and powder kegs to destroy irreplaceable structures related to the story of America's civilization. We use the corrosion of neglect or the thrust of bulldozers.
> (U.S. Conference of Mayors 1966)

The report *With Heritage So Rich* had the advantage, which many reports lack, of appearing at precisely the right time for a positive political response. It was cogently argued, dramatically illustrated, and persuasive. It consisted of a series of essays and a concluding set of recommendations. Some of these were immediately implemented by the National Historic Preservation Act of 1966: for example, the establishment of an Advisory Council on Historic Preservation (ACHP)[4] and of a National Register of Historic Places. A remarkable change of policy in relation to highway construction was introduced in a provision of the Transportation Act of 1966 which requires the Secretary of State for Transportation to refuse approval for projects which would involve damaging or demolishing historic sites unless there is "no prudent and feasible alternative."

A similar provision was included in the Model Cities Act 1966 in relation to urban renewal plans. Later amendments extended this policy to all federal departments. Changes in taxation were made by other legislation, such as the Tax Reform Act 1976 and the Economic Recovery Tax Act 1981, to encourage historic preservation (for example, by way of tax deductions for rehabilitation). A separate Act, the National Environmental Policy Act (NEPA), included additional provisions for preserving "important historic, cultural, and natural aspects of our national heritage."

As this brief summary demonstrates, the fifteen years following the publication of *With Heritage So Rich* witnessed a veritable orgy of legislative activity. In the following pages, the more important features of this are discussed.

The National Register of Historic Places

In thus marking an extensive collection of properties for special attention (including economic) and special protection (including from the government's own activities), the Register helps broaden – or even create – Americans' sensitivity to the historic value of the built environment. At the same time, the Register serves as the pivot upon which the economic and legal supports for preservation turn.

(Keune 1984)

The National Register of Historic Places is maintained by the Keeper of the National Register in the National Park Service of the Department of the Interior. It lists districts, sites, buildings, structures, and objects which are significant on a national, state, or local level in American history, architecture, archeology, engineering, and culture: in short America's cultural resources.[5] These are provided with a degree of protection from the harmful effects of federal action. The federal government is committed, by law, to protect these resources: agencies are required to follow a statutory process of review and consultation with the ACHP in connection with any undertaking affecting properties included in the list. Additionally, and at first sight curiously, this requirement extends to properties which, though not listed, are *eligible* for listing.[6] Though both listed and eligible properties are subject to the review process, only listed properties are qualified to receive grant aid or tax advantages (discussed below).

The statutory requirements relating to the process which reviews whether a federal action will have an adverse effect (popularly known as the section 106 process) read as follows:

The head of any federal agency having direct or indirect jurisdiction over a proposed federal or federally assisted undertaking in any state and the head of any federal department or independent agency having authority to license any undertaking shall prior to the approval of the expenditure of any federal funds on the undertaking or prior to the issue of any license, as the case may be, take into account the effect of the undertaking on any district, site, building, structure, or object that is included in or eligible for inclusion in the National Register. The head of any such federal agency shall afford the Advisory Council on Historic Preservation . . . a reasonable opportunity to comment with regard to such undertaking.

Detailed guidelines on the review and consultation process have been issued by the Council.[7]

The "section 106" process is not a mere

formality: all federal actions and federally funded projects are monitored or reviewed by preservationists. This usually occurs at the State Historic Preservation Office level. Indeed, "review and compliance," as it is called, now occupies a dominant position in the state programs. However, as always, much depends upon the quality of the local administration.

Statewide Comprehensive Historic Preservation Planning

If it is the role of the planner concerned with land use patterns to understand them in relationship to the dynamics of the contemporary land market and its interplay with social and cultural values, then it is the task of the historic preservation planner to understand the evolution of those patterns over time and to assess the significance of remaining fragments. Historic preservation planning is one of several perspectives on, and public interests in, land.

(Ames, *et al.* 1989)

Though historic preservation is very much a local matter, it is more than this: as with all local plans, relationships with wider plans have to be forged. (The imperative is misleading since, in practice, as has already been stressed, there are so few *plans* – as distinct from zoning provisions.) Ideally these would include such functional elements as transportation planning, economic development planning, and environmental planning. The most promising approach is where different planning agencies integrate (or at least cooperate in) their planning processes. In the words of the Delaware Comprehensive Historic Preservation Plan:

It is very difficult, if not impossible, to integrate complete plans that can translate the recommendations of one plan or functional area into terms relevant to

another. Plans must be integrated, or information exchanged, at the points in the planning process when problems and alternative goals are defined and analyzed and decisions made.

(Ames *et al.* 1989: 9)

Moreover, without coordination, historic preservation policies may conflict with land use policies. Duerksen quotes the case where

preservationists have struggled to enact an ordinance to control design details or forbid demolition by private developers in a historic neighborhood, only to discover that the real threat in the area is a city zoning policy encouraging high-rise development. In short, preservationists have focussed on design issues and on saving threatened buildings when the key issue is more often how landmarks and their surrounding areas will be developed according to local zoning classifications and redevelopment programs.

(Duerksen 1983: 44)

Coordination has other advantages, not the least being that it impresses courts that the municipality has a comprehensive plan, and is working to this rather than making a series of ad hoc decisions. It also facilitates the use of sophisticated zoning techniques such as incentives, bonuses, and the transfer of development rights.

Coordination is also desirable between the policies of the municipality and the state. A well-known example is Oregon's statewide planning goals, which are mandatory on municipalities. One of these goals includes the requirement that local programs shall be provided which will "protect scenic and historic areas and natural resources for future generations, and promote healthy and visually attractive environments in harmony with the natural lanscape character." Inventories are required of historic areas, sites, structures, and objects; and cultural areas. An historic area is defined as "lands with sites, structures, and objects that have local, regional, statewide, or national historical significance." A cultural area is "an area characterized by evidence of ethnic, religious or social group with distinctive traits, beliefs, and social forms" (Rohse 1987: 260). Local comprehensive plans and land use regulations are required by statute to comply with these goals.

Highways and Historic Preservation

> It is time that Congress took a look at the highway program, because it is presently being operated by barbarians, and we ought to have some civilized understanding of just what we do to spots of historic interest and great beauty by the building of eight-lane highways through the middle of our cities.
>
> (Senator Joseph S. Clark 1966)

The ravages of highway construction constituted one of the major reasons for the swell of public opinion against "the federal bulldozer." It is therefore perhaps fitting that the strongest federal provision is to be found in a transportation act. The Department of Transportation Act declares that it is a matter of national policy that a "special effort" shall be made to preserve and enhance the natural beauty of lands crossed by transportation lines.[8] As indicated earlier, approval to any program or project cannot be given for the use of:

> any publicly owned land from a public park, recreation area, or wildlife and waterfowl refuge of national, state or local significance as determined by the federal, state or local officials having jurisdiction thereof, or any land from an historic site of national, state, or local significance as so determined by such officials unless (1) there is no feasible and prudent alternative to the use of such land,

and (2) such program includes all possible planning to minimize harm to such park, recreational area, wildlife and waterfowl refuge, or historic site resulting from such use.

The scope of this requirement is much broader than that provided in the NHPA: it gives protection to any site considered by officials as being of historic significance – not only those listed, or eligible to be listed, in the National Register. Moreover, the "no feasible and prudent alternative" is more stringent than the NHPA which provides only for ACHP "comment." By contrast, the Transportation Act permits a harmful use only if a) no feasible and prudent alternative exists, and b) all possible planning is carried out to minimize harm. The courts have held that there must be "truly unusual factors" of "extraordinary magnitudes" for this high standard to be met.[9]

Section 4(f) has been used in relation to a wide variety of historic sites, buildings, and objects, from the French Quarter in New Orleans – the cancellation of an expressway (Baumbach and Borah 1981) – and the childhood home of Thomas Jefferson, to Hawaiian petroglyphic rocks, a truss steel bridge (Riznik 1989), Indian archeological sites, and many others. "Unlike parklands, recreational areas and wildlife refuges, section 4(f) also applies to privately owned historic sites as well as those in public ownership. This extension recognizes the breadth of historically or culturally significant properties and the role private ownership plays in historic preservation. Most of the properties listed in the National Register of Historic Places are privately owned" (Wilburn 1983: 2018).

Other legislation dealing with specific modes of transportation have similar provisions, e.g. the Airport and Airway Development Act 1970, the Federal-Aid Highway Act 1968, and the Urban Mass Transit Act 1976.[10]

The National Environmental Policy Act

Federal preservation laws are weak sisters to federal environment laws.

(Dworsky *et al.* 1983)

The National Environmental Policy Act of 1976 (NEPA) establishes a national policy of environmental protection. The historic preservation element of this refers to the preservation of "important historic, cultural, and natural aspects of our national heritage" and the maintenance, wherever possible of "an environment which supports diversity and variety of individual choice." The legislation requires every "major federal action" which "significantly affects the environment" to be preceded by an environmental impact statement (Duerksen 1983: 270–81). This must contain a detailed analysis of the environmental impact of the proposed action, any adverse environmental effects that cannot be avoided if the proposal is implemented, and alternatives to the proposed action.

There is some overlap between NHPA and NEPA (and the environmental protection Acts passed by several states), and regulations have been issued by the Council on Environmental Quality in relation to coordination.[11] The two Acts can be seen as reinforcing each other. Dworsky *et al.* have written:

> The two laws reinforce each other and can be used effectively in tandem: if NHPA does not apply to a historic resource, NEPA might. While some courts may hold that agencies need not continue to comply with NHPA after a federal project has commenced, courts have generally agreed that NEPA does apply in such situations. If NHPA is weakened through funding cuts and revisions to the federal regulations to the ACHP, NEPA can still be used to compel agencies to consider historic properties.

(Dworsky *et al.* 1983: 305)

Nevertheless, the federal Acts provide no guarantee that cultural resources will be protected: the only means which guarantees protection is acquisition.

Finance for Historic Preservation

The use of federal tax incentives to encourage historic rehabilitation continues to be one of the most successful urban revitalization tools ever implemented by Congress, even though the Tax Reform Act of 1986 has significantly reduced the attractiveness of tax incentives as an investment strategy.

(Chittenden 1988)

Taxation provisions often work against sectoral policies: typically they are, not surprisingly, designed to raise revenue, not to further public policies. So it has been with historic preservation. Prior to 1976, the tax laws actually discouraged the preservation and rehabilitation of historic properties. (Tax deductions were allowable for the costs of demolition.) The 1976 Tax Reform Act created a number of preservation incentives: tax credits for certain rehabilitation expenditures and (as a disincentive to the demolition of historic buildings) an increase in the "tax cost" of demolition.

The use of preservation tax incentives increased enormously after the passing of the Economic Recovery Act of 1981 which introduced a new, and highly attractive, system of tax credits.[12] By the mid-1980s, the program was running at an annual rate of 3,000 projects and an investment of $2 billion. Reagan's Tax Reform Act of 1986 drastically cut these incentives but, even so, some 1,000 certified rehabilitations a year (involving $900 million of investment) were undertaken in the late 1980s. Between 1976 and 1989, a total of some 21,000 historic buildings were rehabilitated with the aid of tax incentives, representing private sector investment of almost $14 billion (Blumenthal and Siler 1990: 1).

In addition to the tax incentives program, the NHPA of 1966 provided matching grants to the states for historic preservation survey, planning, acquisition, and development. With the funding cutbacks of the 1980s, little acquisition and development is being carried out, but most states continue to use grant funds for "survey and planning" – which includes the preparation of nominations to the National Register, and developing technical preservation information (Dworsky et al. 1983: 226).

In addition to federal tax incentives, there are state incentives. These vary considerably among the states, and take many forms. There are, however, basically six taxation methods used to encourage historic preservation: exemption, credit or abatement for rehabilitation, special assessment for property tax, income tax deductions, sales tax relief, and tax levies. A report of the NTHP's "State Legislation Project" gives details for each state (Davis 1985).

State and Local Programs

The council finds that many improvements . . . and landscape features . . . having a special character or a special historical or aesthetic interest or value and many improvements representing the finest architectural products of distinct periods in the history of the city, have been uprooted, notwithstanding the feasibility of preserving and continuing the use of such improvements and landscape features, and without adequate consideration of the irreplaceable loss to the people of the city of the aesthetic, cultural, and historic values

represented by such improvements and landscape features.
(New York City Landmarks Law)

As in so many areas of land use planning, it is at the local level that most of the real action takes place. All states have a State Historic Preservation Officer (SHPO). This is a federal requirement, and the Secretary for the Interior has the responsibility of approving state programs that provide for the designation of a SHPO, a state historic preservation review board, and a scheme for adequate public participation in the state program. Each SHPO is required to identify and inventory historic properties in the state; nominate eligible properties to the National Register; prepare and implement a statewide historic preservation plan; serve as a liaison with federal agencies on preservation matters; and provide public information, education, and technical assistance (ACIR: 1985: 7).

In 1956, New York State became the first to pass legislation enabling municipalities to enact an ordinance for individual landmark buildings (as distinct from historic areas). New York City was the first to take advantage of this (in 1965). A New York City Landmarks Commission was established and empowered to designate properties of significant historic or aesthetic value. Designated properties cannot be demolished or altered without the approval of the commission. This is given only if the commission decides that the proposed works will have no effect on the protected architectural features, is otherwise consistent with the purposes of the landmarks law, or is necessary to secure a reasonable return to the owner (assessed at 6 percent). The "purposes" of the landmarks law are as follows:

to (a) effect and accomplish the protection, enhancement and perpetuation of such improvements and landscape features and of districts which represent or reflect elements of the city's cultural, social, economic, political, and architectural history;

(b) safeguard the city's historic, aesthetic, and cultural heritage, as embodied and reflected in such improvements, landscape features and districts; (c) stabilize and improve property values in such districts; (d) foster civic pride in the beauty and noble accomplishments of the past; (e) protect and enhance the city's attractions to tourists and visitors and the support and stimulus to business and industry thereby provided; (f) strengthen the economy of the city; and (g) promote the use of historic districts, landmarks, interior landmarks, and scenic landmarks for the education, pleasure, and welfare of the people of the city.

(New York City Landmarks Law)

Until the *Penn Central* case was settled by the Supreme Court in 1978, there was some doubt as to the constitutionality of such legislation. This case involved a proposal for the erection of a 55-story office building atop the city's beaux-arts masterpiece, the Penn Central Railroad's Grand Central Terminal: a building which the city had designated as a landmark. The New York City Landmarks Commission rejected the proposal, and the owners took the matter to court with two complaints. Firstly, they argued, there had been, in effect, a taking of their property without just compensation. Secondly, by designating the terminal, the city had discriminated against the owners in requiring them to bear a financial burden which neighboring owners did not have to shoulder.

In a six-justice majority, the Supreme Court upheld the action of the Landmarks Commission. Though previous decisions had provided no clear rule for determining whether a taking had taken place, in this case it was determined that there was no taking: the owners had been left with a reasonable return, and the restrictions imposed were within the police powers of the city. It was explicitly stated that "states and cities may enact land use restrictions or controls to enhance the quality of life by preserving the character and desirable aesthetic features of a city."[13]

The importance of the *Penn Central* case is underlined by Duerksen: "*Penn Central* made it clear that localities could forbid demolition or stop new construction for preservation purposes. Thus, a landowner who did not understand local preservation law could face serious economic consequences." Together with the federal tax incentives introduced in 1976 (which provided landowners with significant benefits), historic preservation law suddenly emerged as a subject "worth studying and practicing, just as environmental law had almost a decade earlier" (Duerksen 1983: 19).

The result was a major increase in historic preservation activity. Though there was some setback with the financial cuts imposed by the Reagan administration, historic preservation was clearly at the stage of becoming established as a significant land use control.

An additional note about the New York Landmarks Commission is appropriate. A major factor in its establishment was the widespread concern about the loss of a railway terminal – Pennsylvania Station, which was destroyed in 1963. It was a curious twist of fate that made another terminal (Penn Central) the subject of a case which confirmed the legitimacy of the commission and its functions. There have, however, been constant rumblings about the way in which these functions have been carried out. In particular, the commission has been accused of as acting "as a kind of planning commission of last resort, stepping in to prevent or slow the pace of development in circumstances in which the planning commission had failed to act" (Goldberger 1990). Paul Goldberger has noted that though this was not what the commission had been created for, "over the last decade it generally did the right thing, even if it did overreach its mandate from time to time."[14]

New York City is only one of forty-five local governments in New York State to have a preservation commission, and of course, New York State is only one of the fifty states.[15] Its experience is therefore not necessarily represen-

III   116   The quality of development

tative, however newsworthy it may be. A contrasting case is that of the city of Roanoke, Virginia.

The Roanoke Vision

> Assisted by a grant from the National Trust for Historic Preservation's Critical Issues Fund, Roanoke revised its existing ordinance in conjunction with a comprehensive planning process called Roanoke vision.
> (Roanoke City Planning Commission 1986)

While only very few illustrations of the experience of particular towns can be provided in this book, cases have been selected with a view to illustrating interesting developments. Roanoke is a good case in point. It exemplifies a trend toward a greater coordination of historic preservation and land use controls (J.H. Miller 1987).

The approach was a truly comprehensive one, encompassing the comprehensive plan, zoning, and historic preservation. The objective was to design a policy which could promote neighborhood conservation and improved design quality on a citywide basis. The city had a strong overlay district (a set of zoning specifications that is imposed on a map as an addition to the underlying district requirements)[16] limiting demolition and providing design controls which had protected two downtown historic districts, but neighborhood areas had remained largely unprotected — to the detriment of the neighborhoods and the historic districts. As a result, there was considerable destruction of the residential fabric of many neighborhoods, as well as the loss of many fine older structures.

> While the development focus and economic realities [in Roanoke] have changed, the regulatory tools which influence investment and set development and land use standards have remained relatively constant. The city

did enact a historic district overlay zone to protect its City Market and Warehouse districts as part of the city's recent downtown revitalization, but the overall zoning ordinance continued to be based on development trends of the 1950s and 1960s. It did not recognize the growing values placed on preserving the scale, style and character of the city's past. In fact, the existing ordinance actually encouraged the destruction of many of the city's older and low income neighborhoods.
> (Roanoke City Planning Commission 1986)

Following extensive public participation, a new comprehensive development plan and a new zoning ordinance was prepared. Of particular interest is the introduction of Neighborhood Preservation Districts, in addition to Historic Districts. These are designed to encourage the conservation and revitalization of older neighborhoods. (They require, for instance, the issuance of a certificate of appropriateness before any building can be erected, demolished, moved or structurally enlarged or reduced in floor area.)

In a break from traditional zoning, new provisions for special uses in historic structures have also been included in the new ordinance. These include arts and crafts studios; art galleries; antique shops/rare book, coin or stamp shops; community centers; professional offices (not to exceed four employees); and multifamily apartments (not to exceed four units within a building).[17]

Historic Preservation and Tourism

> cultural tourism, by creating the conditions for a new humanism, must henceforth be one of the fundamental means, on a universal level, of insuring man's equilibrium

and the enrichment of his personality, in a civilization which, owing to the ever more rapid development of technical progress, may now be daily directed towards the intelligent use of its leisure.

(International Council of Monuments and Sites 1969)[18]

One of the objectives of historic preservation is often the promotion of tourism. Sadly, success here can bring its own problems. Too many people seeking a particular experience can result in its destruction. Fitch notes that many popular places – Kyoto, New Orleans, Paris, Leningrad (St. Petersburg) – are facing threats to their actual physical fabric. He continues:

In many famous individual monuments, tourist traffic has reached its absolute limits: at Mount Vernon, George Washington's residence, stairs and floors have had to be reinforced to carry the weight of visitors; and the abrasion of flooring surfaces is so severe that protective membranes must be replaced in a matter of weeks. Faced with the noise, confusion, and downright squalor which such overcrowding often produces it would be all too easy to reject the whole concept of mass tourism and yearn for a return to the good old days of aristocratic travel.

(Fitch 1982: 78)

More generally, National Park Service officials have commented that Americans are in danger of loving their national parks and historic sites to death (Hunt 1988: II.43). This problem is not, however, shared by most places of historic interest. On the contrary, the economic development importance of historic resources is underlined by the use of the term "heritage tourism." In the words of the ACHP:

Heritage tourism is just one way in which the preservation and maintenance of historic towns and urban areas may contribute to overall economic improvement. Innovative programs initiated at all levels, public and private, illustrate how preservation efforts can support and complement economic and social developments in urban areas.

(ACHP 1989: 35)

Under the heading "economic revitalization" the 1989 ACHP *Report to the President and Congress* summarizes three initiatives in heritage (or "cultural") tourism. In Lockport, Illinois, the Gaylord Building rehabilitation project is the first in the National Heritage Corridor, a 120-mile historic district designated in 1984, which stretches along inland waterways from Chicago to La Salle-Peru. This "blend of natural and historic resources" is attracting tourist dollars to the Lockport area.

In the town of Port Townsend, Washington, tourism tripled between 1983 and 1988 as a result of promoting Victorian-era neighborhoods and downtown facade rehabilitation. And in Georgia, the Antebellum Trail is being promoted (using hotel room taxes) as a tour of sites which Sherman missed on his march to the sea. Many other examples are given in the annual repors of the ACHP. In addition, a publication of the NTHP gives eloquent details of historic preservation and downtown revitalization.[19]

The Widening Scope of Historic Preservation

One of the most characteristic aspects of historic preservation today is that its domain is being constantly extended in two distinct ways. On the one hand, the *scale* of the artifact being considered as requiring preservation is being pushed upward to include very large ones (e.g. the entire island of Nantucket) as well as downward, to include very small ones (e.g. historic rooms or fragments thereof installed in art museums).

On the other hand, the domain is being enlarged by a radical increase in the *type* of artifacts being considered worthy of preservation. Thus in addition to monumental high-style architecture – traditionally the concern of the preservationist – whole new categories of structures are now being recognized as equally meritorious: vernacular, folkloristic, and industrial structures. In a parallel fashion, the time scale of historicity is being extended to include pre-Columbian settlements at one end and Art Deco skyscrapers at the other.

(Fitch 1982)

The lengthy quotation from Fitch nearly says it all! The boundaries of "historic preservation" are being stretched in such a way that the term is now a misnomer. One has only to peruse the volumes of Perspectives in Vernacular Architecture (e.g. Wells 1986; Carter and Herman 1989) to see the way in which interests are broadening. At the same time, new problems are arising, and old problems are taking on new dimensions. Reference has been made above to the criticism of the New York City Landmarks Commission that it had, on occasion, exceeded its mandate. With heightened concern for "historic" areas of the twentieth century, this may become more common. The issue is complicated by an overlapping concern to preserve low-income housing from redevelopment. In April 1990, for instance, the Landmarks Commission gave landmark status to a complex of fourteen buildings in the Yorkville section of Manhattan which were originally constructed as a privately financed experiment to provide housing for the poor. Here is a nice mixture of historic, architectural, social, and economic issues. Some idea of the flavor of the debate can be gleaned from the following quotation from the New York Times:

Paul Selver of the law firm of Brown and Wood, which represents the owner, said there was "nothing special" about the property to warrant landmarking ...

The commission, in its resolution, noted that the projects represented an attempt by a group of prominent New Yorkers "to address the housing needs of the working poor." Investors agreed to voluntarily limit their profits, and the apartments provided occupants with interior plumbing, more window space and more light and air than typical tenement apartments of the time.

(New York Times April 15, 1990)

The commission also praised the "distinction" of the architecture, and maintained that designation would help to protect an area which represents "an important slice of history of the Upper West Side."

Clearly a host of different interests and values are at stake here. In such cases, the matter is settled (perhaps after recourse to the courts) by determining which interest – or interest group – is to prevail. And so we get into what Bishir (1989), in a stimulating and entertaining paper has termed "the politics of culture." She ends this by stressing that, since preservationists are participants in the politics of culture, it is necessary for them to be aware of the impact on their decisions of their value system. Whether self-knowledge is sufficient is an open question.

One final (and significant) point needs to be made.[20] It was earlier stated that "historic and associative" elements remain as important today as they were a century ago. It is, however, important to note the emergence (particularly in the western United States) of scholarly studies of archaeology which have profoundly affected our view of material history. The traditional approaches to the field varied. As Torma notes:

At the onset of the national preservation program, the field was divided into at least two distinct camps – the archaeologists were on one side and the architectural historians and historians were on the other. While the orientation of the archaeologists was cultural, the orientation of the architectural historians and historians was traditional history and history of aesthetics. One group was trained in the social sciences (and some would say the sciences) and the other in the humanities. While the archaeologists looked at all aspects of the "cultural picture" – economic base, diet, foodways, architecture and seasonal migration patterns (to name a few) – those working in the historic sites program were generally concerned with only two issues: is this structure aesthetically beautiful and/or does it have already demonstrated historic value?

(Torma 1987)

The coming together of these different approaches has proved fruitful, and new perceptions of "historic preservation" are emerging.[21] In Stipe's words (1987: 274), the subject matter of historic preservation has become "thoroughly democratized", and topics such as vernacular architecture, and industrial and commercial archaeology are now common and popular topics. The very term "historic preservation" is being replaced by broader concepts of "heritage."[22] It is an exciting time for students and practitioners in this field.

Part IV
Urban growth and urban policy

Chapter 10

Growth management and local government

Introduction

> After twenty years of experience with growth management, it must be said that we know very little about how to manage urban development.
>
> (D.R. Porter 1989a)

A European observer would expect that zoning had a great deal to do with the management of development on the urban fringe (growth management), but he would be sadly wrong. American zoning largely proceeds on the basis of decisions regarding individual lots. What is typically ignored is the cumulative effect of an enormous number of "lot decisions." This is partly because the zoning machine usually operates without the advantage of a guiding plan; partly because zoning has traditionally been unconcerned with the timing of development (or its relationship to the provision of infrastructure); and partly because the normal presumption of municipalities is in favor of development – the more, the better. The last

point goes deep: instead of asking "is the proposed development desirable in the public interest at this place at this point in time?," the typical municipality starts from the presumption that any development is good and, in any case, it is unfair to penalize a particular owner with a refusal: if one farmer's land has been approved for development, why shouldn't his neighbor get equal treatment?

Traditional zoning therefore has difficulty in even attempting to relate development decisions to wider questions of planning. It is essentially reactive and "timeless." The difficulties to which this may be expected to give rise are exacerbated by the fact that zoning maps usually have a similar timeless quality. They show the use to which individual lots of land may – in isolation – reasonably be put, but they do not take into account the effect of the timing of development applications or the effect of a number (and certainly not all) of the proposals emerging at a particular time. The availability of public services (from sewers to roads to schools) does not enter into the political calculus. Development patterns can therefore be haphazard, inefficient, and wasteful, costly to service, and cumulatively disastrous – with inadequate public services, "gridlock" and the like.

Added to the political predispositions are a number of other complicating factors. The dictates of the Constitution are one – particularly the requirement for equal treatment (how does a political body – typically consisting of a very small number of members – defend unequal treatment to landowners on some fuzzy basis of the public interest?). Another is the division of responsibility between different agencies. Transportation planning is frequently the responsibility of an agency different from the one concerned with zoning; schools always are. As a result, zoning is the major discretionary function of municipalities – and sometimes the dominating issue at local elections (Ellickson 1977: 405).

By a curious twist of the tale (common on the American scene), action to promote coordinated planning may be interpreted (not always unjustly) as an underhand means of excluding minority groups from an area – what Bosselman (1973: 249) has characterized as "the wolf of exclusionary zoning under the environmental sheepskin worn by the stop-growth movement." All these considerations help to explain the widespread popularity of large lot zoning: it results in development which makes the minimum demands on public services (and on the demand for an expansion of them); it pays for itself in the narrow terms of a municipal budget; and it excludes minorities from the area. It is, of course, very inefficient. To quote from the still-valid Douglas Commission's report:

> At the metropolitan scale, the present techniques of development guidance have not effectively controlled the timing and location of development. Under traditional zoning, jurisdictions are theoretically called upon to determine in advance the sites needed for various types of development. . . .

the difficulty of predicting has turned many governments to the "wait and see" approach. In doing so, however, they have continued to rely on techniques which were never designed as timing devices and which do not function well in controlling timing. The attempt to use large-lot zoning, for example, to control timing has all too often resulted in scattered development on large lots, prematurely establishing the character of much later development – the very effect sought to be avoided. New types of control are needed if the basic metropolitan scale problems are to be solved.

(Douglas Report 1968: 245)

These ideas will be explored more deeply in later chapters: they are introduced here to provide a reference point for the ensuing discussion of the interesting, and largely unsuccessful, attempts to plan the location and timing of urbanization.[1]

Considerable ingenuity has been displayed in devising techniques of growth management. They range from restrictive subdivision and zoning regulations, "permits" to begin development, "caps" on the number of new dwellings (either annually or over a period of years), phasing development along with the provision of infrastructure, urban growth limit lines, and the preservation of land for agricultural or other highly restricted uses. A complete list of all the possible measures would be a very long one. The massive Urban Land Institute study of 1975 listed eighty-six![2] Indeed, most planning techniques can be utilized for growth management purposes. Some of the discussion of the subject in this book is therefore scattered among the various chapters.

Any account of this subject must include two machinations in mathematical probity which assumed fame in the early 1970s – the growth control programs of *Ramapo* and *Petaluma*.

Ramapo and *Petaluma*

> Phased growth is well within the ambit of existing enabling legislation.
>
> (*Golden v Planning Board of the Town of Ramapo* 1972)

> The concept of the public welfare is sufficiently broad to uphold Petaluma's desire to preserve its small town character, its open spaces and low density of population, and to grow at an orderly and deliberate manner.
>
> (*Construction Industry Association of Sonoma County v City of Petaluma* 1975)

Ramapo is a town in Rockland County, New York, about thirty-five miles from downtown Manhattan. At the end of the 1960s, it had a population of around 76,000, and was growing rapidly. As a result, there was an increasing strain on public services and infrastructure. A master plan had been adopted in 1966, followed by a comprehensive zoning ordinance, and then by a capital improvements program and a phased growth plan. The latter provided for the control of residential development in phase with the provision of adequate municipal facilities and services. The various plans covered a period of eighteen years.

The timed growth plan did not rezone any land: the restraint on property use was regarded as being of a temporary nature. This restraint took the form of a requirement that a special permit be obtained for residential development.

> The standards for the issuance of special permits are framed in terms of the availability to the proposed subdivision plat of five essential facilities or services: specifically (1) public sanitary sewers or approved substitutes; (2) drainage facilities; (3) improved public parks or recreation facilities, including public schools; (4) state, county or town roads – major, secondary or collector; and (5) firehouses. No special permit shall issue unless the proposed residential development has accumulated fifteen development points, to be computed on a sliding scale of values assigned to the specified improvements under the statute.[3]

Thus where the required municipal services were readily available a special permit would be granted, but where a proposed development was located further away, development could not begin until the programmed services reached the location – unless the developer installed the services himself.

The court held that "where it is clear that the existing physical and financial resources of the community are inadequate to furnish the essential services and facilities which a substantial increase in population requires, there is a rational basis for 'phased growth' and, hence, the challenged ordinance is not violative of the federal and state constitutions."

The Ramapo plan implied an annual quota on residential development. This was not a predetermined figure: the actual number of dwellings built was dependent upon capital improvements and the ability of developers to acquire points under the point system.[4] The Petaluma plan, by contrast, operated by way of a fixed quota.

Petaluma lies some forty miles north of San Francisco. In the 1950s and 1960s it experienced a steady population growth, from 10,000 in 1950 to 25,000 in 1970. By the latter date, however, this self-sufficient town had been drawn into the Bay Area metropolitan housing market, and development boomed. Whereas only 358 dwellings had been built in 1969, the number rose to 591 in 1970, and 891 in 1971. Alarmed at this rate of growth, the city introduced a temporary freeze on development. This provided a breathing space during which a growth management plan could be prepared. The plan, adopted in 1972, fixed development at a maximum rate of 500 dwellings a year (excluding projects of four or fewer units). To give effect to this control mechanism it was necessary to have a system which would choose between competing claimants. The instrument devised for this purpose was an annual competition among rival plans in which points were awarded for access to existing services which had spare capacity, for excellence of design, for the provision of open space, for the inclusion of low-cost housing, and for the provision of needed public services. (The policy allocated 8–12 percent of the annual quota to low- and moderate-income housing.)

Not surprisingly, the development interests in the area were highly alarmed, and a case was brought against the city. The district court declared the plan to be unconstitutional, but the court of appeals reversed. Though it accepted the view that the plan was to some extent exclusionary, it noted that "practically all zoning restrictions have as a purpose and effect the exclusion of some activity or type of structure or a certain density of inhabitants." The court's review did not cease upon a finding that there was an exclusionary purpose: what was important was to determine whether the exclusion bore any relationship to a legitimate state interest. The court held that the Petaluma plan did in fact serve such an interest. (See the quotation at the beginning of this section.) Moreover, the plan was certainly not exclusionary in the sense of keeping out low-income households. On the contrary, it was "inclusionary to the extent that it offers new opportunities, previously unavailable, to minorities and low and moderate income persons."

Ramapo and *Petaluma* are only two of many growth management schemes which have been introduced since the late 1960s. They are particularly notable because of the blessing bestowed upon them by the courts (and their prominence in standard texts). However, not all schemes have been approved by the courts. For instance, the attempt by Boca Raton, Florida, to place a cap (of 40,000) on the number of dwellings ultimately to be built in the city (agreed by a public referendum after a superficial review of urbanization trends) was declared unconstitutional.[5] Perhaps Boca Raton was just unlucky, though its action was not backed up by the supportive planning studies which courts like to see. Yet – a point which needs to be constantly borne in mind – most schemes are not in fact challenged.[6] For example, there has been no challenge to the Californian City of Napa's Residential Urban Limit Line, which is intended to limit the city's population to 75,000. This Line represents the boundary beyond which essential public services will not be provided. It is accompanied by the Napa Residential Development Management Plan, which imposes an annual ceiling on new residential construction (Dowall 1984: 83).

The reasons for introducing growth management policies vary. Thus, while Napa's Residential Urban Limit Line is intended to "cap" the population growth of the area, nearby Santa Rosa has an urban boundary designed to permit all the development which is envisaged for the foreseeable future. Should further land be required, the boundary can be extended: the objective is not to prevent growth but to ensure that it takes place in a desirable manner. It is aimed at the problems of "scatteration" and the "unnecessary use" of agricultural land. It also attempts to preserve environmental quality and enhance the aesthetic quality of new housing (Dowall 1984: 84). Other California cities (such as San Rafael and Novato) have also been concerned essentially with the preservation of open space and the establishment of green belts. Similarly, the comprehensive planning program in operation in Montgomery County, Maryland, is "intended to accommodate growth, and to manage it only to the extent needed to moderate its ill effects" (Christeller 1986: 82).

Growth Management and Infrastructure

The growth management movement is a response to large-scale national demographic and settlement pattern shifts, which have put pressures for new types of development on those local areas that, for the most part, are least equipped to deal with them.

(Godschalk *et al.* 1979)

The use of infrastructure planning as a major element in land use controls has become a popular one. A common method of coordinating urban growth and the provision of infrastructure is to require developers to hold back until the necessary provision can be made (as in Petaluma). This also allows developers to proceed if they themselves provide the infrastructure, or the finance for it. (Impact fees, discussed in Chapter 6, may form an element of such a scheme, whether or not the overall intention is the limitation or the management of growth.)

Another permutation of this school of controls is impact zoning. This has been particularly popular in the towns of Massachusetts. "Drawing on NEPA and the lawyer's continuing faith in procedural solutions, these towns have amended their bylaws to require a statement of the impact of proposed subdivisions on town services and the local environment." The amendment to the zoning bylaw may take a form such as the following:

In order to evaluate the impact of the proposed development on Town services and the welfare of the community, there shall be submitted an Impact Statement which describes the impact of the proposed development on (1) all applicable town services, including but not limited to schools, sewer system, protection; (2) the projected generation of traffic on the roads of and in the vicinity of the proposed development; (3) the subterranean water

table, including the effect of proposed septic systems; and (4) the ecology of the vicinity of the proposed development. The Impact Statement shall also indicate the means by which Town or private services required by the proposed development will be provided, such as by private contract, extension of municipal services by a warrant approved at Town Meeting, recorded covenant, or by contract with homeowner's association.

(Haar and Wolf 1989: 592)

Another example of the use of infrastructure planning in land use controls is provided by the experience of Boulder, Colorado. Boulder has had a long history of planning to preserve its dramatic natural surroundings. Frederick Law Olmsted, Jr., produced a report in 1910 on *The Improvement of Boulder, Colorado*, and in 1928 the city became one of the first western cities to introduce a zoning ordinance. Not surprisingly, pressures continued on the peripheral areas and in particular on the mountain foothills. To stem this, a 1959 charter amendment established an elevation of 5,750 feet along the mountainsides, beyond which utility service could not be extended (Cooper 1986: 35). In the 1960s, 40 percent of a one-cent city sales tax was earmarked for open space acquisition (the balance went to road improvements). By the end of the 1980s, the city had spent $53 million on the acquisition of 17,500 acres of open space most of which lies outside the city limits (S. Lewis 1990a: 16).

The city has been able to exercise some control over development beyond its boundaries by virtue of its utility functions: in a part of the country where water is in short supply, Boulder had virtually complete control of the water in its area (Godschalk *et al.* 1979: 258); but it received a setback in 1976 when the courts ruled that it was unconstitutional to use the powers of a public utility for planning purposes.[7] However, by cooperating with the surrounding Boulder County, a comprehensive plan for the larger area was agreed, and this enables the two governments to coordinate planning and annexations. Boulder uses a system of phased-in development in which the area is divided into three sections. The first consists of the nineteen square miles now within the city limits, and has a full range of public services. The second, 7.5 square miles under county jurisdiction, is targeted to be annexed and serviced within three to fifteen years. The third, some fifty-nine square miles, is not projected for servicing after fifteen years — if ever.

An interesting feature of Boulder's planning policy is its purchase of development rights, which keeps land in agriculturally productive use but prevents development upon it.[8] Boulder is by no means alone in its concern for protecting agricultural land from urbanization, as the following discussion demonstrates.

Safeguarding Agricultural Land

> Opinions vary on how much farmland is being lost to urbanization and other uses — and the impact such losses might have on our Nation and the world in the future. As world population increases, total food production must also increase. We would be better able to meet future food requirements if the Nation's best farmland is preserved for agricultural use.
>
> (U.S. General Accounting Office 1979)

> Although a broad, governmentally sponsored program of planning and farmland preservation has little to objectively recommend it, the idea has had a persistence that is hard to explain.
>
> (Delogu 1986)

The safeguarding of agricultural land has a strangely captivating and persistent appeal. It is readily accepted that food-producing land is "under threat," that its loss is irreversible, and

that it is folly to reduce national self-sufficiency in food supplies. This may be expected in countries such as the Netherlands, which have a scarcity of land; but at first sight it seems inappropriate to a country of the vastness of the United States. Nevertheless, there is considerable controversy on precisely this point.[9] The issue achieved salience in the 1960s with concern about environmental degradation, urban sprawl, and the pressure for national land use policies. The federal reaction was to mount the National Agricultural Lands Study (U.S. Department of Agriculture 1981). The accuracy of the data presented in this report has been subject to intense debate, but little consensus has appeared. Action at the federal level has been minimal. The most significant legislation has been the Farmland Protection Policy Act of 1981. This requires the Department of Agriculture to "develop criteria for identifying the effects of federal programs on the conversion of land to nonagricultural uses." It also requires federal agencies to use these criteria:

> to identify and take into account the adverse effects of federal programs on the preservation of farmland; consider alternative actions, as appropriate, that could lessen such adverse effects; and assure that such federal programs to the extent practicable are compatible with state, units of local government, and private programs and policies to protect farmland.
>
> (Farmland Protection Policy Act 1981)

The regulations appeared in draft form in 1982, but became the subject of much controversy (Schnidman, Smiley and Woodbury 1990: 8). While some argued that they did not go far enough, others argued the opposite. When the final version of the regulations appeared in 1984, they did little more than require that federal agencies consider the impact of their activities on the conversion of farmland: there is no requirement that the activities should be changed as a result of the impacts.

While federal action has been less than dramatic, there has been much action at state and local levels. Indeed, the position has not changed since the 1981 National Agricultural Lands Study noted that state and local governments are the prime instigators of agricultural preservation.

At the state level,[10] the most common program is some type of favorable tax treatment such as assessment at existing use (farming) value rather than market value (which may include potential development value). This can apply to both property and inheritance taxes. However, it is doubtful whether such tax benefits are very effective on their own: owners may simply enjoy reduced taxes until the time comes when they want to sell. They are, of course, popular with farmers, and therefore they enter into the arena of state politics.

Also popular are "right to farm" laws: these protect farmers from local ordinances (and private nuisance suits) that restrict normal farming operations.[11] There is little analysis of the effectiveness of such laws (Popp 1988: 525), though one study concluded that, while they reduced the number of private actions for farm-related nuisance, they had no effect on the loss of farmland to other uses, especially in peri-urban areas (Lapping and Leutwiler 1987). They may be of greater help to farmers, at least in the short run, than the strategy employed by one Delaware farmer who placed a huge notice on the boundary between his mushroom farm and a new housing development warning prospective buyers of the unpleasant environment into which they were being enticed to move. (The reader may wish to be reminded of the *Hadacheck* case, summarized in Chapter 2.)

State programs tend to be rather blunt instruments: the real action is at the local level (though in fact it is rather modest). Here, there are three main approaches: agricultural zoning, the purchase of development rights, and the transfer of development rights. Agricultural zoning, as its name suggests, restricts use in the defined zone to agriculture. It is a simple tech-

nique, but it is open to constitutional challenge (Popp 1988: 528) and, not surprisingly, it faces strong political opposition from farmers. However, it can be useful when coupled with other measures such as tax incentives or the transfer of development rights.

The most effective way of safeguarding agricultural land (other than outright purchase at market value) is by the acquisition of the development rights. All the states have passed legislation enabling such acquisitions (usually at local level), but the costs are so high that few local governments can contemplate a program on any significant scale. There are, however, various devices for overcoming this difficulty by the transfer of development rights (TDR). This is a relative newcomer to the armory of planning techniques. It is simple in concept but complex in its details. In essence, it separates the development value of land from its existing use, and "transfers" that development value to another site. The owners of the land in the area to be preserved can sell their development rights to developers in designated "receiving" areas who are thereby allowed to build at an increased density reflecting the value of the transferred rights. Unlike traditional zoning techniques, TDR gives farmers an incentive to retain their land in agricultural use. Few TDR programs have been implemented, though they have attracted considerable interest.[12] The program in Montgomery County, Maryland, is one of the best known, and perhaps the most dramatic.[13] This designates "preservation areas" where downzoning has reduced development density on about a third of the 500 square mile county to one house per twenty-five acres. In addition, there is a transferable development right of one house per five acres – the density which the land had before designation. This can be sold to developers in receiving areas.

> Receiving areas are the designated sites to which development rights can be transferred. They must be specifically described in an approved and adopted

master plan, a process by which areas are screened to assure the adequacy of public facilities to serve them and to assure compatibility with surrounding development. Each receiving area is assigned a base density. Developers can build to this density as a matter of right. To achieve the greater density permitted under the TDR option, the developer must purchase development rights. No rezoning is necessary, but a preliminary subdivision plan, site plan, and record plat must be approved by the Montgomery County Planning Board.

(Banach and Canavan 1987: 259)[14]

Another scheme of great complexity operates in the New Jersey Pinelands (Pizor 1986).

All the evidence shows that the preservation of agricultural land can involve very large costs. In assessing whether these are justifiable, it is necessary to ask not only what the objectives are, but also who actually benefits.[15] In their review of agricultural land protection policies in New England, Schnidman, Smiley, and Woodbury (1990: 321) conclude that there are four interrelated concerns behind the adoption and implementation of such policies in this region. These are the difficult-to-define but easily recognizable quality of the rural landscape; environmental degradation (pollution in all its forms); the quality and regional availability of food products; and the various economic benefits of agriculture (such as its beneficial impact on the economy, and the avoidance of the problems of land speculation and rising land prices).

Interestingly, it is the "aesthetic" concerns which predominate. The term is used here in a very wide sense to mean the general quality of the environment.

> State-by-state review of New England farmland protection efforts reveals that in every state one of the most important concerns was the desire to preserve certain aesthetic qualities which agricultural lands

provide. The determination to protect open space and local community character evolved primarily from intangible motivations such as the value of farming as a lifestyle that is pleasing to the eye. The traditional Yankee farm, with its small fields surrounded by stone walls, woodlands, and rural architecture, has given the landscape a unique visual character that a majority of New Englanders, both urban and rural, want to protect.

(Schnidman, Smiley, and Woodbury 1990: 322)

It is noteworthy that these issues are not only interrelated but also somewhat elusive;[16] but clearly the major issue is a vague unease and concern about the way in which a familiar and friendly environment is changing. This, of course, is a common feature of the operation of the planning and zoning system. Fischel (1982: 257) argues that "the real beneficiaries...and the real force behind the farmland preservation movement, are local antidevelopment interests." By contrast, the American Farmland Trust (1988), in its survey of schemes for the purchase of development rights in Massachusetts and Connecticut, underlines the benefits to individual farmers and to the farming industry generally. A more balanced viewpoint is expressed in a monograph emanating from the long-term research program on farmland protection programs carried out by the Florida Joint Center for Environmental and Urban Problems. As with all good research projects, the conclusions raise as many questions as are answered:

> the Center has concluded that urban conversion of agricultural lands, while not posing an immediate threat to America's food supply or its strategic position in international affairs, does warrant concern on other grounds. Certainly, other things being equal, it makes little sense for a society to shift agriculture from better to worse lands if planning and management would allow

more efficient uses of land resources. Nor is it prudent to convert agricultural land to urban uses if the urban development in question is itself wasteful and socially expensive. The challenge is to develop farmland protection programs that distinguish inefficient from efficient land uses and promote objectives more complicated than simply indiscriminately saving all agricultural land.

(Hiemstra and Bushwick 1989: xi)

Conclusion

The slow-growth movement has proved that it can win elections. What it has not proved, however, is that it can stop growth.

(Fulton 1990)

In this chapter a brief account has been given of the urban growth controls operated by a number of local governments, together with a summary of some policies relating to the safe-guarding of agricultural land. The policies discussed are interesting, and they are certainly popular with local residents (Dowall 1980; Baldassare 1986) – that is those who have not been prevented by the controls from living in the area. There is no lack of critics,[17] some of whom are very sure of themselves. Ellickson (1977), for example, describes growth controls as a type of "homeowner cartel," while Frieden (1979) slates "the defense of privilege." However, the evidence is variable in reliability, and equivocal or contradictory in its results. Thus, while Schwartz *et al.* (1979) conclude that the Petaluma policy resulted in higher house prices, Logan and Zhou's study found that (both in *Petaluma* and *Ramapo*) "despite the sound and fury of legal battles, [their poli-cies] had little demonstrable effect on subse-quent development." On the other hand, another study concluded that "one of the impacts of Ramapo's conscious efforts to slow

its rate of growth has been to shift development to the incorporated villages in the township."[18]

Of course, there is no reason why all local governments should be equally successful (or unsuccessful) in controlling urban growth. Conditions, policies, and administration will vary greatly among different areas. The issues are well treated in a valuable paper by Logan and Zhou, who conclude that:

> the planning mechanisms afforded by governmental autonomy should be evaluated cautiously. The formal ability of a community to control its own growth, and even the formal adoption of growth controls and environmental limitations, do not guarantee their implemention. . . .
>
> Policies may be adopted for symbolic as well as practical reasons, and to appease diffuse public opinion as well as to respond to a well-organized political force. What appears to be a restrictive policy may on close inspection turn out to be a license for more intensive development. Even the most carefully drawn legislation is dependent upon subsequent enforcement. And often the legislation is reactive to a pattern of events that have already run their course, too late to make a difference.
>
> (Logan and Zhou 1989: 469)

However, it may sometimes be that, though policy is expressed in terms of urban growth management, the real purpose is to secure leverage in the planning process to obtain benefits for the locality. Fulton illustrates the point by reference to one city where, in order to obtain permission to proceed with develop-ment, developers are "offering amenities that they would not otherwise offer and that the city would not be legally permitted to ask for" (Fulton 1990: 29).

Moreover, what happens to growth pressures which are stemmed in one area? Do they neces-sarily move to another area where development is in the public interest? How can any rational assessment be made of such matters without a proper land use plan? Interestingly, Judge Choy made a similar point in the *Petaluma* case, where he noted that:

> If the present system of delegated zoning power does not effectively serve the state interest in furthering the general welfare of the region or entire state, it is the state legislature's and not the federal courts' role to intervene and adjust the system. . . . The federal court is not a super zoning board and should not be called on to mark the point at which legitimate local interests in promoting the welfare of the community are outweighed by legitimate regional interests.
> (*Construction Industry Association of Sonoma County v City of Petaluma* 1976)

In short, policies relating to growth management cannot be adequately designed and implemented on a local basis: a regional or state outlook is required. A few states have realized this, and are making attempts to create a new intergovernmental system of land use control. This is the subject of the next chapter.

Chapter 11

Urban growth management and the states

Introduction

This country is in the midst of a revolution in the way we regulate the use of our land. It is a peaceful revolution, conducted entirely within the law. It is a quiet revolution, and its supporters include both conservatives and liberals. It is a disorganized revolution, with no central cadre of leaders, but it is a revolution nonetheless.

The ancient regime being overthrown is the feudal system under which the entire pattern of land development has been controlled by thousands of individual local governments, each seeking to maximize its tax base and minimize its social problems, and caring less what happens to all the others.

The tools of the revolution are new laws taking a wide variety of forms but each sharing a common theme – the need to provide some degree of state or regional participation in the major decisions that affect the use of our increasingly limited supply of land.

(Bosselman and Callies 1972)

This now famous quotation was written in the heady days of the 1970s, and its promise has not been fulfilled. Only a few states have become involved in land use planning (as distinct from environmental planning), and there are marked differences among them in the purpose and scope of their involvement. The collapse of the campaign for a National Land Use Policy Act in the mid-1970s (discussed in the next chapter) led to a general loss of interest in land use planning at the state level. On the other hand, many state Acts have been passed to deal specifically with environmental matters, particularly wetlands and, to a smaller extent, coastal areas. Only a dozen or so states have become involved with land use controls.[1]

Callies and Freilich (1986: 902) have suggested that there are three principal reasons for the introduction of state systems of land management and control:

a) a perception that local governments were unwilling or incapable of making decisions of supra-local impact except on narrow, parochial grounds;

b) a crisis of sorts regarding a particularly large development or a particularly critical feature or resource threatened by actual or pending land development; and

c) the threatened imposition of federal land use controls.

In this chapter, six illustrative types of state land use planning are discussed: Hawaii, Vermont, Florida, California, Oregon, and New Jersey. (In fact, there are not many others, though – as discussed at the end of this chapter – more states have recently introduced or considered new legislation.) Each of these states faced one or more problems which was seen as requiring state action. (That action meant taking back some of the state powers which had earlier been delegated to local governments to enable them to administer zoning.) Hawaii was troubled by the rapid urbanization of its valuable agricultural land; Vermont faced a sudden large increase in development pressures; Florida experienced phenomenal growth; California was witnessing a large loss in public access to the coast; Oregon had similar problems with its coastline, and also problems of urban development and speculation; and New Jersey (the most urbanized state in the nation) was facing massive urbanization.

There are clearly some similarities among these states: all have had to devise ways of dealing with growth. But, as will become apparent in the following account, each has its distinctive set of problems, goals, constraints, and plans. Thus Hawaii introduced statewide zoning, Vermont set up a system of citizen district commissions to administer a development plan system, Florida introduced state controls in "areas of critical concern" and

"developments of regional impact," California established a coastal planning system, Oregon set up a comprehensive system of local planning which had to conform to a long list of state goals, and New Jersey battled with the introduction of a state plan intended to guide growth and conservation throughout the state. Many of the initial provisions have been revised, for a variety of reasons ranging from a recognition that the early provisions were inadequate to changes in political control. The story continues to unfold, of course.

Hawaii

Regulating Paradise: State Planning in Hawaii[2]

It all began in Hawaii.
(Bosselman and Callies 1972)

Hawaii's approach to land use control is, as elsewhere, a product of its history; but this history is very different from that of the other forty-nine states. Indeed, the land use planning system that has emerged is unique. It is briefly discussed here, however, since it was the first: as the quotation indicates, state land use planning "all began in Hawaii." Though the systems devised in other states are different, Hawaii is important partly because it was the first, but also because it is of intrinsic interest.

The particular history of Hawaii led to a concentration of land ownership which lasted until a combination of events (including the advent of a young reform-minded state administration) brought into being the 1961 Land Use Law (Cooper and Daws 1985). The concern was not simply with land ownership, but with the effects of the policies being operated by the landowners. In short, an increasing pace of development (particularly by way of premature subdivisions) threatened Hawaii's agriculture-based prosperity. Thus, unlike the situation typical of other states, controls were introduced, not to retard or control growth but "to promote and to enhance the economic prosperity of the islands" (DeGrove 1984: 12). Moreover, the 1961 law was not motivated (as was the case in the 1970s on the mainland) by environmental considerations, though these arose at a later date: "the prime motivating force was the fear that agriculture, especially the pineapple and sugar plantations, would be wiped out by careless and unplanned urban development" (DeGrove 1984: 15).

Another distinctive feature of Hawaii is its governmental structure (Mandelker 1976a: 269). This is highly centralized, with the state having responsibility for major services such as education, welfare, and housing, which elsewhere are usually delegated to local governments. There is no tradition of autonomous local government: the islands are governed by four counties, and there are no lower levels of local government.

Thus Hawaii is different in significant ways from the other states, and the story of the development of its land use planning is therefore different.

The 1961 Act established a Land Use Commission which was charged with designating all land in (currently) four "districts": urban (5 percent), agriculture (47 percent), conservation (47 percent), and "rural" (1 percent) (Callies 1984: 7).

The urban districts cover land which is in urban use or which will be required for urban purposes in the foreseeable future. The administration of zoning in these districts is the responsibility of the counties. The designation provides no rights to urban development: it merely signifies that the county *may* zone the land for urban development under its zoning code. Thus the counties can impose more restrictive conditions, but they cannot relax the commission's regulations.

Agricultural districts cover land used not only for agricultural purposes but also for a range of other uses, including "open area recreational facilities." In establishing agricultural districts, the commission is required to give the "greatest possible protection" to land which has a high capacity for intensive cultivation.

Conservation districts are primarily forest and water reserve zones, but also include historic sites, mountains, and offshore outlying islands. The administration of planning in the conservation districts lies directly with the state government, operating through the Land Board of the Department of Land and Natural Resources.

The final small category, of "rural districts" was added in 1963 to permit low density residential lots. These are principally small farms and rural subdivisions which are inappropriate for either the agricultural or urban designations. Administration lies with the Land Commission.

The Commission was originally required to undertake district boundary reviews every five years, but now it does so only on petitions from landowners. It operates within the framework of the state plan, which was approved by the Legislature in 1978.

The Hawaii State Plan

The Legislature finds that there is a need to improve the planning process in the State, to increase the effectiveness of public and private actions, to improve coordination among different agencies and levels of government, to provide for wise use of Hawaii's resources and to guide the future development of the State.

(Hawaii Act 100, 1978)

Hawaii was the first state to enact a comprehensive plan. It is a short document which sets out a series of "themes" and policies for the state covering a wide range of issues including health, culture, education, and public safety, as well as land use, population, and the environment (Healy and Rosenberg 1979: 192). Its provisions are of a general rather than detailed and specific character. Nevertheless, "there are many goals, policies, and priorities that relate to land and growth management that are sufficiently specific to be directive in an important way for the commission, state agencies, and the four city-counties in Hawaii" (DeGrove 1984: 33).

The provisions of the state plan relate to a number of important matters concerning population growth and distribution. These include:

a) The carrying capacity of each geographical area;
b) The rehabilitation of appropriate urban areas;
c) Directing urban growth primarily to existing urban areas;
d) Directing urban growth away from areas where other important benefits are present, such as protection of valuable agricultural land;
e) the preservation of "green belts" through a variety of techniques;
f) the identification of critical environmental areas where urban growth should be excluded; and
g) the encouragement of new industrial development in existing and planned urban areas.

All of this has the cumulative policy thrust of directing urban growth toward compact urban centers, and so far as possible existing urban centers, with the related policy of protecting agricultural lands, open space, and critical environmental areas.

(DeGrove 1984: 34)

These extracts give some indication of the character of Hawaii's comprehensive state plan. However, it is one thing to prepare a plan: it is another to make a plan work. And here there is considerable, if not overwhelming, difficulty. Those who have tried to assess the effectiveness of Hawaii's land planning endeavors have come up with only partial answers, such as that things could have gone better if the state had been able to elaborate clearer guidelines and more specific standards to guide its regulatory system.[3]

Hawaii is the only state to operate a centralized statewide system of land use controls. As already explained, the particular history and governmental system of Hawaii accounts for this. In 1978, Hawaii's four counties had only rudimentary land use and zoning schemes. However, matters have changed greatly since then:

Now [1986] both are well developed in all four, and the city and county of Honolulu has one of the most sophisticated land use management systems in the country, with detailed development plans consisting of common provisions text, land use maps, and public facilities maps covering every inch of the island of Oahu, and a county charter which forbids even the "initiation" of a subdivision or zoning change unless it conforms to the development plans.

(Callies and Freilich 1986: 915)[4]

This change raises a question about the continued relevance of the statewide system of controls:

Given this increase in sophistication and capability, together with the regional nature of local government (roughly one county government for each of Hawaii's major islands) is there still a significant need for statewide land use controls as set out in the Land Use Law? Indeed, is there any development which could be construed as truly statewide in importance and effect, given that the boundaries of each county stop at the water's edge?

(Callies and Freilich 1986: 915)

Vermont

Almost Getting it Together in Vermont[5]

With a sense of crisis firmly established, focused largely on a fear that Vermont's way of life, centered on rural and small town areas with great environmental beauty, was about to be destroyed, decisions were made over a very short period of time that moved Vermont into the front ranks of states in the land and growth management area.

(DeGrove 1984)

Like Hawaii, Vermont has some very special features which set it apart from other states. It is a largely rural state (one of the most rural in the United States) in which the pressures for development come mainly from outside. In terms of population (some 563,000) it ranks 48th in the 1990 census. As Heeter (1976: 325) has nicely commented, until the early 1960s, the number of cows in the state exceeded the number of humans. It has a highly decentralized local government system of small New England communities with nine cities and 237 "organized towns." The towns cover all of the land outside the cities, and therefore are of a large area (averaging some thirty-six square miles), though with small populations – averaging less than 1,000 in 1970. There are also five un-organized towns (in sparsely populated areas) and fifty-seven incorporated villages which are urban in character, and have responsibility for some of their own services (such as police, fire protection, water supply). Counties largely exist only on paper. "Thus, between the cities and towns and the state there is a gap in the governmental structure which the state has attempted to bridge by administering many of its programs on a regional basis" (Heeter 1976: 326). Vermont has no tradition of state planning;[6] yet it passed in 1970 a growth management measure which in fact introduced comprehensive statewide planning.

The major reason for this was a transformation of Vermont from a state whose young people traditionally left for better opportunities elsewhere into a state beset by the problems of unprecedented growth. This was caused by a number of factors. The extension of the interstate highway system brought Vermont within easy travel distance of the 40 million inhabitants of the urbanized areas to the south. Several economic changes also took place within the state, some of which were related to this new accessibility, including the expansion of the ski industry and the growth of second homes. Almost suddenly, Vermont changed from a remote area to an easily accessible vacation, second home, and commuters' haven. It is, of course, a beautiful state.

The resultant growth in population led to development pressures and increased land costs (and taxes), which were alarming to the conservative Vermonters. A 14 percent growth in population during the 1960s, though modest by the standards of California or Florida, was greater than the increase over the previous half-century. It was, moreover, concentrated in particular areas, and therefore its impact was larger than the overall figure suggests.

The local governments of Vermont were quite unable to deal with this unprecedented situation. None of the towns had a capital budget program, and few had a zoning ordinance. The desperate situation is described by Healy and Rosenberg:

The lack of full time town executives or town planners frequently meant that the developer was far better prepared technically than the town zoning board, if any, had to approve his proposal. One environmentalist remarked that it was just about impossible to hire a lawyer in southern Vermont who was not on some developer's payroll. In almost every town where there was a development, he said, the first move made by the developer was to put a selectman [councilman] – sometimes more than one – on the payroll.

(Healy and Rosenberg 1979: 43)

In a remarkably short space of time, the state government acted, and legislation (Act 250) was passed in 1970. This introduced a development permit system administered by an appointed environmental board and district environmental commissions. It also provided for the preparation of three statewide plans: an interim land capability plan (an inventory of physical data); a land capability plan (to guide "a coordinated, efficient, and economic development of the state"); and a final land use plan. The development permit system has worked reasonably well, but the plans have given rise to a number of difficulties, and the land use plan never emerged.

The Vermont Development Permit Scheme

From a political perspective, there is a widespread consensus that the local lay citizen approach embodied in the district environmental commissions is one of the outstanding strong points of the law.

(DeGrove 1984)

Establishing new agencies of government is always problematic. In particular, there is the perennial issue of decentralized versus centralized control. In Vermont, it was clear that the local government system could not administer the development permit scheme, but there was little enthusiasm for "putting any more power in the hands of the state's bureaucracy" (DeGrove 1984: 69). The solution adopted placed the major responsibility for administering the development permit system on lay citizen district commissions – with the right of appeal to a lay state board. Thus the process "is strongly decentralized, though decentralized in a way which bypasses existing local govern-

ments" (DeGrove 1984: 74). Ideally, of course, a plan should have preceded the introduction of this system, but there was no time for this:

With developers starting now to subdivide thousands of new acres of land, much of it on mountains and in areas where subdivision regulations are not in existence, towns cannot wait upon the completion of comprehensive plans.

(Myers 1974: 12)

In the absence of a plan, Act 250 provided a list of ten criteria against which development applications are to be judged. No permit can be issued unless a commission finds that the proposed development:

1 Will not result in undue water or air pollution.
2 Has sufficient water for its reasonably foreseeable needs.
3 Will not cause an unreasonable burden on an existing water supply, if one is to be used.
4 Will not cause unreasonable soil erosion or reduction in the ability of the land to hold water.
5 Will not cause unreasonable congestion or unsafe conditions on highways or other transportation facilities.
6 Will not cause an unreasonable burden on the ability of a municipality to provide educational services.
7 Will not cause an unreasonable burden on the ability of the local government to provide governmental services.
8 Will not have an undue adverse effect on the scenic or natural beauty of the area, aesthetics, historic sites, or rare and irreplaceable natural areas.
9 Is in conformance with statewide plans required by Act 250.
10 Is in conformance with any duly adopted local or regional plan or capital program.

Permit hearings are thorough but informal, with an emphasis on the settlement of disputes by negotiation (Myers 1974: 19). Permits are required for (1) housing developments of ten or more units by the same applicant within a five-mile radius; (2) developments involving the construction of improvements for commercial or industrial purposes on a tract of more than one acre in towns without permanent zoning and subdivision bylaws and on a tract of more than ten acres in towns with such controls; (3) developments involving the construction of improvements for state or municipal purposes of a size of more than ten acres; and (4) all developments above an elevation of 2,500 feet (Heeter 1976: 333). Daniels and Lapping (1984: 502) have argued that since these provisions are aimed at large scale developments, many environmentally damaging small scale developments subdivisions have escaped review. Indeed, land sales and development activity is purposely designed with this objective in mind.[7]

The Vermont Plans

The controversy over whether the state land use plan does or does not involve state zoning has been a continuing and bitter one in Vermont, and lies at the heart of the difficulties that have been encountered in adopting the plan.

(DeGrove 1984)

The interim plan (the first of the three required by Act 250) raised relatively little controversy, though signs of future difficulties were apparent. These mushroomed during the course of the preparation of the second plan, the Land Capability and Development Plan (1974). Basic to the difficulties was the lack of clarity in the legislation as to the purposes and scope of the plan. Much of the controversy stemmed from a confusion between two planning concepts: environmental protection, and the coordination of planning across the state. There were also problems of relationship between board members and planning staff.[8] The important outcome was that the final plan has never surfaced. Indeed, the popular distrust of the concept of a state plan led to the repeal of the relevant statutory provisions (Sinclair 1988: 18).

Of course, plans typically have multiple objectives, some of which may be difficult to harmonize, while others may be contradictory. It seems that the Vermont planning system has had successes in improving the quality of large scale developments (Environmental Board of Vermont 1981). On the other hand, land market activity by out-of-state residents, particularly vacation property, appears to have remained significant (Daniels and Lapping 1984: 507). The Vermont planning approach is essentially "reactive": it evaluates planning proposals which are submitted for approval; it does not direct growth to areas which are considered by planners to be suitable for growth. In short, as a growth management system, a lot remains to be desired. A 1988 report by a commission appointed by the Governor underlined the perceived weaknesses, and stressed the fact that "a consequence of the failure to adopt comprehensive local and regional plans is that basic planning decisions are left to the regulatory process." This was inadequate: there was an urgent need "to introduce planning into the regulatory process" (Vermont 1988a).

The 1988 Growth Management Legislation

It is the traditional settlement pattern (village, town, and countryside) that reflects the essence of Vermont. In order to maintain the essential character and ethic of Vermont's built environment, there should be a clear delineation between town and countryside through effective planning and supportive land development.

(Vermont 1988a)

The Governor's Commission on Vermont's Future, from whose report the above quotation is taken, reported in 1988: legislation giving effect to the recommendations was passed in the same year. The new planning system is based on the commission's analysis of the wishes of Vermonters:

1   Vermonters want decisions made at the lowest possible level of government.
2   Vermonters do not want a new layer of government.
3   Vermonters recognize that what is done locally may affect more than one community.
4   Most towns do not have adequate resources to manage growth (financial, technical, and legal).
5   Vermont does not have a clearly defined process for dealing with growth impacts that affect several towns.
6   Effective decision making through local and regional planning will help guide public investment decisions and reduce burdens on the regulatory process.
7   State agencies have conflicting mission statements that interfere with their ability to plan effectively.

Against this background, the 1988 legislation retains the existing regional planning commissions with wider powers, and subject to a requirement that they cooperate with other agencies and levels of government (Vermont 1988b). All regional commissions and state agencies are required to ensure that their planning is consistent with twelve broad state planning goals. In this revised system the regional commissions become the vital force in growth management: they are assigned the responsibility for approving town plans, commenting on state agency plans, and reviewing proposed state capital expenditures.

A new agency, the Council of Regional Commissions, has been created to review regional and state agency plans to ensure its compatibility with state goals. Additionally, a Municipal and Regional Planning Fund has been established to assist municipal and regional planning commissions. A geographic information system (to which all commissions and agencies are to contribute data) is being financed from this fund.

The revised Vermont system is a neat balancing act between the requirements of area-wide planning and the strong proclivity of Vermonters for local control. But, in essence the planning process works from the "bottom up," though within the framework of state policies.

Florida

Florida: Growth Facilitator to Growth Manager[9]

Deep-rooted love affairs are always difficult to terminate, and Florida's love affair with growth has been no exception.

(DeGrove 1984)

Florida is an example of how the state can inject rational policy by centralizing authority, creating a uniform process, and supporting an agenda of utilitarian benefits through government intervention.

(Turner 1990a)

Florida's growth in post World War II years has been phenomenal – a result of its attractive environment, its warm climate, and its low taxes. In 1950 the state had a population of 2.8 million. This increased to 5 million in 1960, 6.8 million in 1970, 9.7 million in 1980, and 12.9 million in 1990. It is now the fourth largest state in the Union.

Such a growth would have presented problems in any state, but the problems in Florida are compounded by its unique, fragile, and complex natural environment. These are the most difficult in precisely the areas of the greatest growth – in the southern part of the state. If ever a situation cried out for strong planning measures, this is it.

It took some time for Floridians to appreciate and acknowledge this, but a number of incidents in the 1960s and early 1970s brought about a new environmental awareness which formed the political basis for a remarkably quick response (how far this was effective is another matter which will be addressed later in this discussion). The story is a complex and fascinating one which has been chronicled in brief by DeGrove (1984), and at length by Carter (1974). The elements of this story include the abandonment of the building of a barge canal across the northern part of the peninsular (cancelled by Nixon); the guaranteeing of an adequate water supply to the Everglades National Park; the decision not to go ahead with the Jetport in the Big Cypress Swamp; and the trauma of the 1971 water crisis.[10]

> The point was driven home as never before that indiscriminate drainage and headlong growth in South Florida had combined to produce an environmental crisis of truly major proportions. The response to this crisis brought about a coalition of political and environmental forces that thrust Florida from being among the least growth management oriented states to becoming a state with advanced land and water management tools with which to attempt the management of growth
>
> (De Grove 1984: 106)

The outcome was a series of Acts to control development. Of particular importance was the Environmental Land and Water Management Act of 1972 which (modeled on the American Law Institute's *Model Land Development Code*) provided for the designation of "areas of critical state concern" (ACSC) and for special measures for dealing with "developments of regional impact" (DRI). The appeal of these techniques (in addition to the fact that the legal

expert F. Bosselman was involved in both the preparation of the ALI Code and in the design of the Florida legislation) was that they furnished a nice balance between state, regional, and local interests. Development normally remains the responsibility of the municipalities, but in the case of an ACSC or DRI, higher levels of government are involved.

Areas of critical state concern can be recommended by the state planning agency[11] in three types of area:

(1) places which are of special environmental, historical, archaeological, or other significant regional or state importance (such as aquifer recharge areas, wetlands, critical habitats);
(2) areas significantly impacted by an existing or proposed major public facility (such as the South Florida Jetport); and
(3) areas of major development potential designated in a state development plan (such as a development under the New Communities Act).

Only four areas have been designated: the Big Cypress Area, the Green Swamp, the Florida Keys, and the Apalachicola Bay Area.

While areas of critical state concern are designated by the state, developments of regional impact are a matter for local governments, subject to review by the regional planning council and the state. As Healy and Rosenberg put it:

> The process does not start when an area is thought ripe for development, nor when new infrastructure makes building more attractive, nor when a state plan indicates that a particular area should grow more quickly. A DRI begins only when a developer proposes to build a particular project in a particular place.
>
> (Healy and Rosenberg 1979: 145)

The system is therefore a reactive one, and it was made more difficult initially because of the absence of a comprehensive state plan. It was, however, an improvement on the previous system in that it brought into the development approval procedure the regional level of government. All of the state is now covered by eleven regional planning agencies (which are essentially multicounty councils of government).

This whole system was characterized by persuasion: persuasion of one level of government by another, and persuasion of developers by the municipalities. In the first five years of the scheme, thirty-eight local decisions were appealed, but negotiation, changes in the proposal, withdrawal etc., resulted in only four cases reaching the adjudicatory commission: in only one case did the commission reverse local approval: "negotiation and compromise, rather than forceful cabinet level decision making, seem to characterize the DRI process" (Healy and Rosenberg 1979: 149). The obvious weaknesses in the system (particularly the absence of a state plan, lack of funding for local planning, and the inadequacy of review of plans by the state) led eventually to the introduction of major changes in the planning system.

The Omnibus Growth Management Act of 1985

The Omnibus Growth Management Act of 1985[12] represents Florida's most ambitious, comprehensive and far reaching effort to date at controlling the adverse physical, economic, and environmental consequence of rapid population and economic growth.
(DeGrove and Juergensmeyer 1986)

In the mid-1980s, Florida overhauled its planning system at state, regional, and local levels (DeGrove and Stroud 1989). The revised system is in essence one of growth management. A hierarchy of plans features a comprehensive state plan with which the plans of state agencies ("functional plans") and regions

(regional plans) must be consistent; similarly with local plans. The state plan adopts twenty-five goals and policies, covering education, children, families, the elderly, housing, health, public safety, water resources, coastal and marine resources, natural systems and recreational lands, air quality, energy, hazardous and non-hazardous materials and waste, mining, property rights, land use, public facilities, cultural and historic resources, transportation, governmental efficiency, the economy, agriculture, tourism, employment, and plan implementation (DeGrove and Juergensmeyer 1986: 23).

The new system transforms Florida planning. In place of the "bottom up" character of the earlier legislation, it is now unequivocally "top down." All municipalities and counties are obliged to prepare and adopt comprehensive plans. The requirement that these be consistent with the state plan is not mere rhetoric. If the plan is not "compatible with the goals" of the state plan, the state can impose some severe sanctions, particularly the withholding of funds.

A remarkable new provision requires local governments to coordinate the provision of infrastructure with urban growth. Development can be permitted only to the extent that the infrastructure can support it:

The most powerful new policy (of the comprehensive Planning Act) was the provision that it would be unlawful for a local government to approve new development unless the infrastructure was in place concurrent with the development. Put simply, the state and all participants must stop the deficit financing of growth caused by the build up of large infrastructure backlogs, and begin paying the cost of growth as it occurs.
(DeGrove 1988: 9)

The Act provides that "it is the intention of the legislature that public facilities and services needed to support development shall be available concurrent with the impact of such development." A document issued by the Florida Department of Community Affairs notes that local governments must design adequate and realistic "level of service" (LOS) standards for roads, sewers, drainage, water, recreation, and (if applicable) mass transit (Florida 1989b). Development which would fail to maintain LOS standards cannot be permitted unless the deficiency will be made good by the provisions of the capital investment plan.

The lynch pin here is the finance for infrastructure. How this will be provided is not yet clear.

"Keys to Florida's Future"

The total cost over the next ten years of implementing the State Comprehensive Plan will be $52.9 billion.

(Florida 1987)

The problem of financing infrastructure and other public services loomed large in the report of the State Comprehensive Plan Committee (Florida 1987). This committee was charged with calculating the costs of implementing the state comprehensive plan, and recommending specific ways of paying for those costs. It showed that the state had a major problem ahead of it:

Our failure to provide the transportation, wastewater, solid waste, corrections, schools, and other public facilities we need threatens to have a chilling effect on the economy.

Our continued balanced growth is endangered by threats to the air, water, beaches, wetlands, and other natural resources that are essential to lasting prosperity for the state.

Primary, secondary, and higher education in Florida fall short of the standards we need to compete successfully.

Florida ranks last in the nation – 50th among the 50 states – in per capita spending on basic human services. And Florida ranks 47th in state and local taxes as a percentage of personal income.[13]

Our failure to pay for growth leads to needless regulation that hinders private enterprise, stifles initiative and incentive, and diminishes our quality of life.

And our low tax rates and our undue reliance on a narrow based sales tax keep us from having the stable and reliable flow of governmental revenues that is needed to attract and accommodate quality growth.

(Florida 1987)

After this dismal litany, the committee set out a wide-ranging set of proposals for reforming the tax system, increasing taxes, and introducing user fees.[14]

This has not proved to be easy. The same people who support urban growth management may well refuse to agree to the levels of taxation necessary to implement it. However, some progress has been made: an increasing number of local governments have imposed road impact fees, and some have introduced one-cent sale taxes (with some of the income being earmarked for transportation). In 1990, the state legislature increased gasoline tax by four cents a gallon and added other charges which are expected to generate an additional $3 billion over five years for transportation (Koenig 1990: 6). Unfortunately this is far short of what is required for transportation, and there is little evidence that the shortfall will be met by other taxes. In particular there was the firm opposition of Governor Martinez to most tax increases. He maintained a strong Republican line of "no new taxes" ever since he was forced to rescind a 5 percent sales tax, which he introduced in 1987 to cover professional services.

Voter resistance to taxes is, of course, nothing new, but the vociferousness of the opposition can outweigh benefits which when presented without a cost tag seem highly desirable. Several Florida communities have passed tax referendums, but others have rejected them. Koenig (1990: 8) instances Broward (Fort Lauderdale) and Hillsborough (Tampa), where the voters rejected a one cent local-option sales tax intended, in part, to pay for more roads. "With the failure of their referendum, planners in Broward project a $217 million shortfall in meeting transportation needs" over the following two years. Given such situations, local governments can be forced to impose moratoria on new development. Naturally, this draws the wrath not only of developers, but also of construction industry workers. Both have shown themselves to be effective in opposition to such action though, unless the system collapses, there are no further alternatives. The choices are limited to raising taxes to pay for the needed infrastructure, putting new building on hold, or abandoning the planners' hope of dealing with the problems. Curiously, another problem has emerged out of all this: the concurrency requirement has tended to drive developers to those areas which have spare road capacity, and the result is urban sprawl in rural areas.

> Broward County's map of areas where plat approvals are deferred illustrates the problem. The over-capacity roads are primarily those in the urban core and suburban areas, while the rural areas remain open for development. As long as the over-capacity roads are not improved, development will be driven into the very areas where, according to the growth management law, it is supposed to be discouraged.... As a result, local governments are scrambling to come up with anti-sprawl measures to combat the very trend that concurrency encourages.
>
> (Koenig 1990: 8)

Whether a future governor of Florida will be able to deal with these problems remains to be seen.[15]

California

The California Coast

> The independent California Coastal Commission ... has survived several hundred bills introduced in the state legislature to kill or cripple California's coastal program, has engaged in a pitched battle with the US Department of the Interior over the federal offshore oil leasing program, and has scored planning successes in providing public access to the shoreline and creating transfer-of-development-credit programs in Big Sur and the Santa Monica Mountains.
>
> (Fischer 1985)

Strictly speaking, California's coastal program is not a statewide comprehensive planning endeavor: as its name suggests, it is concerned only with the coast. But that coast is 1,100 miles long (and the coastal planning area is up to five miles wide in rural areas). It is therefore very much akin to a statewide planning area. However, in many ways it is unique. Even its birth was unusual. It was established, not by state action but as a result of the use of California's extraordinary ballot initiative (well known through the 1978 "tax revolt" of Proposition 13).[16] For some time, attempts had been made to get the legislature to act against the boom in condominiums, subdivisions, and above all, developments which sealed off the beach from the public.[17] Eventually the voters took matters into their own hands and, in 1972, after a tempestuous campaign, succeeded in getting a coastal protection initiative placed on the ballot. Despite a well-funded aggressive counter-campaign by developers, oil companies, and the like (and the opposition of Governor Ronald Reagan) the initiative received a 55 percent affirmative vote. In the ensuing legislation, an interim California Coastal Zone Conservation Commission was established. The commission began regulating coastal development by permit in February 1973.

The commission was in fact seven bodies: the state commission and six regional commissions (which, as will be noted, were later abolished). The mandate of the state commission was to regulate all development in the permit area, to prepare a coastal plan, and to make recommendations for the permanent management of the coastal zone. The coastal plan and recommendations were completed at the end of 1975 and (though they had no legal standing in themselves) greatly influenced the Coastal Act of 1976.

The California Coastal Plan

The legislature further finds and declares that, notwithstanding the fact electrical generating facilities, refineries, and coastal-dependent developments, including ports and commercial fishing facilities, offshore petroleum and gas development, and liquefied natural gas facilities, may have significant adverse effects on coastal resources or coastal access, it may be necessary to locate such developments in the coastal zone in order to ensure that inland as well as coastal resources are preserved and that orderly economic development proceeds within the state.

(N. Williams 1985–90)

As the quotation indicates, the coastal plan is not concerned solely with environmental protection: it seeks to ensure that the coastline is used intelligently and sensitively, with due regard to both the environment and the needs of coastal-related development. However, the plan is highly restrictive in respect to the preservation of wetlands, historic, scenic, agricultural, and forest lands. The basic goals set out in the legislation are as follows:

1   To protect, maintain, and, where feasible, enhance and restore the overall quality of the coastal zone environment and its natural and artificial resources.
2   To assure orderly, balanced utilization and conservation of coastal zone resources taking into account the social and economic needs of the people of the state.
3   To maximize public access to and along the coast and maximize public recreational opportunities in the coastal zone consistent with sound resources conservation principles and constitutionally protected rights of private property owners.
4   To assure priority for coastal-dependent and coastal-related development over other development on the coast.
5   To encourage state and local initiatives and cooperation in preparing procedures to implement coordinated planning and development for mutually beneficial uses, including educational uses, in the coastal zone.

The interim planning commissions were independent of local government, and they operated that way: they made very little effort to develop any collaborative relationships with their constituent municipalities (Fischer 1985: 315). This changed dramatically after 1976 when, subject to conditions, plan making and regulatory responsibility was returned to local government.[18] The conditions were several. The statute mandated development of local coastal plans (LCPs), with regulatory authority over most development to be transferred back to local government only after the commission had certified that the LCP was in conformity with the policies of the Coastal Act. Further, the commission retains some important planning and regulatory responsibilities, including permanent jurisdiction in some areas such as tidelands, submerged lands and trust lands; reviewing and acting upon appeals from local permit decisions; reviewing and authorizing amendments to LCPs; implementing public access programs; reviewing federal activities pursuant to the federal consistency provisions of the federal Coastal Zone Management Act; and periodically reviewing the implementation of certified LCPs to determine if the plans are being implemented in conformity with provisions of the Coastal Act, and making recommendations to local governments or the legislature. Thus the legislation clearly establishes a "shared" responsibility between the commission and local governments.

The California Coastal Conservancy

The Commission did not start with a blank slate. There are dozens of poorly placed or badly designed pre-existing subdivisions – with lots now owned by individuals – that no amount of creative regulation can undo. There are wetlands that need restoration and public accessways to and along the beach that, once required to be dedicated as a condition of development approval, must be improved and managed by a public agency.

(Fischer 1985)

Just such an agency was established in 1976 as a companion to the Coastal Commission: the State Coastal Conservancy.[19] Among its many functions, the conservancy helps to carry out coastal improvement and restoration projects to implement policy established through the plans and regulations of the commission and local governments. It is empowered to buy land, and restore or resubdivide it, or sell or transfer it to others (whether at a profit or a loss). The conservancy is funded by the state, by bond issues, by federal grants under the Coastal Zone Management Act, and miscellaneous other sources. With these funds it carries out a wide range of functions in furtherance of the policies and regulations of the commission and the local governments. Four of these are summarized here.

Vast areas of California's wetlands have been destroyed during the past century: two-thirds over the state as a whole, and 90 percent in the southern part of the state. The preservation of the remaining wetlands (and the restoration of degraded wetlands) therefore has a high priority. Wetlands are not a source of private profit: their value lies in the habitat they provide for wildlife. Individual owners cannot be expected to safeguard them, and there is a limit to the conditions which can be imposed by the Coastal Commission. Hence the importance of the conservancy. Fischer has outlined how the conservancy operates:

Working with the development community, other public agencies (US Fish and Wildlife Service, California Department of Fish and Game, etc) and the owners of degraded wetland acreage, the conservancy first would prepare a resource enhancement plan. In a typical case, the Coastal Commission then might issue one or more permits that, for instance, would allow the filling of on-site acreage. The developers then would pay (as a permit condition) into an in-lieu mitigation fund established by the conservancy. In some cases, the amount would be set at a level sufficient to restore off-site acreage at a ratio of as much as four to one. . . . The conservancy typically assumes the responsibility for acquiring the off-site property, performing the restoration, and holding the land until a permanent, funded (public or, in many cases, private nonprofit) agency can accept title and management responsibility for the wetland property.

(Fischer 1985: 319)

Another priority of the coastal policy is the maximization of public access to and along the shoreline. The commission requires, as a condition for the granting of a permit, that public access be provided. (This is a condition which achieved national publicity in the planning world with the 1987 case of *Nollan v California*

Coastal Commission, which is briefly discussed in Chapter 6.)

This involves the dedication of an easement to a public agency that is willing and able to accept responsibility for maintenance and liability. Since huge numbers of conditional permits have been issued (1,800 in the twelve years up to 1985: a potential of more than fifty miles of additional shoreline access), this is no small task; and it is one which financially hard-pressed local governments are none too happy to accept.

However, a range of public and private agencies have accepted responsibility for some 200 new access facilities (Fischer 1985: 319). The transfer of development credits is a modern technique for compensating owners of land which is required for public purposes. One such use of this has been in the Santa Monica Mountains, where about 10,000 very small, unserviced lots were formed in the first half of the century. The development capacity of the area is very limited, and if all the lots were developed the public services would be over-stretched, "and it would look terrible" (Fischer 1985: 320). The solution adopted has been the development by the commission of a transfer of credits (TDC) program through which conditions are applied to coastal permits. The program, which is administered by the coastal conservancy, provides a means "to combine existing lots or otherwise retire existing lots so that new land divisions would not result in a net increase in the amount of development which could eventually occur." An official report of the California Coastal Commission, from which the foregoing quotation is taken, explains:

> Despite the implication of its title, the program does not actually transfer an individual applicant's proposed development from one location to another; only the development *potential* is "transferred." In other words, the program extinguishes (usually through the creation of an open space easement) the development potential of lots in one area, while allowing the creation of new residential building sites in other areas which are better suited to support development. . . .
>
> For a new single-family residential parcel which is proposed to be created (above and beyond those parcels which already exist), an applicant must acquire one transfer of development credit. That is, the applicant must demonstrate that the development potential on the equivalent of one developable parcel has been extinguished. This is normally done by recording an offer to dedicate an open space easement over the subject property.
>
> (California Coastal Commission 1989)

Several hundred such transactions have been successfully concluded, "making this the most successful TDC program in the country" (Fischer 1985: 320).

Many of the development rights extinguished in the TDC program are of no value for recreational purposes (though they may have stunning scenic value, as in Big Sur). They are purchased by the conservancy to prevent development. In this and other cases (for example, undeveloped coastal subdivisions) the conservancy acts as buyer of last resort.

The future of the coastal program is uncertain. It has aroused a great deal of opposition (see the quotation at the beginning of this section). The opposition continues, and its budget is under constant attack.

Oregon

Comprehensive Statewide Planning

The form of planning in Oregon is not so much different from that in other states, but the substance is. In most states, the cities and counties may plan and zone; in Oregon they must. In most states, standards for local planning are not uniform from one jurisdiction to another, are not particularly high, and are not enforced by any state agency; in Oregon, general planning standards (the goals) are the same throughout the state, they are high, and they are administered by an agency with clout.

(Rohse 1987)

Oregon has had a good long track record for state planning initiatives. For example, between 1969 and 1971 five laws were passed (the so-called "B" laws) which provided for public access to the beaches (not merely to the mean high water line, but to the line of vegetation); issued bonds for pollution abatement; banned billboards; earmarked funds for bicycle paths; and mandated returnable bottles (Little 1974: 8; DeGrove 1984: 238). Other laws established the Oregon Coastal Conservation and Development Commission, and mandated local governments to prepare comprehensive land use plans and develop land use controls. And, in 1973 came the Land Conservation and Development Act which greatly increased the powers and responsibilities of (mandatory) local planning, and provided for a set of state planning guidelines which local plans are required to follow.

As usual, there is no single factor which explains why Oregon acted as and when it did.[20] Certainly, a catalyst was political – in the form of Governor Tom McCall and Senator Hector Macpherson who, in promoting new legislation in 1971, started what DeGrove has called "a kind of blitz, in which powerful forces allied themselves on both sides of the issue, and a hard-fought series of compromises had to be worked out to obtain the bill's ultimate passage." But there was a popular base on which this blitz was waged. Oregonians have a particular pride in the beauty of their state: they see it as a precious heritage which demands to be preserved. Bolstering this is a strong and vocal conviction that "Oregon must not

become another California." (This conviction has been fueled in recent years by the influx of Californians seeking an environment similar to that of California, but with much lower house prices.)

The concern emanates from visible pressures on the land: urban encroachment in the Willamette Valley, land speculation in the fragile landscape of the eastern part of the state, and degradation of the marvelous shoreline.

Oregon's land use law is comprehensive only in the sense that the planning guidelines apply to the whole of the state. The actual preparation and implementation of local plans is the responsibility of the 241 cities and 36 counties. Thus there is in no real sense a "state plan": there are 277 local plans that have been developed in accordance with state standards and have been reviewed and approved by the state.

It should also be noted that though the Oregon planning system can be accurately described as dedicated to urban growth management and the preservation of agricultural land, it does not seek to limit or retard growth. Moreover, the statewide planning guidelines are not wholly concerned with the conservation of resources (Rohse 1987: 4). Indeed, the first goal – to promote public participation – is wide open in this respect. Several goals (8–12) require local governments to provide for certain types of development: recreational facilities, diversification of the economy, housing, public facilities and services, and transportation systems. The purposes of the nineteen goals include:

1 *Citizen Involvement*: To develop a citizen involvement program that insures the opportunity for citizens to be involved in all phases of the planning process.
2 *Land Use Planning*: To establish a land use planning process and policy framework as a basis for all decisions and actions related to the use of land. . . .
3 *Agricultural Lands*: To preserve and maintain agricultural lands.

4 *Forest Lands*: To conserve forest lands by maintaining the forest land base and to protect the state's forest economy. . . .

5 *Open Spaces, Scenic and Historic Areas, and Natural Resources*: To conserve open space and protect natural and scenic resources.

6 *Air, Water, and Land Resources Quality*: To maintain and improve the quality of the air, water and land resources of the state.

7 *Areas Subject to Natural Disasters and Hazards*: To protect life and property from natural disasters and hazards.

8 *Recreational Needs*: To satisfy the recreational needs of the citizens of the state and visitors. . . .

9 *Economic Development*: To provide adequate opportunities throughout the state for a variety of economic activities vital to the health, welfare, and prosperity of Oregon's citizens.

10 *Housing*: To provide for the housing needs of citizens of the state.

11 *Public Facilities and Services*: To plan and develop a timely, orderly and efficient arrangement of public facilities and services to serve as a framework for urban and rural development.

12 *Transportation*: To provide and encourage a safe, convenient and economic transportation system.

13 *Energy Conservation*: To conserve energy.

14 *Urbanization*: To provide for an orderly and efficient transition from rural to urban land use.

15 *Willamette River Greenway*: To protect, conserve, enhance, and maintain the natural, scenic, historical, agricultural, economic and recreational qualities of lands along the Willamette River as the Willamette Greenway.

16 *Estuarine Resources*: To recognize and protect the unique environmental, economic, and social values of each estuary and associated wetlands. . . .

17 *Coastal Shorelands*: To conserve, protect, where appropriate develop, and where appropriate restore the resources and benefits of all coastal shorelands. . . .

18 *Beaches and Dunes*: To conserve, protect, where appropriate develop, and where appropriate restore the resources and benefits of coastal beach and dune areas. . . .

19 *Ocean Resources*: To conserve the long-term values, benefits, and natural resources of the nearshore ocean and the continental shelf.

(Oregon 1990)

The Administrative Structure

As the inability of local governments to adequately cope with all aspects of land use becomes increasingly evident, there is increased need for greater state involvement in the management of land to ensure the preservation of an acceptable living environment for this and future generations.
(Rubino and Wagner 1972)

A state planning agency, the Land Conservation and Development Commission (LCDC), was established to ensure that state policy is implemented. Its seven members are appointed by the governor, and confirmed by the senate. Its first tasks were to adopt the statewide planning goals and to review and approve (technically termed the "acknowledgement" of) local plans. The review of amendments (of which there are several thousand every year) is dealt with by the commission's administrative arm, the Department of Land Conservation and Development (DLCD). However, DLCD has no power to prevent a municipality from adopting an amendment to which it objects. In such cases, DLCD would normally appeal to a body known as the Land Use Board of Appeals

(LUBA). This board was established to provide a simple means for settling land use disputes without the need to go through the state circuit courts. (But LUBA's decisions can be reviewed and enforced by the higher courts – the court of appeals and the Oregon supreme court.)

The important lubricant in this system is the wide provision for citizen involvement (the first of the state goals). This is in the political tradition of Oregon: it is common for local governments to establish citizen advisory committees (CACs) for every city neighborhood and county district. "The groups typically meet monthly. Their advice and concerns are given to the planning commission or governing body. CAC meetings are quite informal, and are open to all, without dues or formalities of membership. CAC meetings are often attended by members of the planning department, who can answer technical questions or keep a record of comments" (Rohse 1987: 57). Local governments are also required (by Goal 1) to establish and support "an officially recognized citizen advisory committee or committees broadly representative of geographic areas and interests related to land use and land use decisions." At the state level, there is the Citizen Involvement Advisory Committee, which has several functions: to advise the commission on matters of citizen involvement, to promote "public participation in the adoption and amendment of the goals and guidelines," and "to assure widespread citizen involvement in all phases of the planning process."

Intergovernmental relations are often characterized by a heavy measure of bluff. A state law may mandate a local government to do something, but if there is no machinery for ensuring compliance or no financial incentive, nothing may happen. The Oregon system has teeth in it. Rohse lists a number of reasons why local governments seek to get "acknowledgement" of their plans:

> First, state law requires it. Second, there often is a financial incentive: state maintenance grants are given to jurisdictions with acknowledged plans. Third, there is the threat of an enforcement order by LCDC if local progress toward acknowledgement occurs too slowly. Finally, the state may withhold cigarette, liquor, and gas tax revenues from jurisdictions that do not complete their planning in a timely manner.
> (Rohse 1987: 6)

But probably more important than any of these, he suggests, is that once a local government's plan and land use regulations have been approved by the commission, the state's role in local planning is greatly reduced: there is no longer any need for it: the goals have been incorporated in the plan. "Acknowledgement returns a large measure of planning power to a local government and facilitates local decision making."

It is, however, the exceptions that prove the rule. The first enforcement action was taken in 1978 after a lengthy dispute between the city and the county of Hood River. The issue concerned the establishment of an "urban growth boundary." (Under Goal 14, "urban growth boundaries shall be established to identify and separate urbanizable land from rural land.") The city and the county were unable to reach agreement, and the commission ordered the approval of the city's proposal (DeGrove 1984: 263).

Urban growth boundaries (UGBs) are, understandably, a source of contention: in more senses than one they are the cutting edge of planning implementation; and wherever they are drawn someone will be upset. Nevertheless, "the requirement for drawing an urban growth boundary around every city in Oregon has gone forward with considerable dispatch, and has not generated as much controversy as might have been expected" (DeGrove 1984: 271). Each of Oregon's 241 cities has adopted an urban growth boundary which has been reviewed and approved by the commission.

In spite of Oregon's well-publicized leadership in growth management, all is not well. A 1991 review complained that urban sprawl within the urban growth boundaries, and large scale development outside them, is threatening the livability of the state's urban areas (Oregon 1991a). Among the proposals being considered are the establishment of "focused growth plans" which would provide for the concentration of public and private investment within UGBs, the creation of "urban reserves" outside UGBs, and the setting up of a new state agency to aid local governments with infrastructure funding.

New Jersey

Toward a Comprehensive Plan

Statewide comprehensive planning is no longer simply desirable, it is a necessity.
(*Mount Laurel I* 1975)

New Jersey has had a series of important regional planning initiatives. The most famous is the Pinelands Commission, established in 1979 (Collins and Russell 1988) which is responsible for the planning of some one million acres in the southern part of the state. Even earlier, though more modest, is the Hackensack Meadowlands Development Corporation, established in 1968 as a development agency in the northeastern part of the state. In 1973, the Coastal Area Facility Review Act of 1973 created a regional commission to regulate large developments in the coastal area.

In 1980, the Democratic administration of Governor Brendan Byrne created a State Development Guide Plan. Though this was short-lived (Republican Governor Thomas Kean abolished it in the following year), the courts continued to use it, in the implementation of the Mount Laurel policy, to identify growth areas where municipalities were required to set aside some 20 percent of their new housing for lower income families.[21]

During the early to mid-1980s, the New Jersey economy boomed, migration (of both people and jobs) into the state grew, and political pressures for more effective planning increased (Lawrence 1988: 19). In response, a State Planning Act was passed in 1985, establishing a State Planning Commission and its staff arm, the Office of State Planning.

The commission is charged with preparing the primary instrument for coordinating planning and growth management in the state – the State Development and Redevelopment Plan (New Jersey 1988a: 16). The statute provides that the plan shall protect the natural resources and qualities of the state, while promoting development in locations where infrastructure can be provided. It also establishes statewide objectives in a variety of areas including land use, housing, and economic development. The plan is intended to be used to guide the state's capital expenditure.

The Act spells out the way in which the plan is prepared. First, a preliminary plan is approved by the commission which is then used in an interactive planning process called "cross acceptance" which is intended to integrate municipal, county, regional, and state land use plans as well as the capital facility plans needed to assure efficient services.

The Preliminary State Development and Redevelopment Plan

The most significant attempt nationally to apply growth management on the state level.
(Callies and Freilich 1988)

The preliminary state plan was completed in January 1989. In accordance with the provisions of the legislation, it establishes statewide goals and objectives for land use, housing, economic development, transportation, natural resource conservation, agriculture and farmland retention, recreation, urban and suburban redevelopment, historic preservation, public facilities and services, and intergovernmental coordination. For example, Goal 2 is "to provide adequate public services at a reasonable cost." This has six "component objectives":

1   To ensure an equitable distribution between the public and private sectors of the costs of capital facilities and services generated by development and redevelopment.
2   To maintain and improve public capital facilities and services to support areas designated for growth.
3   To maximize the use and efficiency of existing infrastructure to furnish public services for development and redevelopment.
4   To provide an efficient and integrated public transportation system.
5   To increase the availability of public transportation.
6   To prevent urban sprawl.

The plan divides the state into seven "tiers" (New Jersey 1988a: 76). These tiers define the areas where the public interest is best served by encouraging development, redevelopment, and economic growth, and those where it is best served by discouraging development that impairs natural resources or environmental qualities. Four of the seven tiers are designated "growth areas": *redeveloping cities and suburbs*, where the primary goal is revitalization; *stable sites and suburbs*, where some growth can be sustained while preserving community character; *suburban and rural towns*, where development which would otherwise be scattered in surrounding areas is to be concentrated; and *suburbanizing areas*, in which the existing or planned infrastructure is to be optimized by developing "corridor centers" and residential uses together with related commercial activities.

The other three areas are designated "limited growth areas": *future suburbanizing areas*, which do not have the necessary infrastructure to support intensive development, but where limited growth can be "managed" until the areas become designated growth areas; *agricultural areas*, where it is intended that any new development is supportive of agriculture; and *environmentally sensitive areas* where any development must be compatible with natural resources.

Three areas of critical concern were the subject of earlier legislation: the Pinelands, the Meadowlands, and the coastal zone. The planning legislation directs that the state development plan will not "affect the plans and regulations" of the agencies established for these areas.

Much of the recent growth in New Jersey has been along transportation corridors, and this pattern is likely to continue in the future. The plan takes this fact as a basis for a major strategy of developing centers in the prosperous corridors. These centers are not envisaged as an elongation of the corridors: on the contrary, they are to be high-density consolidations around existing development. Their attraction is that of good transportation (which an elongation of a corridor would jeopardize). The development of centers provides the opportunity for enhancing the transportation advantages:

> The encouragement of high density development within the corridor is a key element of the overall transportation corridor concept. By promoting high density development adjacent to transportation facilities within the corridor, the use of mass transportation and multimodal transportation facilities is encouraged and a sufficient client base is established to help make the massive expenditures required to construct high speed or rapid transit more feasible. High speed and mass transit reduce the dependency of the automobile as a source of travel, effectively reducing energy needs and despoliation of the environment.
> (Callies and Freilich 1988: 115; Freilich and Chinn 1987)

"Cross Acceptance"

There appears to be a general agreement among the counties with the overall concept and aims of the SDRDP statewide goals. A principal concern is how to implement the plan's strategies and policies without adversely affecting the communities. . . . There is major concern about state interference with Home Rule. Municipal jurisdiction over land use decision making is of paramount importance to local communities.

(Doyle 1990)

It is one thing for a state to prepare a plan: it is another to get that plan accepted by the local governments whose cooperation is needed if it is to be implemented. New Jersey has a strong tradition of home rule, and the 567 municipalities are very suspicious of state action in the land use field. It was therefore essential that the plan preparation process should involve the active participation of the municipalities. For this purpose the Act provides for a procedure known as "cross acceptance":

The term cross acceptance means a process of comparison of planning policies among governmental levels with the purpose of attaining compatibility between local, county, and state plans. The process is designed to result in a written statement specifying areas of agreement or disagreement and areas requiring modification by parties to the cross acceptance.

(State Planning Act 1985)

In general, cross acceptance involves comparing the provisions and maps of local, county, and regional plans and regulations with the goals, objectives, strategies, policies, standards, and maps of the preliminary state plan. The process is an involved one; there is even a "Cross Acceptance Manual" prepared by the Office of State Planning. However, in essence the idea is a simple one: the authorities that need to coordinate their activities are given a mechanism by which they can talk until agreement or compromise is reached (Gilbert 1990). As the quotation at the head of this section illustrates, not all is plain sailing. Interests vary at the different levels of government, and it is not to be expected that agreement will be reached easily. It remains to be seen whether there will be sufficient accord to enable the plan to be finalized – and implemented.

Conclusion

Stafford Hansell, who heads Oregon's land use commission, asks planning opponents, "How would you like it if I moved my hog operation in next door to you?".

(Fulton 1989a)

Interest in state involvement in growth management appears to be increasing, though it is still too early to declare the "revolution" suggested by Callies and Bosselman.[22] In addition to the six states discussed here, several more have passed or proposed legislation. For example, Rhode Island passed a Comprehensive Planning and Land Use Regulation Act in 1988 (Rhode Island 1988 and 1989). This requires consistency between every local government's comprehensive plan and the state's comprehensive plan. Local plans are reviewed by the Department of Administration for consistency with the Act. Any disagreements are decided by a Comprehensive Appeals Board, which can if necessary substitute a plan of its own. Maine also passed a Comprehensive Planning and Land Use Regulation Act in the same year which is very similar to that of Rhode Island. Washington followed in 1990 with its Growth Management Act which requires comprehensive plans for populous and other fast-growing local governments. The legislation was supported by builders and environmentalists alike, but there was considerable opposition from the rural areas who successfully campaigned for exempting smaller counties (*Planning* May 1990: 31).

Georgia is the most surprising newcomer to the small band of states involving themselves with growth management.[23] This is a state in which two-thirds of the counties have had no zoning. Yet, in 1987, Governor Joe Frank Harris appointed a Growth Strategies Commission and, two years later signed a growth strategy bill. The background to this was the rapid growth in the state, particularly in the Atlanta region, and the clear inadequacy of the local government system to cope with the problems of traffic, urban sprawl, water and similar issues. Georgia's constitution expressly prohibits the state legislature from involvement in local zoning, leaving planning solely as an option for local government (Whorton 1989: 16). The commission's report (Georgia 1988) outlined the difficulties and proposed a "Quality Growth Partnership" to solve Georgia's problems. The resulting legislation includes mandatory comprehensive local plans which are to be reviewed by newly established regional planning agencies.

It was once thought that urban growth "not only paid for itself, but also produced surplus revenues for state and local governments" (DeGrove 1989: 1). However true that might have been in the past, it is clearly not so today.[24] It is the realization of this which has led to increased involvement by a number of states in growth management. The key issue is infrastructure, as the experience in Florida demonstrates:

> Florida is unique among the states in its formal requirement that infrastructure be in place up front to accommodate the impact of growth. Oregon's law did not place great emphasis on the issue, nor did the other states which adopted planning legislation in the 1970s. Newer state initiatives do have that emphasis, as is illustrated by the New Jersey State Planning Law's provisions regarding infrastructure. As other states address the issue of state planning and growth management, infrastructure and especially transportation issues tend to be the force that drives these growth management initiatives. States such as Oregon that acted earlier have returned to the issue to give it added emphasis, with Florida being the clearest example.
>
> (DeGrove and Stroud 1989: 66)

Of course, this resolves into an issue of finance. The battle over taxation in Florida shows how difficult this can be.

Another matter of importance is state control over local governments – though such terminology is typically eschewed in favor of cooperation, partnership or "cross acceptance." But whatever the terminology, the state has to have some mechanism to ensure that local governments do implement state-approved plans. Local officials can only be expected to try to keep land use decisions in their own hands (Fulton 1989a: 42).

Of particular interest is the California coastal planning system, which operates through a partnership of the coastal commission, the conservancy (an agency of coastal management), and the local governments. The commission has the responsibility "from time to time, but at least once every five years" of reviewing the way in which local governments are implementing their coastal programs, and ensuring that they are in conformity with the policies of the state plan. If there is a lack of conformity, the commission is empowered to make recommendations on the "corrective actions" that should be taken.[25] Should this point be reached, the matter falls clearly in the political arena, to be settled by the thinking of the time, and the political complexion of the governmental bodies concerned.

It is too early to judge the "success" of state involvement in land use planning. Nevertheless, it seems clear that "the level of consciousness of the environmental and other impacts of land development" has increased, and it is also apparent that stringent controls can work (Healy and Rosenberg 1979: 273). The states discussed in this chapter are, of course, exceptional: otherwise it would not be interesting to write about them. There seems to be cause for mild optimism. But, as Audirac, Shermyen, and Smith have argued in relation to Florida:

Until citizens are willing to tackle the difficult political issue of new taxes, growth management efforts in the state will remain stymied by lack of revenues to effectively fund regulatory efforts, acquire environmentally sensitive lands and development rights, and provide the urban infrastructure and services needed to preserve the quality of life of fast-growing communities.

(Audirac, Shermyen, and Smith 1990: 478)

Chapter 12

The federal government and urban policy

Federal: from *foedus*, meaning league, covenant or compact.

Introduction

> Every time Treasury changes the Tax Code,
> every time Congress alters a welfare
> program, every time the Defense
> Department awards a military contract,
> urban policy is being made.
>
> (Donna Shalala)[1]

The federal government is by far the largest
land owner in the United States (it owns a
third of the land area),[2] and its policies in relation to these lands (land sales, control of much
mining and reclamation, national parks,
national forests, Indian reservations, etc.) have
been explicit and frequently subject to intense
debate and controversy (Clawson 1983). There
is no urban equivalent to these "land policies."
Indeed, it might be argued that the federal
government has neither had nor (with one
major short-lived exception) attempted to
devise an urban policy. But this is to take a very

narrow view of "urban policy." Urban policies
can be explicitly devised and expressed as such,
or they can be the indirect effect of "nonurban"
policies. Policies directed at expanding welfare,
or at promoting equality of opportunity, or
improving communications, or safeguarding
agricultural land (the list is almost endless) can
have a significant effect upon cities.

In this chapter, a number of these policies is
discussed. There are several objectives in so
doing. Firstly, many of the policies do in fact have
an important impact on urban land use, sometimes of great significance. Secondly, the fate of
these policies is relevant to the question of
whether the federal government should, and
could, play a bigger role in land use planning
policies.

Implicit National Urban Policies

> The federal government has virtually ordained the expansion and form of suburbia, and now of exurbia, through FHA rules and supports, and the funding of highways in preference to mass transit. Federal spending has discriminated against the urban areas that need it most.
>
> (*New York Times*)[3]

The most dramatic of the federal "nonurban" policies have been those which have involved large public investments in the development of canals, railroads, airports, and interstate highways as well, of course, as the long litany of pork barrel projects. These were not designed as urban policies, but their impact on urban areas has been immense. A major example, of course, is the interstate highway program:

> The Interstate Highway Act of 1956 was undertaken largely at the bidding of a well-organized pressure group. This Act committed the federal government to spending $33,500,000,000 in fourteen years on building a national network of motor-roads. It was to do much more to shape the lives of the American people than any other law passed since 1945. It reinforces the ascendancy of the private car over all other forms of passenger transport; it made continental bus services fully competitive with the already declining railroads; it boosted freight carrying by truck; it gave a great impetus for black migration from the South, and a huge boost to the automobile, engineering and building industries, thus helping to stimulate the prosperity of the sixties; by encouraging car ownership it encouraged car utilization, thus stimulating the spread of the population into vast sprawling suburbs, where only the car could get you to work, to the shops, entertainment and voting booths; and this change in turn would soon be reflected in political behavior. Yet it can hardly be contended that the administration foresaw or desired these results, any more than it did the widespread corruption and faulty construction that went with the hasty building of the highways.
>
> (Brogan 1986: 631)

Many cities had their hearts ripped out by the interstates and also by urban renewal. But perhaps the most important impact of these policies (and others such as housing and taxation policies) was on the development of the suburbs[4] – which took place at the cost of the inner cities. Suburbs had previously been largely dormitories; new highways increased their attraction not only for house purchasers but also for office and retail functions and even for manufacturing industry, sometimes on such a scale that they became new "downtowns" (Cervero 1986; Garreau 1991).

Strange though it might at first sight appear, a major implicit national urban policy has emerged with agricultural development policies: over a very long period these have provided the United States with "the most highly capitalised and most productive agricultural establishment in the world, while transferring perhaps forty million people, one-fifth of the current [1970] national population from the nation's rural areas to its cities and suburbs" (Wingo 1972: 5). Dramatic in a different way have been "the civil rights policies and the war on poverty, which made the nation's great urban centers the foci of these political struggles and the economic meccas of the deprived and disadvantaged groups in American society" (Wingo 1972: 5).

Another area in which there has been significant urban impact has been with the location of defense contracts and Research and Development expenditure. Location decisions have favored the newer "sunbelt" areas more than those of the "snowbelt" (or "rustbelt") – though there is considerable controversy over how great their impact has been.[5] It might have been hoped that while some of these policies or policy impacts would exacerbate urban problems, others would ameliorate them. Unfortunately, according to a Rand study, the predominant effect has been negative: "though inadvertent, the effects have mostly been adverse" (Vaughan, Pascal, and Vaiana 1980: v).

Clearly, whatever these policies might have been called, they constitute a confused multiplicity of national urban policies. But "confused" is the right word: in no sense do the policies add up to anything systematic.[6] Indeed, the federal government has had extreme difficulty in devising any urban policies (even when it has tried!). There is, of course, an abundance of statutes concerning such matters as housing and transportation which have important urban dimensions; and there are, as already noted, policies relating to other matters which have had major impacts on urban areas. But there is nothing which attempts to deal with the interlocking problems which face urban areas. It might be objected that an explicit federal "urban policy" is inconceivable and impracticable. This may well be so, and Canadian experience suggests that it is (Cullingworth 1987: 32), but the fact remains that the federal government has *attempted* to devise urban policies and is committed by the National Urban Policy and New Community Development Act of 1970 to develop a national urban policy.[7] In this chapter, some of the more important post-World War II federal efforts in urban policy are discussed, and reasons for their typical failure are outlined.

Urban Renewal

The broad opportunities for urban planning possible under the bill may be lost in the scramble for isolated and unrelated projects.
(Perkins 1949)[8]

"America's slum districts had been a national disgrace for over a century before the federal government began to show an interest" (Gelfand 1975: 60). Even then, the action was modest (Wood 1934). The field of housing was generally considered to be beyond the proper role of government in general, and the federal government in particular. (One particular political difficulty was the association of slums with the city of New York). Not only were sacred property rights involved: it was also feared that government action would severely harm the private market. Tentative initiatives were made by the back door, in the New Deal; and the 1937 Housing Act signalled the recognition of both slum clearance and public housing as legitimate areas of public policy.[9] The substance, however, was thin, mainly because of the bitter opposition of the National Association of Real Estate Boards and kindred spirits.

The opposition continued throughout the 1940s (indeed, it has never ceased). The industry argued that the private market could meet all the nation's housing needs without the intervention of government (though "aids to private enterprise" such as those provided by the Federal Housing Administration were championed). The outlook for postwar housing policy was therefore bleak. A housing bill was introduced in 1945, but was killed by vociferous opposition first in 1946, and again in 1948. It eventually passed as the 1949 Housing Act – the single Fair Deal piece of legislation which Truman managed to get through Congress.[10]

The Act embraced the national goal of "a decent home and a suitable living environment for every American family," but the means to achieve this were effectively denied by Congress.

The legislation authorized the building of 810,000 units of public housing over a period of six years, though the program was slow in starting, and it took two decades before this target was reached. Part of the reason for this was the opposition to public sector activities which, following the passage of the Act, moved from Congress to local urban renewal areas. But the task was inherently complex: areas had to be selected and designated for acquisition; sites had to be cleared; complicated negotiations were required for federal funding; and arrangements had to be concluded with private investors and developers for redevelopment. The important role given to private enterprise was part of the political price which had to be paid to secure the passage of the legislation. "Conservatives acquiesced to a continued public housing program *in exchange* for an affirmation of the principle that private enterprise would be the main vehicle for redevelopment."[11] One arm of this was provision for urban redevelopment by private enterprise with local government supervision and federal-local subsidies which bridged the gap between the market value of land and the (much higher) actual cost of acquisition and clearance.[12]

Another problem arose with the concept of linking redevelopment to the politically unpopular provision of public housing.[13] This was of particular importance since redevelopment was to be "predominantly residential." But private enterprise was not interested in low-income housing (whether subsidized or not). Profits lay in other directions, particularly downtown shopping and commercial centers. These were also popular with local political and business elites. As a result, pressure was put on the Eisenhower administration to alter the rules. A major change came with the 1954 Housing Act, when the term "urban renewal" was introduced, indicating that, in addition to redevelopment, the policy now embraced rehabilitation, conservation, and "the renewal of cities." The Act provided that 10 percent of project grants could be used for nonresidential development. The rationale behind this "was that there were nonresidential areas around central business districts, universities, hospitals, and other institutional settings that certain city interests wished to clear and redevelop for nonresidential purposes" (Weiss 1980: 267). In Mollenkopf's (1983: 117) words, "the 1954

Housing Act shifted urban renewal from a nationally directed program focusing on housing to a locally directed program which allowed downtown businesses, developers, and their political allies, who had little interest in housing, to use federal power to advance their own ends."

The proportion allowable for nonresidential purposes was increased to 20 percent in 1959 (together with the needs of dominant institutions such as hospitals and universities), and later to 35 percent. With the administrative latitude allowed, the eventual result of this was to increase the commercial part of urban renewal to one half of the total. Indeed, by manipulating definitions and procedures "local authorities could allocate as much as two-thirds of their funds for commercial development, but still remain within federal guidelines." Thus, to quote Judd:

> Urban renewal was turned into a political pawn for commercial and industrial interests. Though the national legislation emphasized the deleterious effects of a slum environment on its residents, local political and economic elites focused on the economic decline of central business districts. Business leaders and politicians were convinced that urban decay was caused by the loss of economic vitality at the urban core.
>
> (Judd 1988: 269)

Despite amendments to the legislation and some notable achievements, for example in "improving the physical appearance of hundreds of American cities,"[14] urban renewal became subject to increasing criticism. Above all it failed to help the poor: indeed, it made their position worse. The critics were numerous.[15] A damning overview is provided by Barnekov, Boyle, and Rich (1989: 47), on which the following analysis is based.

While urban renewal bolstered central business districts and may even have contributed indirectly to a city's economic vitality, it also dislocated neighborhoods and often created more urban blight than it removed. Friedman (1968: 166) pointed out that "urban renewal takes sides; it uproots and evicts some for the benefit of others." Urban renewal benefited some center city businessman and the upper middle class, who obtained desirable inner city housing at bargain prices, but frequently the real cost was borne by the urban poor. As Hartman (1974: 183) suggests, the aggregate benefits of urban renewal "are private benefits, and accrue to a small select segment of the city's elite "public," while the costs fall on those least able to bear them." Downs (1970: 223) estimated that households displaced by urban renewal suffered an average uncompensated loss amounting to 20–30 percent of one year's income. Moreover, while originally conceived as a means of increasing the supply of low-cost housing, urban renewal actually exacerbated the urban housing problem.[16] Greer (1966: 3) scathingly commented that "at a cost of more than three billion dollars the Urban Renewal Agency has succeeded in materially reducing the supply of low-cost housing in American cities." In subsequent years the pattern was no different: between 1967 and 1971, 538,000 housing units were razed but only 201,000 replaced (Weicher 1972: 6). An additional point was stressed by the Douglas Commission (1968: 165): "the unconscionable amount of time consumed in the process" of effecting an urban renewal scheme. Most schemes took between six and nine years from inception to completion. Fourteen percent took over twelve years, and 3.5 percent took over fifteen years.

Disillusionment with urban renewal led to its decline at the end of the 1960s. It thereby joined public housing as a cause which even its sponsors no longer supported though, of course, its impact has lived on. But the fundamental weakness of urban renewal was that it was conceived in terms of a land use instrument which was to "save" the declining cities. Though there were pockets of success (sometimes monumental) in commercial centers and middle-class residential areas, these typically had little or no wider effects, except of an undesirable nature such as the displacement of low-income households. Their impact on restraining the exodus to the suburbs was minimal. It is likely that a greater force for effective "renewal" lay with the increasing number of new immigrants who, like so many before them, sought out opportunities in the cities – opportunities which urban policy had defined as problems.

The War on Poverty

The purposes of this title are to provide additional financial and technical assistance to enable cities of all sizes [to implement] new and imaginative proposals and rebuild and revitalize large slums and blighted areas; to expand housing, job, and income opportunities; to reduce dependence on welfare payments; to improve educational facilities and programs; to combat disease and ill health; to reduce the incidence of crime and delinquency; to enhance recreational and cultural opportunities; to establish better access between homes and jobs; and generally to improve living in such areas.

(Model Cities Bill 1966)

Prior to the Kennedy and Johnson administrations, explicit federal urban policies had been largely restricted to urban renewal and public housing. Disenchantment with these spread at the same time that urban problems grew – problems which ranged from race, civil rights, poverty, and violence to state and local government finance. A response to these problems emerged as public concern developed, and as the political scene changed – with Kennedy's rediscovery of poverty in the early 1960s, and Johnson's overwhelming election victory in 1964. The era of the "Great Society" was at hand. To an unprecedented extent, federal policies were developed to reach "deeply into the urban social and political fabric" (Mollenkopf 1983: 95).

As with so much in this field, the sequence of events, the policy initiatives and their impact comprise a complex and confused story. Federal programs proliferated on a bewildering scale: 370 new programs of assistance to state and local governments were introduced between 1962 and 1970 (Judd 1988: 305). These covered the whole spectrum of public policy, from food stamps to regional development, from the "War on Poverty" to health services, from education to model cities and the Community Action Program (designed to provide power to inner city residents to improve their neighborhoods).[17] These new programs, "each with its own budget and statutory requirements, generated a massive federal administrative structure and a significant transformation of federal-state-city relationships" (Frieden and Kaplan 1977: 3). Such a change demanded a greater degree of coordination among federal programs (and agencies) than had ever been required before:

> Growth in the number and variety of federal aid programs did not seem to bring with it the relief of urban woes. Proliferating federal aid led to extreme confusion and long delays that threatened to deplete the limited energies of city administrators, who had been seduced into the maze of federal review and approval processes by the lure of "free" federal money. By the mid-1960s, a chorus of complaints sounded in Congress and in the city halls, as well as in the studies of political scientists. Federal officials and local grantsmen alike were caught in the red tape of programs that seemed to be both underfunded and overregulated.
>
> (Frieden and Kaplan 1977: 5)

Two responses emerged. Firstly, the arguments which had been deployed for several years in favor of the establishment of a new federal department for urban policies gained the support that they had previously lacked (Parris 1969); and the Department of Housing and Urban Development (HUD) was created in 1965. Secondly, a task force appointed by President Johnson to advise on the organization and responsibilities of the new department proposed that the coordinative role of HUD should be directed through a model cities program toward the poverty areas of central cities. This, in fact, implied more than the coordination of programs: it explicitly envisaged the redistribution of resources.[18] It was also a clear change from the physical orientation of the urban renewal program.

The way in which this proposal was translated into legislation and policy is dealt with in detail in the studies of Frieden and Kaplan (both of whom were personally involved in the model cities program), and Banfield (1973). Here reference is made to a few key issues.

From its inception, there was debate on how many model cities there should be. One school of thought opted for a very small number[19] (hence the original term "demonstration cities"), while others, as indicated above, envisaged model cities being the major channel for aid to poverty-stricken urban areas. In the event, the need to obtain political support for the legislation led to an increased number of cities – eventually to 150.[20] The idea of "demonstration" cities was thus killed. But, more than this, Congress was not willing to see funds diverted from other programs into model cities; and so the congressional commitment became largely to another categorical program rather than to a mechanism for reforming other grants-in-aid.

Reflecting on the changes made to the model cities program, Frieden and Kaplan (1977: 236) note that while there was support for the idea of better coordination among urban programs, there was also a fear about "a concentration of power within any single executive agency." These two analysts refer to Downs' *Inside Bureaucracy* (1967: 132) with the comment that bureaucracies can be viewed as "a threat to individual liberty, that they are monolithic organizations where a few powerful men at the top concentrate control over a vast range of activities. The implication of this view is that efforts to strengthen coordination among agencies are potentially dangerous, because they may upset the existing balance of power that permits considerable freedom of action for many interest groups" (Frieden and Kaplan 1977: 8). In case this is thought to be an extreme view, the comment is added that "after Watergate and after Vietnam, the dangers of excessive White House power are all too obvious."

Moreover, the redistributive features of the model cities program implied that other federal agencies would be expected to divert some resources from their "traditional" clients. This "went against the grain of normal agency behavior, congressional grant-in-aid policies, and ultimate reliance on established interest groups that benefited from existing programs" (Frieden and Kaplan 1977: 237). They conclude that "if the designers of future urban policies take away any single lesson from model cities, it should be to avoid grand schemes for massive, concerted federal action." (On this, it is interesting to note the comparison between the persistence of the adherence to policies such as defence and the readiness with which urban policies are declared to be misconceived.)

The model cities program (together with urban renewal and other community development programs administered by HUD) was folded into the Community Development Block Grant at the end of 1974.[21]

Regional Planning

The many federal programs affecting metropolitan areas today are initiated from various sources and generally without reference to their impact on the areas to which they are directed.

(Bureau of the Budget 1960)[22]

No attempt has been made in this book to trace the checkered history of regional planning in the United States.[23] It is not a thrilling story: rather it is (generally) a succession of false starts and disappointed hopes. Nevertheless, some progress has been made, particularly with economic planning regions, such as the Tennessee Valley Authority, established in 1933; the Appalachian Regional Commission (1965), and a number of substate, metropolitan, and special purpose regional organizations (McDowell 1985). Also noteworthy has been the prodding of the Advisory Commission on Intergovernmental Relations, and the establishment of planning capabilities in the offices of the governors.

A thrust for creating a means of cooperation between the constituent parts of metropolitan areas came in President Kennedy's 1961 *Housing Message to Congress*. In this, he argued that the old jurisdictional boundaries were no longer adequate:

The city and its suburbs are both interdependent parts of a single community bound together by the web of transportation and other public facilities and by common economic interests.... This requires the establishment of an effective and comprehensive planning process in each metropolitan area embracing all activities, both public and private, which shape the community.

(Shonfield 1965: 354)

Particular progress was made through the Bureau of Roads, which required local governments to cooperate in a regional planning exercise as a condition for highway construction grants. The system was gradually extended and, in 1965, urban areas with a population of more than 50,000 became ineligible for federal grants for highway construction unless they had a "comprehensive transportation process for the urban area as a whole, actively being carried on through cooperative efforts between the states and the local communities" (ACIR 1964: 106).

The Urban Renewal Administration followed suit with its requirement that states and local governments produce comprehensive plans. Then, in 1968, following the Cities and Metropolitan Development Act and the Intergovernmental Cooperation Act, the Bureau of the Budget issued Circular A-95 which sought to establish a "network of state, regional, and metropolitan planning and development clearinghouses" to receive and disseminate information about proposed projects; to coordinate applicants for federal assistance; to act as a liaison between federal agencies contemplating federal development projects; and to perform the "evaluation of the state, regional or metropolitan significance of federal or federally-assisted projects" (Mogulof 1971: 418; Elazar 1984: 186).

Between 1968 and 1970, the number of councils of government (COGs) increased from 100 to 220: almost all the 233 Standard Metropolitan Statistical Areas had regional councils of some type: COGs, economic development districts, regional planning commissions. These varied greatly in the extent to which they became involved in regional planning: many did as little as was possible to meet the federal conditions. Others became actively committed, particularly after HUD introduced yet another regional planning scheme: the comprehensive planning assistance program, popularly known as the 701 program. The influence of this program has been described by Mogulof:

> HUD resources in the 701 planning program have become the institutional support for COGs, and HUD guidelines insist on the representation of a significant percentage of metropolitan area governments on COG policy boards. Additionally, it is HUD which has begun to prod 701 planning agencies with regard to "citizen participation" in their policy structure. And it is HUD which has moved the COG into a new (and sometimes uncomfortable) concern with social problems by requiring that a housing element be a part of the 701 agency's regional planning.
> (Mogulof 1971: 418)

Regional planning was not as effective as its protagonists had hoped. It was mainly a creature of federal initiatives, and frequently did not receive more than nominal support from the member governments, who certainly did not wish to see the growth of an "independent source of regional influence." Instead, they typically saw it as "a service giver, a coordinator, a communications forum, and an insurance device for the continued flow of federal funds to local governments" (Mogulof 1971: 418). Moreover, though one of the major objectives was to ensure that individual federally funded projects were in harmony with metropolitan or regional plans, such plans often did not exist. The importance of the absence of effective machinery for coordination is stressed by Meltzer:

> Governance implies a horizontal capacity to assert command and control across the range of functions at each governmental tier, within a system of hierarchical nested governmental layers. The dichotomy and threat to governance are compounded by the growth of the professions, the functional bureaucracies, and the plethora of organizations and administrative agencies that have been created to deliver the benefits promised by science and technology. Elaborate and massive networks have emerged in connection with each of the major functions – health, education, economic development, transportation, and environment, among others. Each is supported by political, legislative, and beneficiary and citizen constituencies, and by their own captive professional and lay organizations and publications. Each such functional activity system cuts the range of

governments vertically, and commands loyalties to the functional system equaling or exceeding the loyalties to the government of which they are a part. . . . The challenge in government is to affirm political jurisdiction and assert horizontal capacity; the challenge to professional and bureaucratic power is to affirm their functional supremacy and to assert vertical integration. In the case of governance, the question is citizenship; in the case of functional organization, the question is consumerism. The tension between these forces captures the essential conflict posed by government and management control.

(Meltzer 1984: 24)

Meltzer adds that "the metropolitan planning activities were peripheral, and dependent on the very people they sought to influence and oversee; only rarely did they affect private decisions in any demonstrable way."

Weak though the COGs were, they constituted a point from which regional thinking could develop; hopefully, action would have followed. But, in 1982, the Reagan administration rescinded Circular A-95, "and issued regulations that ended the system of federally funded regional clearing houses and encouraged the states and their local governments to develop their own mechanisms to do the job" (Elazar 1984: 187).

Nevertheless, a beginning had been made in regional cooperation, and planning was taken over as a state function.[24] Its character and extent varies greatly, as is illustrated at a number of points in this book. However, the indications are that changing conditions and attitudes in a number of areas are slowly bringing about some renewal of interest in forms of regional planning, though the effectiveness of this is not yet generally significant. We return to this in the final chapter.

The New Federalism

The time has now come in America to reverse the flow of power and resources from the states and communities back to Washington, and start power and resources flowing back from Washington to the states and communities and, more important, to the people all across America.

(President Nixon 1971)[25]

While Johnson's policies embraced a positive role for federal government ("federal activism" or "creative federalism"), Nixon (1969–74) promoted a "new federalism" which he claimed would bring about a return to the original conception of federalism as envisaged by the Founding Fathers. The main feature of this was intended to be the replacement of large numbers of categorical programs (and all the controls which accompanied them) by block grants.

This was, in fact, a reaction against the federalist policies of the previous Democratic administration. One highlight of this was Senator Muskie's extensive congressional hearings launched in 1966. The Senator observed that what had been created was almost "a fourth branch of government, but one which had no direct electorate, operates from no set perspective, is under no specific control, and moves in no particular direction" (U.S. Senate 1966; Haider 1974: 60). Virtually all the new programs were "functionally oriented, with power, money, and decisions being vertically dispersed from program administrators in Washington to program specialists in regional offices to functional heads in state and local governments." From the perspective of the Advisory Commission on Intergovernmental Relations (1970), this left "cabinet ministers, governors, county commissioners, and mayors less and less informed as to what was actually taking place, and [made] effective horizontal coordination increasingly difficult."

The major problems which arose were neatly summarized by Haider:

1   Proliferation of federal grants, agencies, and new programs which were deemed excessive in number and too narrow in functional orientation.
2   Multiplicity of planning requirements and rigid funding arrangements under the federal grant programs which had created rampant confusion over variable matching requirements and involved too much duplication and overlap in structure and purposes.
3   Bypassing of general-purpose units of government which, at times, entailed open discrimination against their participation.
4   The systematic undermining of elected executives and their authority due to unnecessary limitations placed on their involvement in intergovernmental programs as well as the inadequate consultation with them on program structure, funding, administration, and implementation.
5   Inordinate program delays and uncertainties as to eligibility, funding, and actual transference of federal funds, as well as enormous red tape in the total grant acquisition process.

(Haider 1974: 61)

Given this background, there was a great deal of support for Nixon's proposals. In particular, city mayors saw them as a means of obtaining additional assistance with their fiscal problems and of enabling them to recover some of the power they had lost in the Johnson years. (There were, of course, opposing forces, particularly in Congress.)[26]

Nixon's urban aid strategy had two major elements. Firstly, and most innovative, there was "general revenue sharing" which provided federal funds on the basis of a formula encompassing population, incomes, urbanization, and tax effort. The essential policy objective was to allow localities to take spending decisions on the basis of their knowledge and understanding of local needs. Secondly, "block grants" were extended by the merging of groups of categorical grants. The best known of these is the Community Development Block Grant (CDBG) program.

Nixon's ideas were never implemented to the extent which he had envisaged, mainly because of congressional opposition and the political impact of Watergate. General revenue sharing was abolished in 1986. However, the CDBG proved so popular with local political constituencies that it survived constant financial cutbacks – though in an attenuated form. It is appropriate to examine this grant more fully.

Community Development Block Grants

The Housing and Community Development Act of 1974 constituted the most important attempt since the model cities program to formulate a comprehensive national policy for the cities.

(Judd 1988)

The three-year $8.6 billion Community Development Block Grant (CDBG) program was signed into law by President Ford shortly after his inauguration in August, 1974 (Dilger 1989: 159). The Act folded seven categorical programs administered by HUD (including urban renewal and model cities) into this single grant program, which was directly targeted on cities, particularly those showing signs of social and economic distress (Bunce and Glickman 1980: 515). It was intended to achieve a balance between providing maximum flexibility for local decisions and securing the national purpose of developing "viable urban communities by providing decent housing and a suitable living environment, and expanding economic opportunities, principally for persons of low and moderate income."

This has never been an easy balance to attain. At first, there was minimal federal control: eligible local governments simply requested the allotment that was due on a predetermined formula.[27] HUD officials checked entitlement and issued approvals. Any assessment of the value was undertaken later. Local governments took full advantage of their freedom to decide on the allocation of funds and, not surprisingly, there was a number of highly publicized cases of expenditure which *prima facie* seemed inappropriate. Tennis courts took pride of place in these indictments. For instance, Little Rock, Arkansas, used $150,000 from its CDBG to construct a tennis court in a wealthy section of the town. Chicago used $32 million for snow clearance. Other criticized schemes included golf courses, polo fields, and wave-making machines.[28]

There is nothing surprising here: if local governments are given freedom to allocate funds as they wish, they will do precisely this. As Nixon had clearly stated in his message to Congress in 1971:

> Decisions about the development of a local authority should reflect local preferences and meet local needs. No group of remote federal officials – however talented and sincere – can effectively tailor each local program to the wide variety of local conditions which exists in this highly diversified country.

A requirement that "maximum feasible priority" was to be given to projects benefiting low and moderate income families allowed a good deal of leeway. Nevertheless, grants were distributed according to a formula based on population (25 percent), housing overcrowding (25 percent), and poverty (50 percent). The formula was changed in 1977 to direct resources from high-income suburbs and urban counties to needy central cities – though not with complete success (Bunce and Glickman 1980: 533; Dommel and Rich 1987: 561).

However, a HUD study (1982c) showed that 62 percent of benefits went to lower-income groups. The pattern of expenditure, which remained fairly constant through 1987, was: housing-related activities (36 percent); public facilities and improvements (22 percent); economic development (19 percent); acquisition and clearance (6 percent); administration and planning (13 percent); and other activities (3 percent) (U.S. HUD 1988b: 1).

As is not uncommon with public policies, different sources provide different conclusions on the effectiveness of the CDBG; but it does seem that, despite an attempt at targeting needy areas, the CDBG benefits were spread widely, and became even more so after the 1974 legislation gave more discretion to cities in the allocation of funds. The increase in benefits going to wealthier areas was a result of local politicians using their discretion in favor of "important segments of their electoral constituencies" (Robertson and Judd 1989: 312). Local discretion increased still further under the Reagan administration.

Carter's "New Partnership"

> For the first time in its history, this country now has an explicit urban policy.
> (President Carter 1978)[29]

In March, 1978, Carter submitted to Congress proposals for "a comprehensive national urban policy" (White House 1978). Reviewing previous policies, he noted that during the 1960s the federal government had taken "a strong leadership role" in identifying and dealing with the problems of cities. This proved to be inadequate because the federal government alone had neither the resources nor the knowledge required "to solve all urban problems." During the 1970s, federal government "retreated from its responsibilities" leaving state and local government with insufficient resources, interest, or leadership to accomplish all that needed to be done." The lessons had been learned:

> These experiences taught us that a successful urban policy must build a partnership that involves the leadership of the federal government and the participation of all levels of government, the private sector, neighborhood, and voluntary organizations and individual citizens.
>
> (U.S. HUD 1978b: 134)

The "new partnership" thus involved a positive role for the federal government, together with incentives to state and local governments, and to the private sector.

Carter's policy consisted of a large package of existing legislation and new proposals, with an emphasis on the stimulation of private investment.[30] "The vitality of American cities is crucial to maintaining our nation's economic strength and quality of life" (White House 1978). There was no suggestion that the migration from the northern cities should be stemmed, even if this were thought to be desirable:

> The continuing decentralization of the nation's 218 million people and the dispersal of its economic activities are having important consequences, good and bad, for each region and urban area of the country as it approaches 1980. For some this "thinning out" process has created the special pressures of rapid growth, for others the social and fiscal strain of population and employment decline.
>
> (U.S. HUD 1978b: 2)

The stimulation of local economic development was not conceived merely as "first aid" to the northern cities; rather it was part of a wider national economic policy designed to cope with "a new stage of urban development," of which the main features were decentralization and the dispersal of population and economic activity.

Among Carter's specific policy initiatives was a national development bank which was rejected by Congress, and the Urban Development Action Grants (UDAG) program, which passed Congress with relative ease, and proved to be widely popular: its bounties were distributed extensively.

UDAG was aimed at the stimulation of private investment to create jobs in distressed communities by schemes agreed between the private and governmental sectors. The grants were intended to leverage private money, particularly in distressed cities. Unfortunately, this was easier said than done since to target distressed cities was not the same thing as to alleviate the distress. "The contradiction in the UDAG program was that relative city distress was the primary criterion for awarding an Action Grant, but at the same time the extent of private financing was an important factor in deciding which projects received funding. Cities that provided the best investment opportunities – where private funds were more available – were not likely to be severely distressed" (Barnekov, Boyle, and Rich 1989: 79). There has been much controversy over the success (measured in differing ways) of the UDAG program (see, for example, Gist and Hill 1984; Webman 1981). There was, however, no doubt about the popularity of the UDAG scheme with "property developers, construction firms, the urban chambers of commerce, and pro-development mayors. UDAGs were urban renewal in another guise" (Markusen and Wilmoth 1982: 130).

Two unintended results of the policy of promoting private development (through UDAG and other programs) were, firstly, that increased competition among cities developed, and this led to escalating subsidies (Fainstein *et al.* 1986: 267); and, secondly, that federal controls had to be increased, thus reducing local discretion. On the latter point, Kettl noted that "regulation gradually emerged as the key strategy for implementing the new generation of urban aid programs. The creeping growth of the new rules ... gradually shifted power back to the federal government" (Kettl 1981: 123).[31]

The major legacy of the Carter administration was its reorientation of policy toward the stimulation of private investment. Indeed, "in the last year of the Carter administration, attention turned away from urban policy toward reindustrialization of the national economy" (Judd 1988: 355).

An emphasis on economic development as the foundation of federal policy was embraced by Reagan. His administration brought about the dramatic change of raising unfettered economic forces to the mainspring of "policy" – a policy of "do nothing." This had the powerful (but highly controversial) support of the President's Commission on A National Agenda for the Eighties (which Carter found unacceptable).[32]

National Agenda for the Eighties: *Urban America*

Our nation's settlements have not been and will not be significantly dependent on what the federal government does or does not do.
(*Urban America* 1980)

The essential message of *Urban America* was that unfettered market forces would benignly bring about an efficient and equitable urban settlement pattern, with the economy operating at such a high level that many "social" problems would disappear (or at least they would be reduced to a level which the enhanced resources of a liberated economy could meet). Such problems as persisted should be approached directly by "people policies" (as distinct from "places policies"). Above all, policies which tied people to declining areas should be avoided: "urban programs aimed solely at ameliorating poverty where it occurs may not help either the locality or the individual if the net result is to shackle distressed people to distressed places." Such policies are inherently wasteful. By contrast, "a federal policy presence that allows places to transform and assists them in adjusting to difficult circumstances can justify shifting greater explicit emphasis to helping directly those people who are suffering from the transformation process."

Strong criticism was expressed about the concept of a national urban policy: "efforts to revitalize urban areas through a national urban policy concerned primarily with the health of specific places will inevitably conflict with efforts to revitalize the larger economy." They will therefore do more harm than good. The forces underlying urban change are "relatively persistent and immutable," and thus are highly resistant to public policies which try to stem them or harness them to policy goals which are not consistent with wider economic development purposes. Two major policy goals were stressed:

The federal government can best assure the well-being of the nation's people and the vitality of the communities in which they live by striving to create and maintain a vibrant national economy characterized by an attractive investment climate that is conducive to high rates of economic productivity and growth, and defined by low rates of inflation, unemployment, and dependency.

People-oriented national social policies that aim to aid people directly wherever they may live should be accorded priority over place-oriented national urban policies that attempt to aid people indirectly by aiding places directly. ... A national social policy should be based on key cornerstones, including a guaranteed job program for those who can work and a guaranteed cash assistance plan for both the "working poor" and those who cannot work.
(President's Commission on a National Agenda for the Eighties 1980: 101)

The report did not enter into much detail about the translation of principles into practice, but it did list "prime candidates" that should be "scrutinized for eventual reduction or elimination" such as economic development, community development, housing, transportation, and development planning. Also suggested was a scrutiny "for major restructuring or elimination" of such programs as "in-kind benefits for the poor (such as legal aid and Medicaid), the growing inventory of subsidies that indiscriminately aid the nonpoor as well as the poor (such as veterans' benefits), protectionist measures for industry (trade barriers for manufacturers and price supports for farmers), and minimum wage legislation."

Though the philosophy of *Urban America* was very much to the liking of the new President, the report was never explicitly "accepted" by him. Given the number of constituencies which would have been affected by the "scrutiny" list, this is hardly surprising. But, as we shall see, President Reagan moved forcefully to develop policies which bore a strong resemblance to it.[33]

The Reagan Years

Improving the national economy is the single most important program the federal government can take to help urban America; because our economy is predominantly an urban one, what's good for the nation's economy is good for the economies of our cities, although not all cities will benefit equally, and some may not benefit at all.
(Savas 1983)

Reagan's pursuit of privatization was in the tradition of previous administrations, but he gave it a particular twist: so much so in fact that the difference became one of kind rather than of degree.[34] Reagan's policy was of the utmost

simplicity (some would say simple-mindedness): free rein to private forces was the key to economic growth and thus to urban regeneration and the solution of many so-called "social" problems. Government "intervention" was not only inadequate: it was counterproductive. Government action was no solution: it was part of the problem.[35]

Reagan's first major policy statement was made in an address to Congress in February, 1981. This *Program for Economic Recovery* was, as its title suggests, focused on economic matters. Most of the address dealt with general economic policy issues: proposed limitations in the growth of federal expenditure, reductions in tax rates, "an ambitious program of reform" to reduce federal regulatory burdens (including a suspension of "the unprecedented flood of last-minute rulemaking on the part of the previous administration"), and the establishment of a monetary policy "to provide the financial environment consistent with a steady return to sustained growth and price stability." Urban affairs arose only incidentally – which was precisely what was intended. There was no "urban policy," other than cuts in programs. Programs which were regarded as counterproductive were reduced or completely eliminated: the Economic Development Administration, the Urban Development Action Grant, the Community Development Block Grant, and subsidized housing (Glickman 1984). Much of this was, of course, along the lines proposed in the *Urban America* report, though the suggestions for a "national social policy ... including a guaranteed job program ... and a guaranteed cash assistance program" were conspicuously absent.

A striking aspect of the Reagan policy was the rejection of economic development programs: these were regarded as being ill-conceived and misguided. "Studies" had shown that the cost per job created by the Economic Development Administration was as high as "$60,000 to $70,000 for each person-year of employment." Not only was this cost excessive, but "EDA programs have had little effect on local economic development since such development generally would have occurred without EDA investment" (White House 1981). Nevertheless, Congress refused to eliminate EDA, as well as UDAG.

The absence of any urban policy was, of course, intentional. In the words of Barnekov, Boyle, and Rich (1989: 106), "the idea of an urban policy ran counter to the administration's governing philosophy which attributed no lasting value to federal assistance to urban areas." This created some difficulty, however, when the time arrived to fulfill the congressional mandate to present the 1982 *Urban Policy Report.* One school of thought held that since there was no intention of having a policy, the legislation should be repealed; but another argument prevailed: "the Reagan administration had an urban policy whether it recognized it or not" (Ahlbrandt 1984: 479).[36] There was some difficulty with the drafting and, after the rejection of a first draft, E.S. Savas produced another which presented other problems. Barnekov, Boyle, and Rich (1989:106) comment that Savas's "language was too blunt and his economic view of the world too simplistic . . .; but since Savas's views were consistent with the administration's efforts to fundamentally alter the federal government's responsibilities to distressed urban areas, his policy prescriptions remained the same": the first priority was to strengthen the national economy; "states and cities, properly unfettered, can manage themselves more wisely than the federal government"; the private sector should be encouraged to assist urban communities ("the private sector, both corporate and voluntary, contains important sources of strengths and creativity that must be tapped for the Nation to progress"); federal urban programs had not resulted in improving urban conditions; and there would be no significant increases in federal assistance to the poor (U.S. HUD 1982b: 1).

Enterprise Zones

No one knows with any certainty the outcome of the enterprise zone experiment.
(Bollinger 1983)

The one and only urban policy initiative made by the Reagan administration was the introduction of legislation establishing up to seventy-five enterprise zones. This was a policy import from Britain and, like all such imports, its character changed in transit (Butler 1981: 27; Hall 1991: 179). The original (1968) British concept was:

a precise and carefully controlled experiment in non-planning . . . to seize on a few appropriate zones of the country, which are subject to a characteristic range of pressures, and use them as launchpads for "Non-Plan." At the least, one would find out what people want; at the most, one might discover the hidden style of mid-20th century Britain.
(Banham *et al.* 1969)

The idea did not receive an enthusiastic reception; indeed, it was almost totally ignored until the late 1980s, when the Conservatives took it over and remolded it. The British enterprise zones (of which there are twenty-three)[37] are areas in which planning restrictions are eased and property taxes abated. The zones are located in uninhabited derelict industrial areas, and operate in reality (to use American terminology) as industrial parks (Butler 1991: 29). Their success has been modest (far from the promise of the early rhetoric): only a small number of new jobs have been created, at a not insignificant cost (though many more jobs have moved into enterprise zones from elsewhere – mostly within the same region).[38] The latest judgment is that the experiment is over and that the lessons learned from it have been incorporated in new policies (Hall 1991: 189).

The American version of enterprise zones reveals its provenance in its anti-planning ideology of simply creating conditions for private enterprise to flourish. But the resemblance ends there. The distinctively American features are threefold: firstly, the primary aim of enterprise zones is the economic improvement of poor neighborhoods; secondly, community institutions are crucial to economic development; and, thirdly, small businesses are favored over large ones (Butler 1991: 32).

Though the concept was an attractive one, Congress failed, for a variety of reasons, to pass the necessary legislation (though it passed the Senate twice). The reasons were partly procedural, partly technical, and partly political. Above all, one question proved difficult to answer: would enterprise zones create new jobs, or would they merely attract jobs from somewhere else? The importance of this question is shown by the conclusion of the British experience noted above.

At first sight it might seem that the very concept of enterprise zones would be anathema to the Reagan administration, but closer examination reveals its orthodoxy: the underlying principle was that "cities with lagging economies must make substantial monetary and policy concessions to hold and attract business and jobs" (Barnekov, Boyle, and Rich 1989: 119). The provisions of the bill were essentially a package of tax advantages and regulatory reliefs, to be provided by both federal and state governments.

Despite the legislative difficulties, the idea of enterprise zones was a popular one. By 1990, thirty-seven states had adopted the concept, and over 1,400 zones had been designated in more than 675 jurisdictions;[39] but no progress was made with the federal legislation.

Enterprise zones have sometimes been viewed as a political strategy rather than as an urban policy. It was "politically appealing because it combined a simple interpretation of urban problems with an apparently pragmatic and purposeful means of resolution" (Barnekov, Boyle, and Rich 1989: 120). A more cynical view suggests that it provided "a facade of concern for poverty and unemployment while mystifying the urban question and luring critics off the track" (Walton 1982: 14).

Any overall assessment of American enterprise zones is inherently difficult; and the difficulties are exacerbated by the fact that "states are trying very different things, and often are attempting a substantial range of things within these programs" (Green and Brintnall 1991: 253).

National Land Use Policy

An Act to provide for the establishment of a national urban growth policy, to encourage and support the proper growth and development of our states, metropolitan areas, cities, counties, and towns with emphasis upon new community and inner city development, to extend and amend the laws relating to housing and urban development and for other purposes.

(Housing and Urban Development Act 1970)

Running, in a curious way, outside the main stream of federal policies, has been a lengthy and largely fruitless battle over legislation concerning the federal role in urban growth policies. Though its impact was not perceptible, the first was the National Urban Policy and New Community Development Act of 1970. The Act was vigorously opposed by President Nixon, though he did not veto it since "skilful manoeuvering by liberal forces in a Democratic party firmly in control of both houses of a Congress impatient with the Executive branch allowed the Congress to seize the policy initiative from the President and force its will upon him" (Wingo 1972: 4). Nixon pushed forward on his new federalism program and ignored the 1970 Act as well as other pressures for broader federal urban growth policies.[40]

At this point in time, the concern was largely with urban *growth* as distinct from urban *policy*. Lyndon Johnson had expressed alarm at the demographic projections that were then being made: if present trends continued "by 1985 as many people will be crowded into our cities as occupy the entire nation today." This, together with the fear about further riots in the cities, formed a powerful argument for considering a national urban policy (Eisinger 1985: 5).

Nevertheless, the national land use policy legislation was part of a broad concern for environmental matters which resulted in some notable legislation: the National Environmental Policy Act 1969, the Clean Air Act Amendments 1970, the Federal Water Pollution Control Act Amendments 1972, and the Coastal Zone Management Act 1972.

The story of the congressional battles over the land use bill in its various forms is a confusing one. Helpfully, Robertson and Judd give a clear account of the first version of the bill sponsored by Senator Henry M. Jackson in January 1970:

> The bill was designed to encourage the states to supersede the authority of local governments in land use decisions. It would make $100 million in grants available to the states, requiring them to set up statewide land use agencies, which would then initiate a process of comprehensive land use planning. First, these agencies would have to compile data on existing land uses and natural resources and project the land use needs of future economic activity. These studies would provide the information for a second planning stage – a comprehensive, statewide land use plan that would define the states' industrial, commercial, residential, and recreational land use needs for the next fifty years.
>
> Following a three-year period, the state land use agency would be required to show that its state had given it the authority to implement a comprehensive land use plan.
>
> A new federal agency, the Land and Water Resources Planning Council, would oversee the state plans. If implemented, states might end up asserting eminent domain and police powers that would supersede the powers of local governments, even acquiring property, if necessary. If states failed to make progress in defining their land use objectives, they would be subject to losing not only the federal land use assistance funds made available in the legislation, but airport and highway funds as well.
>
> (Robertson and Judd 1989: 298; Plotkin 1987: 169)

The bill attracted surprisingly little controversy in its early days, partly because diverse interest groups were prepared to support it in furtherance of their particular concerns. However, this was the quiet before the storm, and a host of amendments and alternative bills emerged. The administration produced its own bill (in the fear that the earlier version might pass) which inter alia shifted responsibility for "areas of critical concern" from local to state governments.[41]

A new version of the bill, incorporating parts of both the earlier versions, was debated in 1972 and 1973. Several issues proved particularly troublesome:

– Should the bill establish a national policy? If so, what should the substance of such a policy be?
– Should federal review go to the substance of a state plan or simply to assuring that an adequate planning process existed? Didn't substance imply federal interference, which would upset the administration and conservatives generally?
– How should environment be defined? Did it mean only ecological considerations, or other social and economic needs as well?

(Lyday 1976: 37)

Opposition increased during 1973, despite

Nixon's endorsement, to such an extent that in June 1974 the bill was defeated.

The National Land Use Policy Bill was described by one member of the House Interior Committee as a "most misunderstood piece of legislation." This was partly because its provisions were vague, because a range of bodies were concerned about the impact on their operations (including the mining, timber and farm interests, and the construction trades). It was the taking issue, however, that proved most troublesome.

> The taking issue was the most troubling issue of all. In part it rested on a misunderstanding of the historical evolution of land use controls. Controls were rejected on ideological grounds but those who opposed were simply not persuaded that the purposes were essential to the public welfare. Regulation was supported by those who thought it would be employed to further their interest in protecting the natural environment.
>
> Supporters of the bill in both the House and the Senate sought to calm opponents' fears with amendments stating that the bill would in no way alter existing constitutional doctrine. But the opponents, convinced that eager public servants would attempt to extend their control over vast areas of privately owned land to further personal and bureaucratic ends, were not reassured.
>
> (Lyday 1976: 51)

Lyday, in reviewing the experience of more than three years during which the bill staggered around Congress, comments that the complex set of issues involved in the growth/preservation debate could not be effectively dealt with in "a process that is better suited to making compromises where fairly clear interests and ends have been articulated." She concludes:

> In the final analysis, it is probably not wise to search for comprehensive rules to deal with land use conflicts nor to attempt to define comprehensively what land should be preserved and what kinds of development are "needed." These must be evaluated in terms of who benefits and who pays, and the trade-offs will vary according to the specific competing claims being made. The debate over the land use bill revealed that there is no consensus about benefits or losses, and little technical or scientific basis against which such values could be weighed. Until there is some agreement on ends and how they can be compared one with another, neither planning nor institutional reform is apt to alter the status quo.
>
> (Lyday 1976: 53)

Conclusion

> It is not feasible for the highest level of government to design policies that can operate successfully in all parts of the nation.
> (*Report on National Growth 1972*)

There are some lessons to be learned from the various federal policies which have been discussed in this chapter. Here attention is drawn to the more important of them.

Urban renewal had some notable successes, but it also had some disastrous consequences, particularly in its insensitivity both to the people and the communities affected, and to the urban fabric. It also clearly reveals a frightening weakness in the political system. In order to obtain support for the program, it was necessary to give the major role to private enterprise. Not surprisingly, therefore, the search for profits took precedence over matters of public policy. It is nonsensical to blame those involved for this selfishness: it was built into the scheme of things.

The model cities program started with the idea of establishing priorities, but this was shattered by the political process. To obtain political support for a program, benefits have to be widely spread to embrace as many constituencies as possible. The program also laid emphasis on coordination, but this ran up against another problem: congressional fears that effective coordination would mean concentration of power. Frieden and Kaplan (1977: 238) comment that a major lesson of the model cities program was that "grand schemes for massive, concerted federal action" simply will not work.

Less threatening to the federal government and, for a time, more effective, was the thrust toward regional planning. Unfortunately, it was deprived of its raison d'etre and largely faded from the scene. Perhaps it was in advance of its time. Be that as it may, it is now appearing in a different guise with the state initiatives illustrated in Chapter 11.

Most federal programs have a short life – generally not more than the length of two administrations, and frequently less. This is the inevitable result of changing circumstances and the political seesaw, as well as the experience of implemented policy. Some underlying policies, however, seem to be embedded in the system: above all, a commitment to ensuring that the engine of change shall be powered by private enterprise. This is the American way, and it is unlikely to change.

Nevertheless, as experience with regional planning shows, federal promotion and funding can have a modest impact which may develop in response to regional circumstances. There seems little possibility, however, of a federal land use policy (outside the established "environmental" programs). If local governments are to be made to shed their narrow, exclusionary practices, controls will have to be applied by the individual states. Unfortunately, this is the level at which "interstate economic competition and suburban political power promote spatial inequality" (Judd and Robertson 1989: 22). It has to be admitted that

it is not easy to see how the character of American urban policy might be changed: "It is the prisoner of an intergovernmental system that uses public resources to protect inequality and that structures politics in such a manner as to make it extraordinarily difficult to enact policies of redistribution" (Judd and Robertson 1989: 22).

It is also the prisoner of simplistic beliefs in the power of market forces to produce an efficient and, in the long run, equitable urban form. Byrne, Martinez, and Rich (1985) have demolished the intellectual underpinnings of this neoconservative argument (that the underlying forces of societal change are beyond social choice). Moreover, both explicit and, more importantly, implicit urban policies have created an "artificial" market framework which may or may not be desirable in the long run. Current policies do not operate on "a featureless plain"; indeed, to carry further the image, the plain is strewn with incentives and barriers to action.

The urban geography of the United States (as elsewhere) has been dramatically reshaped since World War II by a huge range of public policies, some of which have worked as hoped, some of which have worked in unexpected ways, and some of which have not worked at all. This chapter has highlighted a number of these. Though a coordinated and effective national urban policy seems as remote from reality as an unhampered free market, it is incumbent upon government to try to shape its policy instruments to achieve the maximum effect.[42] Though there are limits to effective governmental action, major patterns of urban growth and urban decline have been significantly influenced by government policies. These governmental actions have been reinforced by *inactions* – above all by "public decisions to allow the economic fate of those cities to be determined in the private sector" (Gurr and King 1987: 190).

Part V
American Planning in Comparative Perspective

Chapter 13

Cross-cultural perspectives

Introduction

> The primary value of foreign exploration lies not so much in the discovery of readily transferable concepts, technologies, or techniques that can be packaged up, carried back into the United States duty-free, and unwrapped to delight policy makers, but rather in the stimulus of insightful reflection of culture and experiences.
>
> (Haar 1984)

In this chapter, American land use planning is considered in "cross-cultural perspective": how and why does it differ from other countries, particularly Britain and Canada.[1] The differences are fascinating, and there is added interest in the question as to whether time is narrowing or widening them.[2]

Whatever the answer to this question, there is no doubt that there is benefit to be gained by a comparative international perspective. This does not mean that there are "lessons to be learned." Far from it: policies are the cultural products of history, time, and place: they are rarely exportable. But, as Haar notes in the quotation at the beginning of this section, comparisons prompt new questions (or at least reformulation of old ones) and, in contemplating these, new policy initiatives may emerge, molded to fit a different political terrain. Thus a mutant of "enterprise zones" traveled from Britain to the United States with the same name masking significant differences. Interestingly, the efficacy of both was similar, and far less than anticipated. There was, however, more "traffic" in the opposite direction, first with the War on Poverty programs, and later with economic development programs – by which the British were impressed because of the perceived success of projects such as Baltimore's Inner Harbor, Boston's Faneuil Hall Marketplace, and Pittsburgh's Golden Triangle. Here, it seemed, was very tangible evidence that free market policies worked! (Apparently the itinerary for visiting British dignitaries excluded the Renaissance Center in Detroit.)

These tangible values of international comparison are somewhat unusual, but there are other benefits, as Wolman notes:

The utility of comparing the United States with foreign countries frequently is seen to lie in the potential for transferring the policies or practices of those countries to the United States. While the possibility of such transference does exist, the difficulty inherent in transplanting across systems of divergent political, social, and economic characteristics should not be underestimated. Perhaps a more valuable benefit to be gained from comparative analysis is that such analysis can both broaden the sense of the possible and provide a framework for better understanding our own behavior. The process of comparative analysis should lead to questioning what frequently is simply assumed without question.

(Wolman 1985: 116)

The similarities and contrasts are interesting, but they are also elusive. Scottish planning is different from that in England and Wales; Northern Ireland is different again. In Canada, the operation of planning in Newfoundland is very different from Ontario, while Quebec and Alberta display markedly individual characteristics. Indeed, to generalize about Canadian planning is almost inevitably a travesty of the truth. Of course, the United States is so diverse that generalization is extremely hazardous. Yet, the internal differences are largely a matter of degree, not of kind, whereas there are essential differences among the countries which reflect their particular histories, cultures, and values.

It is these differences that form the basis for the following discussion. The focus is on land use planning but, of course, this cannot be understood except through an appreciation of broader social, economic, political, and historical factors.

Some Major Points of Comparison

The United States is egalitarian and populist, Great Britain is deferential and elitist, while Canada (and Australia) fall in between.

(Lipset 1963b)

A familiar and rueful joke holds that while Canada had hoped to achieve a synthesis of British governance, French culture, and American know-how, it has been left, instead, with the residue of British know-how, French governance, and American culture.

(Friedenberg 1980)

Comparisons between the United States, Britain, and Canada have an honorable history and, indeed, some may argue that there is as yet little to match the depth of analysis of the classics of Tocqueville's *Democracy in America*, Bagehot's *The English Constitution* and, later, Porter's *The Vertical Mosaic*.

The subject has proved to be of endless fascination, which straddles a wide range of disciplinary fields, from sociology to political science to marketing.[3] It deserves far fuller treatment than is possible here, where the object is the modest one of exploring some ideas which help to explain the differences and similarities among the planning systems of three countries which share something of a common heritage – and something of a common language. The commonalities, however, begin to look less convincing when one begins to examine them. The first and most obvious contrast is in size, both of area and of population.

There are, of course, major differences in the size of the three countries and their populations. In area, Canada is almost a tenth larger, but its population is little more than 10 percent of the United States. Great Britain[4] has more than twice the population of Canada, but a land area about the size of Oregon.

Britain led the way in the industrial revolution and, as a result, produced the first large urban slums and also the first public health and planning legislation, together with a local government system increasingly well-equipped to implement some (though not all) of these new public policies.[5] There was also a marked increase in the standard of living, but the bill for the progress made in those times is far from fully paid, as is clearly evidenced today in cities such as Birmingham, Glasgow, and Liverpool.

Canada, with its staples-producing economy, was very different, though this is not to suggest that urban squalor was absent. There was little pressure for planning, though an abortive movement began in the early years of the twentieth century with a nice compound of British and American influences.[6] Other significant differences arose in the interwar years. The Depression hit all three countries, but curiously gave rise to an unprecedented housebuilding boom in England and Wales (Richardson and Aldcroft 1968). This peaked out in 1938 at 340,000 completions; only slightly lower than the United States figure (406,000 starts). More strikingly, the figures for 1933 were 218,000 completions for England and Wales, and 93,000 starts for the United States. Over the whole interwar period, England and Wales produced 4 million houses (of which one million were public housing) while the United States (then with three times the population) produced 10 million. As a result, suburbanization in England and Wales proceeded relatively more rapidly than in the United States and, given the relatively tiny size of England and Wales, its effect was more dramatic. The embryonic planning system could not cope with the pressures; indeed, by 1937, it had reacted by zoning sufficient land to accommodate nearly 300 million people (Barlow Report 1940: 113). An important factor here was the "compensation bogey": municipalities could not prevent development without paying compen-

sation. This interwar experience was important in the development of support for major changes in the postwar period.

The catalyst of war brought about an abrupt change in the whole climate of public opinion towards planning (Titmuss 1958) and, in Britain, a system was established which dealt decisively with the compensation problem: both development rights and development values were nationalized (as were railways, hospitals, health insurance, and a range of important public utilities). At the same time, local authorities were given extremely strong powers of development control. A major expansion of public housebuilding was set in motion by the Labour government of 1945–51 and, curiously, expanded for a while by the succeeding Conservative government. (The explanation lies in the fact that, at this time, it was easier to expand the public sector – which employed many private building contractors – than the private sector, which could not readjust quickly to an uncontrolled market.) The Conservatives also resumed the prewar slum clearance program, which reached massive proportions.[7] The number of demolitions in England and Wales rose to around 60,000 a year in the early 1960s.

It should be noted that this was not "urban renewal" in the American sense: most of the redevelopment was with public housing, the total amount of which increased to 2.5 million by 1953; this was 18 percent of the total British housing stock. Though the Conservative government gradually increased the private supply of housing, the planning controls remained in force.[8]

In the United States, such provisions were inconceivable but, in any case, there was little reason for them. Land was plentiful: or, at least, appeared to be so. The housing finance system, however, entered into a period of collapse as the interwar Depression deepened. "New Deal" measures were introduced on a scale unknown before, in any of the three countries. Even a modest amount of public housing was built. The postwar period saw little

Table 2 Area and population of the United States, Canada, and the UK

	Area (sq ml)	Population (millions)
USA	3,619,000	249(1989)
Canada	3,850,000	26(1988)
UK	94,000	57(1989)

Urbanization

	Population Density (per sq ml)	Urban Population (%)
USA	68	76
Canada	6	76
UK	601	93

Constituent Countries of the UK

	Area (sq ml)	Population (1989) (thousands)
England	50,363	47,689
Wales	8,018	2,873
Scotland	30,414	5,091
N. Ireland	5,452	1,583
Total	94,247	57,236

Selected Provinces/States

	Area (sq ml)	Population (Canada: 1988; USA: 1990) (thousands)
Quebec	594,855	6,639
Ontario	412,586	9,431
British Columbia	364,944	2,984
Prince Edward Island	2,815	129
Texas	266,807	16,987
Oregon	97,073	2,842
Maryland	10,460	4,781
Rhode Island	1,212	1,003

Sources: *World Almanac 1991, Canada Year Book 1990, Whitaker 1991*, and *Britain 1991 Official Handbook*.

of the British approach of "winning the peace" (which soon had a hollow ring in that country as widespread shortages of the necessities of life developed). There were generous provisions for returning American veterans (the GI Bill of Rights) which, together with other housing incentives, led to a rapid expansion in house-building, from 142,000 starts in 1944; to 326,000 in 1945; 1,023,000 in 1946; and rising to a peak of 1,952,000 in 1950 (U.S. Bureau of the Census 1976).

In Canada, the situation was different again. A mere 800,000 dwellings were added to the stock between 1921 and 1941, and little effective governmental action was taken (Saywell 1975). At the end of World War II, Canada had the worst housing shortage of the three countries. Moreover, it lacked the mechanisms for dealing with this, either through the private sector as was done in the United States, or through the public sector, as in Britain. The situation was transformed in the early postwar years with the establishment of the Central Mortgage and Housing Corporation (CMHC), which provided finance and technical services, enforced construction and layout standards, and promoted the formation of large building firms.[9] With its heavily British bias, CMHC played a major role in the postwar development of the suburbs (Carver 1960 and 1975).

Its impact, nevertheless, was less dramatic than was the case in the United States where the federal road program, tax-aided owner-occupation, and urban renewal had a greater effect on the cities, sometimes of a devastating character.

The federal role in Canadian government is not easy to summarize. This is because of the numerous ways in which responsibilities are shared between the federal and provincial governments. Old age pensions and family allowances, for example, are generally a federal responsibility, while other social security services (such as allowances for the blind, the disabled, and the unemployed) are a joint responsibility, as are health services ("socialized medicine" as it would be called in the United States). Parks may be "national" or "provincial"; water resource management rests on federal-provincial agreements; civil aviation is unequivocally a federal matter (Cullingworth 1987: ch. 2).

There are, of course, many points of contrast with the United States and, as would be expected, even more so with Britain, where the differences in geography and the system of government are so great. Here attention is concentrated first on the contrasts between Canada and the United States.

One significant difference is between the federal highway programs in the two countries. Canada has built only one national road, the Trans-Canada Highway. Though individual provinces have financed many more, the mileage is small in comparison with the massive United States interstate highway system.

Urban highway building (with little federal funding) has also been on a smaller scale, with relatively attractive (sometimes superb) public transit systems as a viable alternative for commuters. There has been a correspondingly smaller impact of road building on inner city neighborhoods. Likewise, urban renewal (during its comparatively short life) was far less widespread and disruptive. This may have been related to the fact that it involved far less federal funding in Canada than in the United States. Moreover, "an interesting contrast with the US is provided by the extent to which private capital was involved in large scale urban renewal in Canada in the absence of government policies and programs" (Goldberg and Mercer 1986: 259, n.6; Collier 1975).

A deterrent to urban destruction in Canada has been the attractiveness of the inner cities to articulate middle-class residents who did not flee to the suburbs to anything like the same extent as in the United States (or Britain). Resistance to change preserved urban living qualities and increased their attractiveness. Public policy reflected this: a short-lived urban renewal program in the latter part of the 1960s aroused bitter controversy, and was replaced by a Neighbourhood Improvement Plan, which placed emphasis on "a safeguarding of the built environment of inner-city neighbourhoods and the opening of the planning process to citizen participation" (Filion 1988: 16).

Clearly, public-sector involvement has been different in scale and in form in the three countries. Nevertheless, both countries are witnessing a withdrawal of federal interest and an attempt to shift responsibilities to the provinces or states. This is partly a matter of political philosophy and partly a matter of financial expediency (reducing federal expenditure).

The situation in Britain is different and more complex.[10] The seesaw of party politics saw the nationalization (under the postwar Labour government) and then the denationalization of major sectors of the economy (under the following Conservative administration). In more recent years, a new concept, "privatization," has been imported from the United States.[11] The vicissitudes of postwar domestic polices have been striking. Housing policy and some aspects of land policy (such as "betterment": the collection of land value increases) have swung violently with changes in the political complexion of the government.

One policy change is particularly noteworthy since it mirrored events in both the United States and Canada. In all three countries, a force for change was concern over the implications of anticipated population increases. In 1955, the population of Britain was around 50 million, and was projected to increase only slightly (by less than 2 million) over the following quarter-century. These projections were raised in 1960 to give an increase of 9 million (to 62 million) by the end of the century, and revised again in 1965 to give a massive increase of over 19 million. Thus, in the mid-1960s it was expected that the British population would increase from 53.1 million to 72.5 million in the year 2000 (Cullingworth 1979a: ch.4).

These figures were alarming and, added to the already existing difficulties of obtaining agreement among local governments for selecting sites for large-scale development, resulted in the designation of thirteen additional new towns between 1961 and 1970. Shortly after this policy had been put into effect, the population projections took a dramatic dip downwards. The commitment to the "new new towns," however, could not easily be changed, and most of them went ahead.

These changes in policy, though significant, were in fact only incremental. Quite different were the dramatic changes which have taken place since Mrs. Thatcher became prime minister in May, 1979. To describe the period as "a swing to the right" is to understate the breadth and depth of the change. The roots were essentially ideological and, of course, bore a strong resemblance to Reagan's similarly ideological stance, though the British policies embraced a high degree of centralism which was quite the opposite of Reagan's policy of transferring power from the federal to the state level.[12]

British urban policy, before Thatcher, was heavily influenced by social concerns. Inner city problems were seen as being essentially social in origin, and policy was therefore designed to assist "disadvantaged" areas to break out of the "cycle of poverty" – a concept embraced by the U.S. War on Poverty (Lewis 1959 and 1966). Increasingly, however (again as in the United States), the emphasis shifted to economic development which, in the small country of

Table 3 Postwar British Governments

Labour 1945–51
Conservative 1951–64
Labour 1964–70
Conservative 1970–74
Labour 1974–79
Conservative 1979 to date

Britain is more of a central than a local responsibility. The Thatcher government also had a mistrust of urban local government (particularly in those areas which were controlled by strong, sometimes belligerent, Labour councils).

In addition, there was an overriding concern to reduce public expenditure. There was little room for reductions in central government programs. The cost of welfare benefits (entirely a central responsibility) was rising as a result of the growth in the numbers of the unemployed and the elderly. Other programs, such as defense, and law and order, could not be cut because of electoral commitments. Local government, therefore, became the prime target for expenditure cuts.

Three main methods were used: the imposition of severe limits on local expenditure; the privatization or deregulation of some local services; and extensive audits of local authorities – "in the hope that better informed local taxpayers would, through the ballot box, curtail local authority spending."

The impact of these measures was dramatic. The Conservatives radically restructured and restricted the way in which central government financed local authorities. They abolished cities' own source of income, the rates (or local property taxes) and replaced it with a community charge (or "poll tax") and a centrally determined and assigned "uniform business rate" on non-domestic properties. In the six largest urban areas in England the government eliminated the metropolitan county councils – a complete tier of elected government. Tenants in public housing and parents of children in state schools were given the right to vote to leave the local government sector and to choose private provision at least partially funded by central government. The government argued that it was transforming an expensive, inefficient bureaucracy into an organization which was more accountable to its citizens and more democratic. Local

government, by contrast, argued that individual consumer choice was being reduced and the center's determination to bypass local authorities left them reduced to mere functionaries of an increasingly powerful and undemocratic state. Whoever was right, an extraordinary change in urban policy took place in the 1980s.
(Parkinson 1989: 427; see also Ascher 1987, and Jones 1988)

It used to be said that while British planning was essentially concerned with "controlling" development, the North American equivalent was overwhelmingly concerned with "promoting" it. Much of that difference has now gone; but not all of it. Indeed, perhaps the most surprising aspect of the British planning scene, despite the changes made by the Thatcher administration, has been the permanence of its essential elements such as the control of urbanization, the nationalization of development rights in land, and the system of discretionary planning controls.

The contrast with zoning is marked, though more so in theory than in practice, since the United States and Canada have evolved new techniques of planning control which are far removed from the original rigid system of zoning. Though the Thatcher administration introduced enterprise zones and "simplified planning zones," where local government controls are reduced, their impact on the overall planning system (or anything else for that matter) has been slight (Anderson 1990). More effective have been the ten urban development corporations (UDCs) which have extraordinarily wide powers of land acquisition (Wood 1986; Church 1988; Anderson 1990). Even more remarkable is the extent of their role (within their designated areas): they have all the planning powers of the local authority in whose area they are situated, and are free of the controls which that authority would normally impose. UDCs are appointed (not elected) bodies "intended to eliminate the political uncertainty produced by local democracy which the government regarded as a major deterrent to private investment" (Parkinson 1989: 435). With massive funding from central government, UDCs are undertaking large-scale physical regeneration, though with little direct benefit to local residents who do not have the skills for the new jobs, or the incomes for the new houses (Church 1988; Goodwin 1991).

There is little by way of Canadian equivalents to these programs. Each province has its own set of policies, though these were initially stimulated by the early postwar role played by the CMHC (noted earlier). Several major initiatives were made by the province of Ontario between the mid-1960s and the end of the 1970s (Cullingworth 1987: ch. 8). The most important of these, called "Design for Development" embraced a regional development policy, the reorganization of much of local government on a regional basis, an urban growth plan for Toronto and other urban centers, the establishment of a commission to control development (with British-style planning techniques) in the Niagara Escarpment, and the development of two new towns (which were never built, though the land was purchased). This remarkable set of initiatives was an exercise in both regional and provincial planning: as originally conceived, provincial capital expenditure was to be "directed to regional needs... thus regional development will be contained within the broader spectrum of provincial development." The grand design eventually faded and lapsed, but the regional machinery and much of the planning strategy remained.

Provincially sponsored regional planning was also undertaken in other provinces such as Alberta, Quebec, and British Columbia (Cullingworth 1987: ch. 9). The latter also instituted in the 1970s a land use policy for the preservation of agricultural land over a huge area surrounding the Vancouver urban region. By 1975, some 4.7 million hectares were "protected." In the words of Hagman and

Misczynski (1978: 283), this "may have represented the most massive downzoning in Canada to that date." Towns in Saskatchewan operated land banking policies which, though typically forced upon them by the large number of tax forfeits which occurred during the Depression, proved to be a successful land planning tool (Ravis 1973; Watson 1974; Spurr 1976).

At the federal level, forecasts of major increases in metropolitan populations (from 8.6 million in 1966 to 24.8 million at the turn of the century) led to a remarkable attempt to fashion a new national urban policy. Based on the eloquent Lithwick Report (1970), a Ministry of State for Urban Affairs (MSUA) was established. The underlying rationale for this federal intervention in urban affairs was essentially that "the major forces influencing cities do not lie within their control, and policy that assumes the opposite – that individual cities can solve their own problems – is destined to be ineffectual." Unfortunately, "ineffectual" was the term which precisely describes the operation of the MSUA. There are many reasons for this, including the fact that the new ministry had no programmatic responsibilities (and therefore no political clout) and that its terms of reference were extraordinarily vague. They were to:

> formulate and develop policies for implementation through measures within fields of federal jurisdiction in respect of
> (a) the most appropriate means by which the Government of Canada may have a beneficial influence on the evolution of the process of urbanization in Canada;
> (b) the integration of urban policy with other policies and programmes of the Government of Canada;
> (c) the fostering of cooperative relationships in respect of urban affairs with the provinces and, through them, their municipalities, and with the public and with private organizations.

Though it lasted for nearly a decade, the

MSUA proved incapable of forging any new policy approach, or effecting any system of coordinating the multiple efforts of the federal government to improve urban conditions.

This rapid review prompts a range of questions about the reasons for the differences. One relevant matter, of course, is the constitutional framework within which each country works. This is the subject of the following section.

The Constitutions

The British North America Act . . . is a document of monumental dullness which enshrines no eternal principles and is devoid of inspirational content. It was not born in a revolutionary, populist context, and it acquired little symbolic aura in its subsequent history. The movement to Confederation was not a rejection of Europe but was rather a pragmatic response to a series of economic, political, military, and technological considerations.

(Cairns 1988)

The United States is a land of constitutions, constitutional law, and constitutional interpretation. Americans have an abiding faith in the doctrine and workings of judicial review for defining or checking a multiplicity of statutes, and they place great reliance upon basic documents like written constitutions, fundamental charters, and organic acts for guidance in carrying on public affairs.

(Grant and Nixon 1968)

Unlike its European partners and competitors, modern Britain has never endured political upheaval. For three centuries it has had no revolution; for nine, no invasion. One unwelcome product of that happy history has been a deep reluctance

among Britain's rulers to criticize the system that sustains them. . . . The system is as flawed as a tone-deaf orchestra.

(*The Economist* May 11, 1991)

The American and Canadian Constitutions differ markedly. Of particular relevance to the subject of this book is the contrast in the allocation of powers and thus the relative strengths of the federal and the state/provincial governments. The United States was concerned to diffuse power generally, and to limit the power of the federal government in particular. This emanated, of course, from a revolutionary war fought to establish popular sovereignty.

The preamble to the United States constitution spoke of "We, the People"; the British North America Act of "the Provinces of Canada, Nova Scotia and New Brunswick." Unlike the American, the Canadian Constitution, as it emerged from the negotiations of 1864–67, was not submitted to ratification by popular conventions. . . . In short, the British North America Act of 1867 was an instrument of governments alone which embodied almost no hint of democracy or popular sovereignty.

(Smiley 1987: 39)

The role of the courts is similarly different:

Throughout the history of the United States, common law and the courts have been perceived and used as a check on the power of the state. American jurisprudence reflects a concern for limiting governmental coercion over individuals. In Canada, the courts have been much more closely identified with the state, and perceived as an arm of the state.

(Rocher 1988: 2)[13]

Even more extraordinary to American eyes is the provision in the Canadian Constitution which enables the federal or a provincial legislature to opt out of many of the constitutional restrictions by providing that a law shall operate "notwithstanding" the provisions of the Charter.[14] This neatly precludes "nine non-elected permanently tenured judges ... setting permanent constitutional policy," and helps to avoid what is seen as "the American dilemma of allowing constitutional supremacy to degenerate into judicial supremacy" (Morton 1987: 54).[15]

It should also be noted that the Canadian Constitution was modeled on that of Britain. The preamble to the 1867 Act states that:

> Whereas the Provinces of Canada, Nova Scotia and New Brunswick have expressed their Desire to be federally united in One Dominion under the Crown of the United Kingdom of Great Britain and Ireland, with a Constitution similar in principle to that of the United Kingdom.

The most important feature taken over was the parliamentary system, together with the convention that the government has to have the support of the House of Commons if it is to remain in office.

Canada, of course, had no revolutionary war and, indeed, provided a home for thousands of "United Empire Loyalists" who had supported the British cause during the American revolution.[16] Canada remained a British colony, and had a succession of constitutional enactments (passed by Britain) culminating in the Constitution Act of 1982.

There is little notion of popular sovereignty in Canadian constitutional history. (The Bill of Rights dates only from 1982.) Debates on constitutional change were more in the nature of bargains struck among elites. The major focus was economic development, not political philosophy. For this purpose, a strong central government was needed. In Watts' words:

> Since a major purpose of confederation was to create the political preconditions for the great project of transcontinental economic development based on interregional railways, it seemed necessary to give the central government all the powers required to stimulate development. Pressures for economic growth, which in nineteenth century United States strengthened state and local governments and private profit making enterprises, in Canada had the contrasting effect of strengthening the federal government at its creation.
>
> (Watts 1987: 112)

Given this pragmatic approach, the central government was assigned the senior role in the federal system, complete with all residual powers (precisely the opposite to the United States).[17] Provincial powers related to:

> "matters of a merely local or private nature in the province." The enumerated powers included property and civil rights, the management and sale of provincially owned public lands, hospitals, municipal institutions, local works and undertakings, the incorporation of companies with provincial objectives, the solemnization of marriage, and the administration of justice.

However, history has a way of falsifying the assumptions of constitution makers, and redirecting their intentions. Provincial subordination to a strong central government in Ottawa has long gone, despite the centralizing effects of two world wars, the Depression of the 1930s, and postwar reconstruction. Though Canadian federalism has fluctuated between the extremes of decentralization and centralization, the outcome has been a major shift of power to the provinces, spearheaded by Quebec's move toward "Sovereignty-Association."[18]

As the Canadian provinces have grown in population and power, centrifugal forces have increased in opposition to the historic "nation building" (and thus centralizing) stance of the federal government. To quote Watts again:

> Many responsibilities that seemed relatively insignificant (or within the purview of private or religious organizations) when they were assigned to the exclusive jurisdiction of

the provinces under the British North America Act [1867] have, with the growth of an industrial, urban, and secular society and of more interventionist governments, become among the most important. Exclusive constitutional responsibilities for property and civil rights, health care, social services, and education has made provincial governments extremely important in the day-to-day life of citizens. The control of lands and natural resources has made provincial governments major agents in economic development and has attracted to them the support of provincially oriented elites based on resources.

(Watts 1987: 114)

The tendencies are not all in one direction, of course, and the federal government has that very important "power of the purse" which can overcome provincial resistance to some federal initiatives. Moreover, as in the United States, much "word energy" is dissipated in debate on issues of federal/provincial responsibility. (There is a popular quip that Canada may be the only country in the world where one can buy a book on federal-provincial relations at an airport.) But the main lines are usually clear and, so far as land use planning is concerned, there is no doubt as to where responsibilities lie.

Britain does not have a written constitution. Though there are several important constitutional Acts,[19] there is no document equivalent to the United States Constitution, which establishes rights and the relationships between the three branches of government. There are no checks and balances: Parliament is supreme, and even the small range of powers transferred to the European Community can be rescinded by a further Act of Parliament. The courts have no power to declare Acts of Parliament unconstitutional as do the American and Canadian courts.

The historical roots of British government are parliamentary, not democratic. The Reform Act of 1832 increased the electorate from 5 to 7 percent of the adult population, and the proportion was increased by steps, but it was not until 1928 that universal suffrage (including women) was achieved. "Undemocratic" institutions such as the House of Lords and the Monarchy survive with limited powers, but also with resplendent trappings which are as adored (or at least as keenly followed) in the United States as they are in Britain. As the *Washington Post* commented in May 1991 on the festive preparations for the visit of the British Queen: "Two centuries ago George III lost the colonies; Queen Elizabeth is in danger of winning them back." In a review of four new books on the British royalty, Florence King has noted that there is a "secret yearning for hierarchy that ostensibly egalitarian America so stoutly denies."[20]

The monarchy is part of what Bagehot termed the "dignified" parts of the Constitution, as distinct from the "efficient" parts which provide the rules by which the Constitution works. The "dignified" parts of the Constitution garnered popular support for the governmental system, even though these parts did not wield any power, while the "efficient" parts did the actual business of governing. "Identify the efficient parts and you identify who governs; identify the dignified parts and you identify how they are enabled to do so" (Dearlove and Saunders 1984: 2). In Bagehot's words:

There are two great objects which every constitution must attain to be successful, which every old and celebrated one must have wonderfully achieved: every constitution must first *gain* authority, and then *use* authority; it must first win the loyalty and confidence of mankind, and then employ that homage in the work of government.... The dignified parts of government are those which bring it force – which attract its motive power. The efficient parts only employ that power. The comely parts of a government *have* need, for they are those upon which its vital strength depends. They may not do anything definite that a simpler polity would not do better; but they are the preliminaries, the needful prerequisites of *all* work. They raise the army, though they do not win the battle.

(Bagehot 1867: 61)[21]

Allied to this is long-established respect for authority: "certain persons are by common consent agreed to be wiser than others, and their opinion is, by consent, to rank for much more than its numerical value" (Bagehot 1867: 171). This is not only a British trait. In Europe generally, the political culture provides a framework which facilitates planning:

Hierarchical social and political systems, where the governing class is accustomed to govern, where other classes are accustomed to acquiesce, and where private interests have relatively less power, can more readily evolve urban and regional growth policies at the national level than systems under the sway of the market, local jurisdictions, or egalitarian political processes.

(Berry 1973: 180)

Of course, Bagehot was writing in the middle of the last century, when social class was more important in the government of Britain than it is today. But no American visitor to Britain can fail to be struck by the class divisions that remain.

Another issue which is of particular interest to the student of constitutions is the importance which Bagehot attached to the cabinet:

The efficient secret of the English Constitution may be described as the close union, the nearly complete fusion, of the executive and legislative powers. No doubt by the traditional theory, as it exists in all the books, the goodness of our Constitution consists in the entire separation of the legislative and executive authorities, but in truth its merit consists in their singular approximation. The connecting link is *the Cabinet*. By that new word we mean a

committee of the legislative body selected to be the executive body.

(Bagehot 1867: 65)

It was, of course, the apparent separation of powers which was taken into the U.S. Constitution, and which is such a strong feature of the land use planning process. Of singular importance in the United States is the significant and independent role played by the courts. It could be argued that, in one sense, judicial controls in the United States take the place of the political and administrative controls of Britain and Canada.

This raises a host of interesting questions which would repay further inquiry. Unfortunately, the legal position in relation to major American planning issues is often unclear, and the U.S. Supreme Court has frequently shied away from giving a much-needed lead. One result has been an inconsistency of decisions between the states. The same has been the case, but to a lesser extent, in Canada, though a provincial court will take account of the decisions of other provincial courts, and even those of British courts. While an aggrieved party in the United States would, after exhausting administrative remedies, naturally turn to the courts for redress, in Britain he would rarely be able to go further than an appeal to the central planning department. On balance, Canada leans to the British model, though some of its appellate bodies are more independent than are the British.[22]

There is no equivalent in Britain to the U.S. Supreme Court, and the unwritten Constitution provides few of the protections of the U.S. Constitution and its Bill of Rights. It should be noted, however, that the House of Lords is the highest court of appeal, and it adheres to the *stare decisis* principle: it is bound by its previous decisions – unless a change is made by the legislature.[23] In this, it is different from the U.S. Supreme Court which "considers itself free to overrule its earlier decisions, to discover, that is, that the constitution which it is interpreting really has all along had a different meaning from what was supposed" (Boorstin 1953: 15).

Though Canada has a Supreme Court (created by an Act of Parliament in 1875), neither it nor the lower courts deal with many land use planning cases, despite the extensive controls operated by Canadian local governments.[24]

Ideas and Images

The American tradition is, in essence, an individualistic tradition which has tended to look upon the state with doubt or suspicion.
(Laski 1948)

Unlike Americans – and, for that matter, unlike the British, who had their share of rebellions and revolutions, some of them glorious – Canadians as such have no tradition identifying government as the source of oppression.
(Friedenberg 1980)

The cultural differences among the three countries are manifest in numerous ways. Porter underlines one:

Canada has no resounding charter myth proclaiming a utopia against which, periodically, progress can be measured. At the most, national goals and dominant values seem to be expressed in geographical terms such as "from sea to sea" rather than in social terms such as "all men are created equal," or "liberty, fraternity and equality." In the United States there is a utopian image which slowly over time bends intractable social patterns in the direction of equality, but a Canadian counterpart of this image is difficult to find.
(Porter 1965: 366)

A British comparison is also difficult to find.

Indeed, the British are coy about making sweeping statements of national sentiment, philosophy, or even policy (except for party political statements with catchy titles such as "Britain Strong and Free"). This may be due to the low profile of "a nation which has lost an empire" (and embarrassed memories of "Land of Hope and Glory"); or it may be the national preference for understatement.

Of particular significance is the way in which ideas are expressed in the United States and the way in which they influence (or do not influence) policy. Tocqueville noted that in America, "ideas are all either extremely minute and clear or extremely general and vague; what lies between is a void." This "Tocquevillian void" continues (Burns 1989: 125 and 665). A contemporary illustration is provided by President Bush's rhetoric on his education proposals announced in April 1991. As reported in the New York Times (April 20, 1991), "Mr Bush envisions a profoundly altered educational landscape.... The best minds in America would invent schools that would 'break the mold' and produce world-class students. Parental apathy would vanish and communities would honor educational excellence." This arresting outcome of a "new education vision" is, however, typically light on concrete proposals for implementation or finance.

Despite the fondness for high-flown rhetoric, which inspires more than it informs, Americans have little time for ideology. There is an operational consensus which embraces conservation and equality (of opportunity, not achievement; what Potter (1954) has termed "parity of competition"), but ideas of socialism and communism (which are typically seen as the same thing) are alien: so are ideas and policies which are, or become, associated with them. Thus major British policies relating to the nationalization of development values, government sponsored new towns, and discretionary planning controls, or British and Canadian health and welfare programs, stand little chance of attracting public acclaim, and still less congressional support. There are, however, policies which, at least at first sight, do not fit into this picture. For example, the 41,000-mile interstate highway system was a remarkable policy of colossal dimensions: one which might be the envy of any proclaimed socialist government. Though dressed up partly in terms of national defense (a response to the intensification of the Cold War) it was a major act of domestic policy. It also was carried out in an untypically American manner, not only with the (necessary) large-scale land acquisition, but also with a draconian overriding of property rights in some inner cities[25] (as well as enormous public expenditure).

Is it inconceivable that other "socialistic" policies might attract similar wide-ranging support? Perhaps it is; and the doubts are increased by the experience of the New Deal. Outside the United States, this is viewed as a remarkable change in political activism, with a new-found commitment to alleviating widespread social distress. Less known is the conflict it created between the "progressive" President and the "conservative" Supreme Court. Still less appreciated is the bitter division between the supporters and the opponents of the New Deal. World War II may have restored the normal consensus (Brock 1987: 124), but the opposing forces remobilized after the war to virtually kill Truman's Fair Deal.

In comparing the American planning system with that of Canada, one is struck by a fundamental difference which might be encapsulated in the suggestion that Canadian deference to authority places "peace, order, and good government" above the American priority for "life, liberty, and the pursuit of happiness" (Friedenberg 1980). Canada here follows the British tradition. As Goldberg and Mercer (1986: 15) comment in their insightful analysis of Canadian culture, there is a "British continuity in Canada, meaning that there is a continuing link and affinity between the United Kingdom and Canada." They add:

Canada demonstrates the Britannic community . . . by establishing powerful and generally well-staffed bureaucracies which are accountable to political masters, be they municipal councils or provincial or federal ministers of the Crown. The relatively greater role played by bureaucrats in Canada would be anathema to most Americans as would be the broad discretionary powers wielded by senior officials.
(Goldberg and Mercer 1986: 129)

All this gives Canadian urban and regional planning a more wide-ranging and acceptable role than is the case in the United States. It stops short, however, of fully accepting the degree of discretionary controls which characterizes the British planning system: an issue which is discussed later.

In comparing differences among the three countries nothing is more striking than the attitudes to government, from the deferential to the mistrusting. By contrast with the United States, Canada and Britain are positively law-abiding. It is interesting to speculate whether this has any connection with the widespread American "suspicion of government in general, and of the full time professional in particular" (Sharpe 1973: 148). One school of thought ascribes this to "that fundamentally American tenet of democracy, that all authority must emanate from the people." In Britain much greater trust in the bureaucracy is evident because "the notion of the independent authority of government under the law has continued to exist side by side with the notion of the political power of the people," whereas in the United States, "the revolutionary experience led to the view that there was no office that did not derive from the citizenry, hence no limit to the exercise of citizen competence" (Almond and Verba 1963; Sharpe 1973: 149).

Sharpe adds that another factor which helps to account for the low status of government in American society is "the sheer abundance of natural resources which have been available in the development of the United States." He continues:

> It is one of the few countries where virtually all natural resources have been available on a gigantic scale at a relatively little cost. This must have had a profound effect on the style of government as compared with other democracies where city government, like all government, is predicated on the scarcity of most resources.... The relatively weak attachment to the concept of the public interest in the United States may also be linked to the existence of an abundance economy. Common interests in government will be seen as being relatively less important where individual citizens are self sufficient. During the formative years of the nascent democracy the proportion of the population that were as farmers virtually self sufficient for all public services except roads perhaps, schooling and police was very large indeed.
> (Sharpe 1973: 157)

This is a very helpful insight into the character of American attitudes to government and, as discussed later, attitudes to land and property. Of course, conditions have changed dramatically since the time when the majority of workers were farmers; and no longer is it the case that "very little government went a long way" (Dahl 1971: 53). Nevertheless, old philosophies persist long after their rationale has passed.

Persistence of old philosophies also characterize British government, whose elitist origins are still apparent. Government is carried on in an elitist atmosphere of determining what actions are in the public interest, as viewed through the eyes of the government. This precludes wide and open debate: participation is concentrated in the circle of elected members who make up the government of the day. In Tant's words:

> Since government is not (directly) accountable to the people, there is little need for the people to be well-informed about the details of public policy; it is for parliament, not the unsophisticated public, to scrutinize government decision making. Indeed, law, order and stability might be threatened by disclosure to those unschooled in responsible judgment: hence parliamentarians are not obliged to be "responsive." Thus the consistency between the basic nature of British government and official secrecy is quite apparent. Official information is "owned" by government "office," the proper custodians being the government of the day. The release of information to the general public must be "authorized," and government may authorize or not as it sees fit.... We can therefore identify an idea central throughout the whole of the British tradition, and upon which official secrecy rests. That is, government – whatever the form or ideological justification – is the sole arbiter of the national interest/public good.
> (Tant 1990: 480)

The reader should note that this was written in 1990!

However, lest it be thought that the British are subservient, note Birch's illustrations of the strength of the British attachment to personal liberty:

> The British would never accept the widespread security checks for bureaucratic posts that are taken for granted in the United States. The British would not agree to a proposal to ban extremist parties in times of peace, as communist and fascist parties are banned in West Germany. If it were revealed that the British police had conducted several hundred illegal break-ins a British prime minister would not feel able to tell parliament that such actions were justified in the campaign against potential terrorists and criminals, as the Canadian prime minister did in 1978. If any British minister were to make such a statement, it would be followed by a storm of public

protest, which simply did not happen in Canada. Equally, it is inconceivable that a British government would instruct the security police to compile files on the political affiliations and activities of all candidates for political office, irrespective of party, as the Canadian government has done. Nor would British citizens accept the peacetime identity cards, and the need to register addresses with the police, that are routine in some continental countries. . . . British motorists would not easily accept the low speed limit that is obeyed in docile fashion all over the United States, together with regulations making it a legal offence for a motorist to adjust the carburetor on his own car. British swimmers would hardly put up with the situation on American beaches, where the provision of life-guards is immediately followed by rules making it an offence to swim anywhere except in a small roped enclosure in front of the life-guard.

(Birch 1990: 13)

A major point arises from this recital (which could easily be lengthened): there are differing ideas, within democratic societies, of the proper role and limits of government action. What one "free" society rejects as an unwarranted invasion of personal freedom, another regards as a sensible rule for the common good. As the following chapters show, this has important implications for land use planning.

Chapter 14

Local government systems

Introduction

Few countries are so principled in their devotion to local autonomy and decentralized democracy as the United States.

(Clark 1985)

Much of land use planning operates through the agency of local government. American practice (with extreme local autonomy) has been discussed at length in previous chapters. In this chapter, the objective is mainly to give a comparative account of the local government systems in Britain and Canada, together with a discussion of the prevalence of corruption in American local government. This provides a necessary background to the discussion of comparative planning systems in the following chapter.

American local planning is characterized by a great variety: it ranges from local governments with no zoning, to those with highly active (or reactive) land use controls. American local governments have a small number of elected members, and operate with a marked sensitivity to local feelings: that, after all, is what they are elected to represent. (As we shall see later, British local governments operate in a significantly different way.) Policies may be directed to serve the wishes of the local landowners (who may take it for granted that they should be able to make the maximum profit from their property), and to promote the growth of the area. Alternatively, they may be directed to implementing the desires of the electors to maintain the character of the area and exclude all new development, or at least all development which does not harmonize with the character of the area.[1] Whatever the policy, it is typically determined by the local government itself, not by a higher level of government. Furthermore, unlike Britain, there is no political party control (as indeed is also the case at the state and federal levels); nor is there any system of control by a higher level of government as there is in Britain and Canada.

A few areas have energetic and strong counties, which follow plans of growth management that local governments implement with varying degrees of enthusiasm; others regard this as neither possible nor desirable. Another contrast is between the small localities with a minimal planning competence and those big cities which have sophisticated planning systems. Some of the major cities have local planning capabilities of considerable breadth. San Francisco is a good illustration, about which an English civil servant has commented:

> San Francisco: where the city has recently moved to prohibit demolition of single family housing, where there is a strict quota on the downtown office space that may be approved each year, and where even developments permitted by the zoning ordinance may nevertheless be called in for review by the planning commission to ensure neighborhood compatibility of what is proposed. An illustration of American development control extending far beyond that in most of Britain.
>
> (Wakeford 1990: 3)

Though outstanding examples of planning initiatives could be cited for cities in Canada and Britain, in those countries the difference between the major local governments and their smaller counterparts would be more a matter of degree than of kind: the reason is the existence of a local government *system*, exhibiting a degree of uniformity, operating within a single legal framework (national in Britain, provincial in Canada), and subject to effective administrative controls by a higher level of government.

There is, nevertheless, one common feature: local governments in all the three countries under discussion derive their powers from a higher level of government (rather than, for instance, the Constitution): the state in the United States, the central government in Britain, and the provinces in Canada. In each case, the higher level of government can add to or subtract from their duties and responsibilities; it can even reorganize or abolish them. For instance, British local government was completely reorganized between 1963 and 1973. This reorganization included the estab-

lishment of the Greater London Council and of metropolitan councils in six provincial conurbations; all these were later abolished by the Thatcher government under a policy termed "streamlining the cities" (GB 1983). In Canada, as already indicated, Ontario reorganized a large part of the province on a regional basis. Other provinces have acted similarly, and "as a result, every major Canadian metropolitan region is governed in part by some form of metropolitan-wide government" (Goldberg and Mercer 1986: 129). There is little of equivalence in the United States, despite the attempts made by the federal government to promote regional planning (discussed in Chapter 10).

Indeed, the major characteristic of American local government, when compared with British and Canadian, is its fragmentation ("Balkanization" might be a better word). This results in parochialism, an extremely narrow view of the public interest, and typically a weak system of government. Gottdiener has written:

> On the one hand, the fragmentation through small, relatively new developments in Privatown prevents residents from engaging in more socially oriented activities of regional concern. On the other hand, the weaknesses of the local political organization and the polity of Privatown produces no political leaders with broad public interests or orchestrated party power. Consequently, the combined activities of builders, speculators, homeowners, politicians, and planners have melded over the years to produce a sprawling, multiproblematic landscape which lacks social comprehensiveness and political mechanisms that would allow for the effective and equitable shaping of submetropolitan growth.
>
> (Gottdiener 1977: 175)

One point is common to all three countries: whatever enthusiasm there might have been for regional government has largely evaporated.[2]

British Local Government

> The electorate has little interest in local government affairs, has little knowledge or contact with locally elected leadership, and casts ballots much more on the basis of the popularity of the two major national parties than on the basis of local issues.
>
> (Scarrow 1971)

British local authorities, as elsewhere, are not responsible for all local services. Though they are sometimes referred to as "omnibus" authorities this is true only in contrast to "single-purpose" authorities. The number of the latter has been on the increase since the 1930s. Before this time, local government was responsible for a very wide range of services, including poor relief (welfare), education, health, and public housing; utilities such as water, town gas, electricity distribution, and public transport; and local services such as refuse collection and disposal, street lighting, sewerage systems, and libraries.

These services were steadily removed to central government, which has administered them either directly or through separate, appointed (not elected) boards – poor relief (1934), major "trunk" roads (1936), hospitals (1946), electricity supply (1947), gas (1948), water supply and sewage treatment (1974), community health care (1974), and provision of sewers (1983). Some of these services were later privatized by the Thatcher government.

Local government today is thus a much less important part of the machinery of government than formerly; but it is still significant, and it has gained greatly expanded powers of land use control.

The system has been affected by numerous reorganizations, and is not uniform over the country as a whole. Since the abolition of the Greater London Council, London has had a "unitary" system of boroughs which are solely responsible for most local government services, though there is a separate board for London Transport; and the metropolitan police force is directly administered by the central government. A similar unitary system operates (following the abolition of the metropolitan county councils) in six major provincial urban areas, such as Greater Manchester and Liverpool.[3] Elsewhere (i.e. over the major part of Britain) there is a two-tier system. At the upper level are the county councils which are responsible for strategic planning, transport planning, highways, traffic regulation, consumer protection, refuse disposal, police and fire services. The second tier consists of district councils which have responsibility for public housing (on average a quarter of the total local housing stock), environmental health, refuse collection, and planning decisions of a local nature. Responsibility for some functions, such as museums, art galleries, and parks, varies according to local agreements.

Such is the system in England and Wales. In Scotland it is similar (with the upper tier authorities being called "regions"); but in Northern Ireland, sectarian strife led to the transfer of most services to the central government which operates partly through area boards. (Land use planning is administered by a divisional office of the Department of the Environment for Northern Ireland.)

Local governments are locally elected, frequently on a party political basis (Widdicombe 1987: 119). They form "the only centre of political power in the whole state system outside of parliament itself which can claim legitimacy for its actions on the basis of a popular vote" (Dearlove and Saunders 1984: 380). This means that they can operate in a way considered inadequate, undesirable, or even intolerable, by central government; and the fact that some have done so is one of the reasons why powers have been so dramatically reduced over the last half-century, by both Conservative and Labour governments.

Council members are called councillors, and there is a sharp distinction between councillors and "officials," or "officers." The term "officials" is used in a much broader sense in the United States where it can encompass professional staff, elected members, and political appointments. By contrast, in Britain the term usually refers to paid, career, civil-servant-type office holders, while councillors, mayors (and secretaries of state at the central government level) are not termed officials even if they hold an "official" post. (The distinction is normally, though not always, between an appointed and an elected position.) Councillors are elected by local citizens every three or four years. In some areas (all county councils and the London boroughs for instance) the council is elected in its entirety every four years. Elsewhere, one third of the councillors are elected every three years. The councils elect their own chairperson who may be called mayor or, in the ancient square mile city of London (population 4,000) and certain large cities, lord mayor. The position of mayor is almost entirely honorific, and it rotates among the senior members of the council. It is therefore very different from the American system, where the mayor can achieve a position of real power and, sometimes, even national recognition.

Councillors receive no payment for their services, but they can claim a flat-rate attendance allowance for time spent on council business. They can also claim traveling and subsistence allowances.[4]

Research undertaken for the Widdicombe Committee (1986: 47) showed that, in most countries, local authorities are based on existing communities rather than on an assessment of the optimum size for efficiency of services. As a result, they are much smaller in terms of population than British local authorities. Thus, the lower tier authorities average 30,000 in Sweden, 12,000 in the United States, and 1,000 in France, compared to 120,000 in Britain.

Councils vary in size, of course, according to the number of electors but, like the central legislatures,[5] they are much larger than American (and Canadian) councils,[6] with an average of forty-eight members. The smallest have around thirty members. For example, the town of Alnwick in the north of England, with a population of 31,000, has twenty-nine members. Among the largest is the city of Birmingham, with a population of nearly a million, and a council of 117 members. The administrative county of Lancashire has a population of 1.4 million, and a council of ninety-nine members.[7] The City of London is an anomaly with the largest number of members (161) for the smallest population (4,300 in 1989).

As already noted, members are generally elected on a party political basis, though some stand as "independent" or "non-party." The members of the city of Birmingham district council, for example, consist of sixty-seven Labour, forty-three Conservative, and seven "others." The political composition can, and often does, change dramatically. Local election results are much more volatile than those of general elections. Thus while Birmingham had a Labour majority in 1989, it had no clear majority in 1982, and alternated between Labour and Conservative between 1963 and 1980 (Butler and Butler 1986: 443).

So far in this section on British local government, the discussion has been confined to easily measurable matters. Important though these may be, they fail to convey the character of the system. This is important because it is easy to be misled by formal statements or simply by a failure to realize the distinctive character of the system.[8] A useful description is given by Peterson and Kantor:

> In English local politics, institutional arrangements impede the regular emergence of either highly visible candidates or controversial issues. First, strong local personalities – English equivalents of Richard Daley, John Lindsay, Samuel Yorty, and Richard Lee – cannot easily come to personify local parties and government

because the mayoralty is an almost entirely honorific position. . . . Power is typically fragmented among committee chairmen with the two centralizing forces – party leaders and the town clerk – almost unknown among the general public. Second, local issues are not easily generated. This is partly due to fairly strict constraints that departments of the central government place on local authorities. . . . But greater centralization of power in England than in the United States is not a sufficient explanation for low levels of local political controversy. . . . More important is the fleeting attention newspapers pay to local affairs, the lack of bond and tax referenda, and the few institutionalized opportunities for groups and organizations to discuss and criticize local policies.

(Peterson and Kantor 1977: 213)

Foreign observers have frequently commented on the lack of interest displayed by local newspapers in local political affairs (unless there is a scandal, which does not happen very often). Elkin, for instance, notes that:

It is possible to follow politics in most cities of the United States by reading a local newspaper. While the information is unlikely to be comprehensive, a fair idea can be gained of the participants, the issues, the stakes, and the alliances. In reading a newspaper in London, one by and large got news about parliament, the cricket matches, and the Queen's functions, but little if anything about local government. The news that appeared was generally of the kind that simply announced a new local authority project.

(Elkin 1974: 105)

Another strand in this is that local authority officers are not expected to speak to the press, unless they are stating the views of the council (i.e. the members). But a matter typically becomes one on which the press is informed only when it has completed its "confidential" stage, that is when it has been considered by the relevant committee(s). Most planning applications, for example, are sub judice until they have been decided. Thus neither the press nor the public are informed of them. (However, there are some types of planning applications which local authorities are required by law to publicize: these are noted later.) The introvert nature of British local government is nicely illustrated by the fact that the central government has found it necessary to pass legislation which secures the right of admittance of the press and the public to meetings of the council and its committees (Byrne 1990: 320). Other legislation gives the public the right of access to certain types of council documents (Byrne 1990: 159).

The British elect their politicians and then expect them to get on with the job: that is what they have been elected to do. In contrast to Americans, the British "believe it is the business of government to govern. The voter may control the government by giving or withholding consent, but he may not participate in its affairs" (Banfield 1961: 64).

To Americans this is a startling statement, and it needs some elaboration, though convincing explanation is not easy. There are several points to be made. First, the political stage in Britain is essentially a national one. Though members of both the central parliament and the local governments are elected on a geographic base, the main electoral issues are generally national. Debates may be peppered with illustrations of local impacts, but the orientation is national. Decisions are taken on the basis of an interpretation of the general public interest. There is little of the American bargaining between locally based interests. It would be difficult to imagine an American local municipal representative saying, as did a London councillor:

There is a real contrast on questions of individual welfare and considerations of policy. One makes great efforts to help individuals. But in matters of development and reorganization, for example, it is the good of the community that counts first above that of any particular neighborhood.... One must have an overall plan for the highways, sewage, refuse disposal, and the like. One can't organize services to suit one area and not another.

(Elkin 1974: 124)

Clearly a system such as Britain's can operate only if it is acceptable to the electorate; and it is acceptable, in the sense that elected members are expected to work for the public interest, not narrower local concerns. Of course, it is not uncommon for local problems to be used as illustrations of the inadequacy of national policy. Furthermore, electors may complain about "the system" in a general way, but they are seldom moved to mount a campaign on the style so common in the United States.[9] They tend to see their councillors or MPs as members of a political party and, at election times, it is the party that is voted for, not the individual. As one eminent local official has commented, if electors "get into the decision-making process after their chosen representatives had been duly elected, this could only mean that the current form of representative democracy had broken down."[10] Election to government, whether local or central, seldom involves personal popularity: it is more likely to be a reward for service to the party. In short, unlike the American situation, members have little in the way of an electoral base, and therefore little reason to cultivate it. They are thus free to search for (their view of) the public interest of the whole community rather than being forced to mediate between conflicting interests.[11]

The relative independence of British local authorities from community control is made even more significant because of the nature of their land use planning responsibilities. Firstly, these responsibilities are not optional: local authorities are required to carry them out in accordance with national legislation. Thus every community has a uniform system of planning controls. Secondly (and dramatically different from the American system), these controls are operated with a high degree of discretion. There is nothing equivalent to zoning and the incredible mass of litigation that accompanies it. If there are difficulties between a developer and a local authority, these are usually settled by discussion.

Comparing the two countries, Asimow stresses the "deep cultural differences":

Americans traditionally distrust officials and favour adversary procedures and judicial interventionism. In Britain, on the other hand, people are comfortable in relying on official discretion to strike compromises and make individualized judgments which are never reviewed by the courts.

(Asimow 1983: 275)

Another factor of crucial importance is the role played by officers (local government staffs and civil servants). Here the difference between Britain and the United States is striking. Officers see themselves, and are seen by members, as independent professionals whose concern is the promotion of the public interest as viewed through their professional eyes. They are highly skilled in their functions, whether these be technical or administrative. Their experience is typically longer and deeper than that of their political masters who, in any case, may be replaced at the next election.

Theoretically, officers make recommendations, while members make decisions. In fact, most proposals, though discussed between officers and committee chairmen, are accepted by the members. The officers are full-time experts; the members are part-time laypersons. One writer has described the role of members as "invigilators" (Scarrow 1971: 18). Indeed, while an American local government would undertake negotiations with, say, citizen groups, their British counterparts are more likely to await the outcome of "negotiations" among officers. In the words of one elected member:

> We [i.e. the elected members] tried to get the various officers involved to agree with each other before they came to Committee. The Committee didn't like to be faced with disagreement. When we did get it, we generally sent the officers back to seek agreement.
>
> (Elkin 1974: 124)

Another striking feature arises here: the small number of people concerned with policy. Unlike the situation in the United States, where a multiplicity of interests have to be catered for, or at least taken into account, British local authorities operate, to a great extent, in an atmosphere of independence. They not only determine the solutions to local problems: they largely define what the "problems" are. Attention is directed toward establishing the public interest. The comparatively large size of British local councils facilitates "specialization" by council members: "most councillors concentrate their energies by serving on specialized subject matter committees and working closely with permanent officials on detailed administrative work, rather than by serving as political leaders in omnicompetent decision-making councils and articulating demands of local constituents or acting as brokers among local interests" (Scarrow 1971: 17).

A major feature of this system is that there is no clearly identifiable executive branch of government (deSmith and Brazier 1990: 399). Decisions, both on policy and on its detailed implementation, are made by the council, or its committees, or its officers acting under delegated powers.

Within the council, apart from the leaders and chairpersons, many members participate in council business only in a very part-time manner. In the words of a London councillor:

> There tends to be only a few members who can spend the time required to be a leader or a chairman. Most ordinary members give only two or three hours a week. They tend to be passengers, frankly. They have confidence in people who have gone into the problems and who have the time, unless they see something outrageous and then they get upset.
>
> (Elkin 1974: 131)

Another political scientist has stressed the differences between the American and the European traditions of government:

> [The American system] presumes that *all interests* are capable of representing themselves with the public administration more or less in the same way; and in any case, that the natural functioning of a civil society tends to increase the occasions for diminishing unequal representation between interested parties. On the other hand, the European-type system assumes that some interested parties are better able to represent themselves and others less able to. Consequently, public administration has a job of bringing equality to the maximum possible level where inequality naturally exists and where the natural functioning of a civilized society does not increase opportunities for the emergence of equality, but perhaps has the opposite effect.
>
> (Pizzorno 1971)

This provides a possible explanation for the relative absence in American local government of an explicit concern for the public interest. As Kaplan (1967: 210) has expressed the matter, policy is established "through the open agitation of issues and the open clash of opposing groups in a free political market place." This has the effect of giving the public administrator and the professional a less salient position than he has in the European context.[12] Could it also explain the lowly position of land use planners? Certainly, there is a marked contrast with Britain, where not only is the "advice" of the professional planner taken very seriously, but a range of decisions is actually delegated to him.[13]

Canadian Local Government[14]

Local government came to Canada slowly and with great difficulty. In general, municipal organization in Canada has been opposed by established elites who feared popular control of these new governments, and by the citizenry in general who viewed local government as simply another source of taxation. The history of local government has been a history of provincial governments imposing municipal organization on people either hostile or, at best, indifferent to it.

(Siegel 1980)

Each of the ten Canadian provinces has a somewhat different system of local government, but they are all derived from the British or the American models, or both. The form that these governments take is determined by the individual provinces. Reference is sometimes made to the fact that local governments are creations, if not "creatures" of the provinces. However true this might be, it does not mean that they are tame pets: they have some political substance. In Crawford's (1954: 18) words, "the protection of the municipalities lies, not in their legal or constitutional position, but rather in the needs of the people which must be met and the difficulty, especially in urban communities, of meeting those needs through the medium of any other level of government." It is therefore misleading to view municipalities as mere agents of the provinces. They have the power which derives from being responsible for a range of services for which they are politically accountable within a given territory (Cameron 1980: 233). Of course, municipalities differ, not just in size and socioeconomic character, but also in influence and political clout. One good

illustration of this is Winnipeg, which has over 50 percent of the population of Manitoba. This inevitably affects the relationship between the city and the province (Feldman and Graham 1979).

The majority of the Canadian population lives in municipalities, of which there are 4,634. However, most of the vast land area is not organized in municipalities (Statistics Canada 1982).

There is an extraordinary variety of nomenclature which reflects the differing histories of the provinces: Ontario's townships, Manitoba's rural communities, British Columbia's district municipalities. In Quebec, the parishes, townships, united townships, and "municipalities without designation" are so similar that all newly created municipalities are called municipalities without designation.

The great majority of municipalities are unitary, but there were, in 1988, 165 authorities in a two-tier system. Of the 4,238 unitary authorities, 3,913 had populations of less than 10,000. Thirty-three had populations of over 100,000. Most have between five and fifteen councillors.

These bare statistics give a picture of the variety of municipal institutions in Canada. However, the figures are of limited value, except in highlighting the small size of the majority. Municipalities with different names may operate the same functions, while those with the same names may have widely differing functions. Moreover, many local functions are operated by ad hoc bodies. Education is a particular case: in all provinces responsibility lies outside municipal government. Police, public health, parks, and libraries are other functions often performed by ad hoc boards. This proliferation of specialized boards stems in part from American influence:

This practice in Canada, more extensive in Ontario and the prairie provinces than in the rest of the Dominion, was in part an imitation of the United States and in part an effort to remove from the sometimes penny-pinching control of councils those services which a council might be disposed to sacrifice in preference to expenditures which would give more immediate and tangible results.

(Crawford 1954: 131)

Another illustration of the difficult birth of local government comes from the province of Newfoundland. Here, though St. John's was created a municipality in 1888, it was not until 1938 that any further ones were formed; thus "the most startling fact about Newfoundland's local government is that, apart from the City of St. John's, there were no local government bodies in the island until 1938" (Crosbie 1956: 332). There still remains the long standing opposition to the establishment of local governments and the property taxes that go with them. Much preferred is direct provision (and financing) by the province. By contrast Ontario, as Upper Canada, fought fiercely for the right to local self-government and, by the time the prairie provinces were opened for settlement, "they benefited by the experience of a quarter of a century of municipal government" in the central provinces; "the question at issue was not whether local self-government should be permitted but rather the basis on which it should be established" (Crawford 1954: 19). In British Columbia, geographical isolation necessitated some form of local government.

It is clear that the development of local government in Canada took place, and continues to do so, pragmatically in response to population growth and the need for local services, particularly in urban areas. But the form which this development takes varies among provinces.

Most of Canadian local government is run on nonparty lines. Alberta is typical: "Alberta's citizenry strongly favor professional management, not local party politics, and the dispersal rather than the concentration on local formal

Power" (Ashton and Lightbody 1990: 511). Indeed, Montreal, Quebec City, and Vancouver are the only major municipalities with well-developed party systems and a governing majority; and it is interesting to note that in Vancouver the strongest of the local "parties," the Non-Partisan Party, claims that it is not a party. This is equivalent to those conservative members in British local government who are self-styled "non-party" (Sancton and Woolner 1990: 500). The comment is prompted that each country, state, and province has its own particular mystique in relation to its political system, even to the extent of denying it!

This nonparty character of Canadian local government is, historically, one of its two salient features: the other is its role as an agency for service delivery. These two features have reinforced each other: local housekeeping does not provide an arena for the politics of controversy. Numerous commentators have stressed this preoccupation with the administration of services in contrast to the formulation of policy. Plunkett, for example, states that:

> Generally speaking, municipal governments were regarded as being concerned primarily with administration and not policy. From this concept emerged the "non-partisan" tradition of Canadian municipal councils whose main concerns were to ensure the prudent administration of municipal services without unduly burdening the property taxpayer.

(Plunkett 1972: 17)

Plunkett uses the past tense because "to this traditional role of municipal government in Canada has now been added a genuinely political dimension."

There are two influences here. Firstly there has been an increase in demand for services which is both quantitative and qualitative. Thus, larger cities require disproportionately larger police forces, and sanitation services have to be bigger and more sophisticated. Secondly, municipal government is now faced with a wide range of decisions on matters which are of economic and social importance. Roads, once considered a purely technical matter, now have to be considered in terms of their environmental and socioeconomic impact. Additionally, the cities are now facing, or being presented with, new problems "for which they may have little or no constitutional or fiscal power." Housing is a good example (Plunkett 1972: 18). The problems facing local governments are thus administrative and political; and difficulties with the latter can preclude solutions to the former.

Curiously a growing awareness of, and experimentation with, a policy-oriented system of municipal government has come about following reorganizations designed essentially to create better "service delivery authorities." It is neither feasible nor appropriate in this volume to recount in detail the progress of local government in each of the Canadian provinces (Tindal and Tindal 1990). The salient features can be noted, however, and the growing importance of planning indicated. First in the field was Ontario with the establishment of Metropolitan Toronto in 1953.

At the outset, it is interesting to note that "reform" has generally proceeded on the basis of mutual agreement between the municipalities concerned. There has been little of the urban-rural conflict which has dogged local government reorganization in Britain. On the contrary, though there was strong opposition to the city of Toronto's original proposals for amalgamation with the surrounding authorities, there was general support for a two tier system (Rose 1972). The political attractiveness of this was such that all future municipal reorganizations in Ontario (twelve in total, covering over a third of the provincial population) followed the same pattern, even in areas of quite different character.

The crucial element in this was the incapability of the existing municipal structures to cope with urban growth on the scale which emerged in the postwar period. The brute

physical planning problems were acute: in Metro's first ten years, the construction program reflected this: three-quarters of spending was on roads, sewers, and water mains, a fifth on schools, and a mere 3 percent on housing and social services.

Metro Toronto's two-tier system has worked effectively in the physical provision of services, but new strains may be growing as the need increases for soft services and social planning, and as the physical area of the urban region continually extends beyond its fixed boundaries.[15] Its success, in both administrative and political terms, made it a model not only for other reorganizations in Ontario but also in other provinces, though with less happy results. The Montreal Metropolitan Corporation, established in 1959, never solved its internal political problems. The Metropolitan Corporation of Greater Winnipeg, established in 1960, faced similar problems and also a lack of support from the provincial government which had created it (Brownstone and Plunkett 1983). Subsequent reforms created the indirectly elected Montreal Urban Community in 1969, and the single-tier Winnipeg "Unicity" in 1972.

The legislation that created the Montreal Urban Community was paralleled by provisions for the creation of similar bodies for the Quebec and Hull regions (Godin 1974). The Winnipeg solution is unique. Neither has operated smoothly, particularly in the field of planning. Part of the difficulty is inherent: planning over a wide area seeks to allocate resources in the interests of the area as a whole. These wider interests can, and do, conflict with local interests. Metropolitan government can also become administratively cumbersome. With a two-tier system the clash between local and metropolitan interests may lead to the development of machinery which, even if it is effective in achieving reconciliation or compromise, can be slow and inefficient. Such problems can arise even in single-tier government, as is illustrated by the extraordinary, convoluted and lengthy planning process in Winnipeg which the Taraska report documents (1976: 31).

Local government reorganization in Ontario, Quebec, and Manitoba was preceded by, and followed by, comprehensive studies. By contrast, British Columbia proceeded stealthily and by what analysis has termed a "strategy of gentle imposition" (Tennant and Zirnhelt 1973). The provincial government gradually introduced a system of regional districts. The best known of these is the Greater Vancouver Regional District, but there are twenty-seven others.

These reforms (and those of other provinces) were undertaken for a variety of reasons, but the efficiency of "service delivery" was a major factor in most. This has had some unfortunate results, as many critics have pointed out. Tindal and Tindal for instance argue:

> this urban service emphasis made the reforms regrettably narrow in scope. There has been little improvement in the fragmentation of responsibilities among municipalities, separate boards, and the senior levels of government. While the present distribution of governmental responsibilities has evolved over a long period and not always according to some rational plan, efforts to delineate a new and more appropriate division of provincial and local responsibilities have been [limited]. . . . Yet if the local government system is to be strengthened it needs not only new and more appropriate boundaries but also a respectable range of responsibilities and some freedom to exercise them.
>
> (Tindal and Tindal 1990: 122)

Similar arguments were prevalent in Britain in the 1960s, but they have now been strangled by the anti-local government approach of the Thatcher government. In the United States, there is little political demand for change, an issue to which we return in the last chapter.

From Convention to Corruption

> Americans expect corruption in government and to a remarkable degree accept it.
>
> (Delafons 1969)

> The real problem is the structure of zoning itself, with its emphasis on very local control of land use by a dizzying multiplicity of local jurisdictions.
>
> (Bosselman and Callies 1972)

The political culture of planning includes the "style" of administration. Any traveler to other countries is struck by the different behavior and attitudes of shopkeepers, taxi drivers, waiters, and customs officials. The same differences in style apply to officials at all levels of government. Here, we are concerned with local government officials (using the term in its American sense to embrace all who occupy "official" positions, whether they are elected, appointed, paid, or unpaid).

An observer of the American governmental scene cannot but be struck by the way in which styles of behavior which would be unacceptable in other countries are regarded as normal in the United States. We are not referring here to "corruption" though, as will be shown, there is a fine line separating this from acceptable forms of behavior. We are referring to such matters as the response anticipated to requests for "favors" – though sometimes these are so firmly expected that the word "favor" hardly seems appropriate. An eloquent example is given by Krumholz and Forester in their description of the administration of the city of Cleveland. It is unusual to find such a clear and authoritative statement, and a lengthy quotation is appropriate.

> Each committee chair, who alone establishes what will be on the agenda for committee hearings, may delay a piece of legislation

while working out a deal with the department – that is extracting some concessions. The concession may have to do with the acquisition and price of a specific parcel of land either in the project or elsewhere, or it may involve an entirely unrelated matter such as the hiring or salary of an individual employee (possibly a client or relative of the councilmember) or the renting of a storefront owned by a favored constituent or some other condition.

When I was appointed community development director in 1977, I spent the first two days of my tenure listening to various councilmembers tell me the names of their relatives and friends, who were liberally sprinkled throughout my 170-person staff and whose jobs I "had to protect" if I didn't want "trouble" with the legislation I had before Council.

The rewards council presidents dispense to their council supporters include a share of the campaign contributions that the president has collected for his or her own re-election campaigns; tacit permission to control all zoning changes and the many permits issued in each ward; assignments to powerful committees that offer other "business" opportunities; jobs for friends or family members; a fair share of the city's capital improvements and street repaving; and contracts for favored social service agencies. If a supportive councilmember is defeated for re-election, the president can usually place him or her in another job.

The most important planning or development initiatives come, not from the city, but from developers, utility companies, the Growth Association, and major law firms. Developers contribute to political campaign chests, employ legal retainers, and provide jobs and other gifts in return for the millions in public development subsidies dispensed by the city. The gas and electric utilities employ "government relations" executives who spend much of their time

with City Council. Much of their apparent work has to do with picking up councilmembers' luncheon checks, buying them gifts, and otherwise making them comfortable. This is perceived to be good business, since the city routinely but ineffectually intervenes whenever the utilities file for a rate increase with the Ohio Public Utility Commission.[16]

(Krumholz and Forester 1991: 9)

What is significant and compelling here is the matter-of-fact manner in which the recital is given.

It could be suggested that the prevalence of this type of behavior, whether or not it is called "corruption," is partly a function of the division of powers in American government and the resultant weakness and lack of accountability of the multitude of individual governmental units. Though the system is supposed to operate in the manner of checks and balances, it has the effect of dispersing responsibility and making control (and coordination) more difficult. In Dahl's (1967) telling revision of Acton's famous aphorism, "power corrupts and the absence of power corrupts absolutely." There is also the simple factor of the small size of American councils. It is much easier to bribe, or influence, a majority of council which has only seven members than it is where, as in Britain, the average council has forty-eight.

Another, rather different illustration of the point, is provided by the action taken by Congress on the retirement of House Speaker Tip O'Neill. It relates to the funding of a new road and harbor tunnel in Boston:

About eighty-five percent of the $5 billion project is to be reimbursed by federal coffers under the Surface Transportation Act, an arrangement secured by former House Speaker Thomas P. O'Neill, Jr. as a kind of retirement gift to his home town. To date, $1.4 billion has been allocated, according to state officials. President Bush included an additional $2.1 billion for the project in a proposal sent to Congress earlier this year.

(*New York Times* June 1, 1991)

Again this is reported without comment, as if channeling a few billion dollars to a retiring senior congressman's home town was the most natural thing in the world! Another example is that of Senator Robert Byrd who, after becoming chairman of the Senate Appropriations Committee, declared that he intended to become a "billion-dollar industry." His plan was "to funnel a billion dollars' worth of federal projects and agencies into his impoverished state" of West Virginia. In September 1991, it was reported that he had nearly succeeded (Ayres 1991). Lest it be thought that this practice is restricted to a few powerful members, note should be taken of the (not untypical) federal transportation bill of 1991 which, in its House version, provided for 460 projects earmarked for 267 congressional districts (*New York Times* October 26, 1991).

At a much lower level, one can quote the "allocation" to each member of a state assembly or a local government of a certain amount of resources for investment in the member's area. In all this, favors, perks, service, courtesy — there are many words that might be used – are implicitly distinguished from "corruption." But the dividing line is certainly not clear, and it is easily crossed. Given that this type of behavior is normal, the threshold to criminal behavior is inevitably low.

Some idea of how rampant corruption is in the land use planning process is given in the following quotation (relating to Fairfax County, Virginia) from the National Advisory Commission on Criminal Justice:

> In the 1960s, some developers and their lawyers apparently began to work together with several of the members of the Board of Supervisors in order to ensure that rezonings needed for high-profit development were approved by the Board of Supervisors. Subsequent investigations during this period indicated that lawyers representing some developers provided money to supervisors to rezone a factory site, approve sites for apartment complexes, and approve a shopping center complex. Money was paid to supervisors involved in land use decisions in the form of "campaign contributions" or "no-interest loans." Some members of the county planning staff were also involved in some deals. In short, the squirearchy that controlled county politics used its position to enrich both the county political machine, through campaign contributions, and individual board members through direct payments. It appears that the loosely run land use regulatory system existing in the county during this time encouraged these abuses: the practices continued until the land use system was overhauled after the scandals surfaced.[17]

The report gives many other examples.

A foreign observer cannot but be amazed at the amount of corruption which plagues so many areas of American public policy. If it were not such a serious matter, it might even be looked upon by the observer (and perhaps sometimes *is* by the players) as some kind of game. Any new federal program is immediately studied to identify loopholes. How can the provisions of the legislation be circumvented or, better still, deployed to unintended advantage? How much goes unreported, and even unnoticed is unclear: certainly some incredible swindles have come to light, of which the most recent is the collapse of the Savings and Loan Industry, described by one writer as "the greatest-ever bank robbery" (Mayer 1990).

There is a widespread feeling in the United States that any governmental body is inherently corrupt. Indeed, it is often held that it is "almost impossible" for a person to enter politics without becoming dishonest: "public service is essentially dishonest and corrupting."[18] The problem is, of course, particularly acute in land use planning where (as in

Canada and, to a much larger extent, in Britain), the opportunities are far greater than in other areas of public policy.

The British tend to be somewhat smug about corruption in local government. The general attitude seems to be that, though there is corruption, it is petty, and not a matter to be concerned about: certainly, it is nothing compared to the situation in the United States. However, there is no room for complacency, as a succession of official reports shows.[19]

It is clear that, if any reform of the American land use system is to be effective, it must take account of the human weaknesses which are all too apparent in its administration.

Chapter 15

Land use planning systems

Introduction

The publicly supported private developmental style that characterizes the American scene, incorporating bargaining among major interest groups, serves mainly to protect developmental interests by reactive or regulatory planning, ensuring that the American urban future will be a continuation of present trends, only changing as a result of the impact of change produced by the exploitive opportunity-seeking planning of American corporations.

(Berry 1973)

The major part of this book has been concerned with the American land use planning system, though "system" hardly seems to be an appropriate word for this multiplicity of different procedures, policies, and practices. Even the courts are unsystematic, with their wide range of differing and sometimes conflicting opinions. The Supreme Court has notably failed to bring order and certainty into those aspects of land use planning which it has considered.[1]

Of course, there are some commonalities, but they do not constitute a "system" in the sense that British and West European planning practices do. Indeed, there is usually not even the pretense of overriding purpose. Nevertheless, too much emphasis should not be placed on a form of words which purport to state a comprehensive planning goal, as did the early English planning legislation which established the post of Minister of Town and Country Planning, whose duty was that of "securing consistency and continuity in the framing and execution of a national policy with respect to the use and development of land throughout England and Wales."[2]

In this chapter, the focus is mainly on the land use planning systems of Britain and Canada. The discussion is intended to highlight the differences between these two countries and the United States.

The British Planning System

> Provision for the right use of land, in accordance with a considered policy, is an essential requirement of the Government's programme of postwar reconstruction.
>
> (GB White Paper 1944)

The British system is the easiest to describe, simply because it is a "unitary" one: national legislation applies to the whole country, and it is mandatory.[3] There are central government departments which are responsible for ensuring that the law *and government land use planning policies* are carried out by local governments. (The departments are the Department of the Environment, for England; the Welsh Office, for Wales; and the Scottish Environment Department, for Scotland.)

The departments operate an extensive system of controls over local government. Many Acts of Parliament require local authorities to submit schemes, proposals, development plans and such like for their services to the appropriate department. The central government also has the power to control local authority borrowing, and even to "cap" the amount that can be levied in local taxation. Local authorities are bombarded by bulletins, circulars and guidelines dealing with a wide range of matters from explanations of legislation to statements of government policy which have to be followed. Local bylaws require central approval before they are legally valid.

An eloquent example of the degree of central control is given by Byrne (1990: 265): "in 1990 a number of councils failed to get Home Office approval of bylaws banning dogs in parks." Though a trite matter, it demonstrates the extent to which central government controls can operate.

Of particular interest to this book, of course, is the control over land use planning. This takes many forms. Local "structure plans," which are broad policy statements of the local authority's policies and proposals for development (or conservation) in the area, require approval by the secretary of state.[4] This may be given in whole, in part and with modifications, or totally rejected.

The functions of a structure plan, as set out in the 1990 consolidating Town and Country Planning Act, are to state the local planning authority's "policy and general proposals in respect of the development and other use of land" in the area concerned, "including measures for the improvement of the physical environment and the management of traffic." It should be limited to policies and general proposals of structural importance, and should take into account:

a)   current policies with respect to the economic planning and development of the region as a whole;

b)   the resources likely to be available for the carrying out of the proposals of the structure plan; and

c)   such other matters as the secretary of state may direct them to take into account.

Structure plans form the framework for detailed operational "local plans." These outline the local planning authority's "proposals for the development or other use of land in their area ... including such measures as the authority think fit for the improvement of the physical environment and the management of traffic."[5]

These plans are prepared and "adopted" by local authorities after a public inquiry has been held. There are no legal provisions relating to the procedures to be followed at such an inquiry. It is required, however, that the inquiry shall be held by an "inspector" appointed by the secretary of state. His report is made to the local authority. At first sight this seems curious, but the rationale (originally) was that a local plan would be prepared within the framework of a structure plan, and since structure plans are approved by the secretary of state, local authorities could safely be left to the detailed elaboration of local plans. This went to the very kernel of the philosophy underlying the legislation, namely that the central department should be concerned only with strategic issues and that local responsibility in local matters should be a reality.

This logic is now flawed in those parts of the country which have no county level of local government, that is where the metropolitan county councils have been abolished. In those areas, "unitary" plans are now required, but these are also "adopted" by the local authorities. Apart from the general powers of the secretary of state to "call in" a plan which he considers to be unsatisfactory, the only central government influence is the statutory requirement that the local authority "shall have regard to" matters such as "strategic guidance given by the secretary of state," and current national and regional policies. These are readily available in circulars and their policy successors: planning policy guidelines, regional policy guidelines, and mineral policy guidance notes. In Scotland, there are national policy guidelines.

In this system there is nothing resembling a zoning bylaw. Planning applications are determined by the local authority within the framework of the development plan (a term which encompasses both the structure plan and the local plan). It is not, however, legally binding. In dealing with planning applications, local authorities must "have regard to" its provisions, but also to "any other material considerations." Expressed in different words, local authorities must take into account all considerations which have a bearing on a planning application, of which the development plan is an important one.

This is so different from the system in most other countries that some further explanation is desirable. British planning is conceived in terms of policies which will advance the public interest. (This is in striking contrast with the United States, where the concern is frequently

with resolving conflicts between private interests in land.) Since circumstances are constantly changing, these policies have to be expressed in broad terms: they certainly can not be translated into a detailed map, particularly at the structure plan level. Concern for individual property rights is, by American standards, minimal. The public interest overrides them, often without any compensation being payable. Though there is statutory provision for what is termed "public participation" this is, by American standards, tightly defined, and bears little or no resemblance to the effective power of American communities. Moreover, in determining the outcome of a planning application, it is common for local wishes to be overridden *by the local authority* because of concerns for broad (often nonlocal) policies related to such matters as the provision of adequate land for new housing, and the safeguarding of agricultural land.

A major feature of this system is that, with certain exceptions, *all* development requires the prior approval of the local authority. This approval can be one of three kinds: unconditional permission, permission subject to such conditions as is thought fit, and refusal. There are also strong powers for enforcement. An owner can be compelled by an "enforcement order" to "undo" unauthorized development, even if this involves the demolition of a new building. Moreover, a "stop notice" can be used in conjunction with an enforcement notice to put a rapid end to the carrying out or continuance of development which is in breach of planning control.

Given this public policy character of British land use planning, the courts have little role to play. Indeed, they become involved only when it is alleged that an authority has acted contrary to the provisions of the planning legislation ("ultra vires"). In place of the courts, stands the central government (technically the secretary of state). An applicant who is refused planning permission (or is granted permission with onerous conditions) can appeal to the secretary of state, and large numbers do so – around 28,000 appeals were decided in the year 1989/90 (GB 1991).

Each case is considered on its merits. This allows a great deal of flexibility, and permits cases of individual hardship to be sympathetically treated. At the same time, however, it can make the planning system seem arbitrary, at least to the unsuccessful applicant. Although broad policies are set out in departmental publications, the general view in the department is that a reliance on precedent could easily give rise to undesirable rigidities. As Mandelker has pointed out:

> Conditions vary so fundamentally from case to case and from one part of the country to another that it would be impossible, if not wrong, to draft rules that would hold good uniformly. The basic problem is that a variety of factors operate in a planning case; the art of making a decision lies in the striking of a proper balance. Under the circumstances, there is little that the [central government] department can do beyond listing those factors which it considers crucial, and expressing rules of thumb which will help select those which should preponderate.
>
> (Mandelker 1962: 117)

Another feature of this system is noteworthy. Appeals can be made only by the original applicant; no one else has "standing" (to use the American term). Moreover, if the decision is an approval, no "third party" has any standing, even if they have a real grievance.

Underlying all these procedural matters is the major feature of British planning: state ownership of development rights. This is a highly complex matter of which only the barest outline can be given here.[6]

Nationalized Development Rights

> It is clear that under a system of well-conceived planning the resolution of competing claims and the allocation of land for the various requirements must proceed on the basis of selecting the most suitable land for the particular purpose, irrespective of the existing values which may attach to the individual parcels of land.
>
> (Uthwatt Report 1942)

Effective planning necessarily controls, limits, or even completely destroys the market value of particular pieces of land. Is the owner to be compensated for this loss in value? If so, how is the compensation to be calculated? And is any "balancing" payment to be extracted from owners whose land appreciates in value as a result of planning measures?

This problem of compensation and betterment arises fundamentally from what the Uthwatt committee (an "expert" committee which was set up to resolve the problem) called "the existing legal position with regard to the use of land, which attempts largely to preserve, in a highly developed economy, the purely individualistic approach to land ownership." This "individualistic approach," however, has been increasingly modified during the past hundred years. The rights of ownership were restricted in the interests of public health: owners had (by law) to ensure, for example, that their properties were in good sanitary condition, that new buildings conformed to certain building standards, that streets were of a minimum width, and so on. It was accepted that these restrictions were necessary in the interests of the community: *salus populi est suprema lex*, and that private owners should be compelled to comply with them even at cost to themselves.

But clearly there is a point beyond which restrictions cannot reasonably be imposed without payment of compensation; and "general considerations of regional or national policy require so great a restriction on the landowner's use of his land as to amount to a taking away from him of a proprietary interest in the land."

This, however, is not the end of the matter. Planning sets out to achieve a selection of the most suitable pieces of land for particular uses. Some land will therefore be zoned for a use which is profitable for the owner, whereas other land will be zoned for a use having a low, or even nil, private value. It is this difficulty of "development value" that raises the compensation problem in its most acute form. The development that may legitimately (or hopefully) be expected by owners is in fact spread over a far larger area than is likely to be developed. This potential development value is therefore speculative, but until the individual owners are proved to be wrong in their assessments (and how can this be done?) all owners of land having a potential value can make a case for compensation on the assumption that their particular pieces of land would in fact be chosen for development if planning restrictions were not imposed. Yet this "floating value" might never have settled on their land, and obviously the aggregate of the values claimed by the individual owners is likely to be greatly in excess of a total valuation of all pieces of land. As Haar (1951: 99) has nicely put it, the situation is akin to that of a sweepstake: a single ticket fetches much more than its mathematically calculated value, for the simple reason that the grand prize may fall to any one holder.

Furthermore, the public control of land use necessarily involves the shifting of land values from certain pieces of land to other pieces: the value of some lands is decreased, while that of other land is increased. Planning controls do not destroy land values: in the words of the Uthwatt Committee, "neither the total demand for development nor its average annual rate is materially affected, if at all, by planning ordinances."[7] Nevertheless, the owner of the land on which development is prohibited will claim

compensation for the full potential development of his land, irrespective of the fact that the value may shift to another site.

In theory, it is logical to balance the compensation paid to aggrieved owners by collecting a betterment charge on owners who benefit from planning controls (Hagman and Misczynski 1978). But previous experience with the collection of betterment had not been encouraging, and the Uthwatt committee concluded that the solution to these problems lay in changing the system of land ownership under which land had a development value dependent upon the prospects of its profitable use. They maintained that no new code for the assessment of compensation or the collection of betterment would be adequate if this individualistic system remained. The system itself had inherent "contradictions provoking a conflict between private and public interest and hindering the proper operation of the planning machinery." A new system was needed to avoid these contradictions and to unify existing rights in land so as to "enable shifts of value to operate within the same ownership."

The logic of this line of reasoning led to a consideration of land nationalization. This the committee rejected on the grounds that it would arouse keen political controversy, would involve insuperable financial problems, and would necessitate the establishment of a complicated national administrative machinery. In their view the solution to the problem lay in the nationalization, not of the land itself, but of all development rights in undeveloped land.

This is essentially what the 1947 Town and Country Planning Act did. Effectively, development rights and their associated values were nationalized. No development was to take place without permission from the local planning authority. If permission were refused, no compensation would be paid (except in a limited range of special cases). If permission were granted, any resulting increase in land value was to be subject to a development charge. The view was taken that "owners who lose development value as a result of the passing of the bill are not on that account entitled to compensation." This cut through the insoluble problem posed in previous attempts to collect betterment values created by public action. Betterment had been conceived as "any increase in the value of land (including the buildings thereon) arising from central or local government action, whether positive, for example by the execution of public works or improvements, or negative, for example by the imposition of restrictions on the other land." The 1947 Act went further: all betterment was created by the community, and it was unreal and undesirable (as well as virtually impossible) to distinguish between values created, for example, by particular planning schemes, and those due to other factors such as the general activities of the community or the general level of prosperity.

If rigorous logic had been followed, no payment at all would have been made for the transfer of development value to the state but this, as the Uthwatt committee had pointed out, would have resulted in considerable hardship in individual cases. A £300 million fund was therefore established for making "payments" (as distinct from "compensation") to owners who could successfully claim that their land had some development value on the "appointed day" – the day on which the provisions of the bill which prevented landowners from realizing development values came into force. Considerable discussion took place during the passage of the bill through Parliament on the sum fixed for the payments, and it was strongly opposed on the ground that it was too small. The truth of the matter was that in the absence of relevant reliable information any global sum had to be determined in a somewhat arbitrary way; but in any case it was not intended that everybody should be paid the full value of their claims. Landowners would submit claims to a centralized agency, the Central Land Board, for "loss of development value," that is, the difference between the "unrestricted value" (the market value without the restric-

tions introduced by the Act) and the "existing use value" (the value subject to these restrictions). When all the claims had been received and examined, the £300 million would be divided between claimants at whatever proportion of their 1948 value that total would allow. (In the event the estimate of £300 million was not as far out as critics feared. The total of all claims finally amounted to £380 million.)

These provisions, of which only the barest summary has been given here, were very complicated and, together with the inevitable uncertainty as to when compensation would be paid and how much it should be, resulted in a general feeling of uncertainty and discontent which did not augur well for the scheme. The principles, however, were clear. To summarize, all development rights and values were vested in the state: no development could take place without permission from the local planning authority and then only on payment of a betterment charge. As a result, landowners only "owned" the existing use rights of their land and it thus followed, firstly, that if permission to develop was refused no compensation was payable, and, secondly, that the price paid by public authorities for the compulsory acquisition of land would be equal to the existing use value, that is, its value excluding any allowance for future development.

Not surprisingly, the scheme ran into difficulties and it was considerably amended by the Conservative government in the early 1950s. Later Labour governments introduced alternative schemes, but these also were abolished by succeeding Conservative governments. The story is a complex one, but the essential point is that there has been no denationalization of development rights. It has come to be accepted that the decision as to whether particular parcels of land are to be developed ought to be a public, not a private, matter.

The British scheme is certainly not readily exportable, though there are features of American zoning and planning which bear some resemblance to its philosophy that land profits should not accrue to private owners. American local governments can, and do, downzone land, which amounts to reducing its development value, within the compass of the law. Moreover, as discussed in Chapters 6 and 7, they can also require developers to provide a wide range of public facilities, or cash in lieu. In this way, some "development value" is captured for the community.[8]

The Canadian Planning System

The Village of Lakefield, Ontario passed noise abatement legislation permitting birds to sing for thirty minutes during the day and fifteen minutes at night.

(Peter 1985)

Judges and lawyers play little more than a peripheral role in the [Alberta] planning process.

(Laux 1990)

In Canada, the planning system varies in its nature and extent among the ten provinces.[9] Most provinces require approval of municipal plans and zoning bylaws, though again there are differences in the character of the approval process, and in some provinces, such as New Brunswick, Quebec, and British Columbia, the role of the provincial government is minimal (Rogers 1991: 7).

Canada displays a wealth, if not a confusion, of instruments for the implementation of planning policies or controls. These range from traditional zoning bylaws to flexible development agreements, and from standard subdivision controls to the transfer of air rights. In this section, some of the main instruments are examined with illustrations from selected provinces.

Population growth and urban development were slower in Canada and there was less pressure for development controls than in the United States. Nevertheless, all the provinces eventually passed legislation empowering municipalities to operate zoning controls. The support for these came from the enfranchised property owners who dominated municipal politics. Van Nus (1979: 237) writes that "the principal basis of political support for zoning was the desire to prohibit the intrusion of uses which could reduce neighboring property values. When they set out to sell zoning to the public, planners appealed above all to the determination to maintain property values. They pitched this appeal in particular to real estate interests."

As in the United States, zoning proved to be an inadequate tool for dealing with changing conditions or for controlling development in the interests of nonlocal factors, such as traffic flows or the preservation of aesthetic views. Methods to deal with planning matters such as these have been superimposed on the zoning system with varying degrees of success. This has constituted an attempt to adapt the static concept of zoning to the realities of urban dynamics (Makuch 1983: 297). At the extreme, zoning could be a straitjacket, but its worst effects have been mitigated by two factors. Firstly, zoning ordinances only gradually spread across Canadian cities. Secondly, ways were increasingly found of giving some flexibility to this inherently inflexible instrument.

At the end of World War II, zoning in Canada existed not as a means of implementing plans but as a legacy of the law of nuisance. Yet statutory planning went much further in words: "to the control of land use in the interests of health, safety, convenience, morals and general well-being." In a 1949 report commissioned by the Central Mortgage and Housing Corporation, Spence-Sales argued that this newer "theory of zoning as an instrument for achieving wide planning purposes" had taken the place of "the theory of zoning as a device for the prevention of nuisance.... A concept of zoning which concentrated mainly upon the fixity of land values by preventing change in the established usages within an area, and regards all such changes as in the nature of nuisances, vitiates the scope and tenor of the purposes of zoning control as a means for attaining a planned use of land as a hole."

Here Spence-Sales was arguing against the bias in favor of the maintenance of property values which was so widespread in the Canadian provinces and which, indeed, had been hallowed in a Model Zoning Bylaw published by the National Research Council in 1939. His view was that "zoning requires to be looked upon as one of the most important instruments in the implementation of an operative planning scheme, whereby a flexible control is maintained over the constantly changing pattern of urban development."

The notion of flexibility, of course, was foreign to the original concepts of zoning. Indeed, the essence of zoning had been the certainty which it had provided for landowners. This was equally the case in Britain and the United States, though the two countries had attempted to deal in different ways with the consequential difficulties. In Britain, the 1932 Planning Act had provided for planning schemes which, in effect, were zoning plans. However, most schemes in fact did little more than accept and ratify existing trends of development, since any attempt at a more radical solution would have involved the planning authority in compensation they could not afford to pay. In most cases the zones were so widely drawn as to place hardly more restriction on the developer than if there had been no scheme at all.

This was tantamount to resolving the issue by ignoring it. In effect, all that a British scheme did was to ensure that if development did take place in a particular area it would be controlled in certain ways. But the controls were weak and, given the extensive areas of land covered, ineffective. Flexibility was therefore not an issue.

Given the slow rate of urban development in Canada and the limited use of zoning up to this time, (as well as more subtle matters such as the less litigious nature of Canadians), the variance device seems to have worked more smoothly than was the case in the United States. The "particularity issue" was very different:

> In the case of the technicalities underlying zoning, one of the most critical factors in nullifying the elastic legal basis upon which Canadian planning law is established has, to a very large extent, been frustrated by the particularity of zoning techniques which have been borrowed from American precedents.... The adoption of American zoning techniques in the provinces of Canada raises the important question of their suitability in a country which may have certain similarities in its urban developments, but in which the legal basis for planning is of a different order.
>
> (Spence-Sales 1949: 83)

Nevertheless, since there was seldom a master plan or any kind of planning policy framework, changes could be made easily simply by amending zoning bylaws. Hulchanski (1981: 42) comments that in Edmonton, "the zoning bylaw could be amended from time to time on most any grounds. There was no set of criteria for judging amendments. Either they were approved or they were not. And most applications for amendment were approved. The city did not want to do anything that would discourage development, especially during the Depression."

There is a burgeoning literature on Canadian urban history, encompassing both zoning and planning,[10] but unfortunately most of this stops around the end of World War I.[11] In fact the period up to the end World War II in Canada was not an eventful one for planning. In Carver's words:

> In both the US and in Britain the foundations of present planning ideas and methods was laid down during the period

between the two wars. In Canada this did not happen. For us the economic Depression of the thirties was a vacuum and a complete break with the past. We had no Frederick Osbornes, Abercrombies and Clarence Steins. We had no public housing programs and none of the adventurous social experiments of the New Deal .... So in 1946 we almost literally started from scratch with no plans or planners and we immediately hit a period of tremendous city growth.

(Carver 1960: 2)

There was inevitably a period of improvisation as provinces battled with inadequacies both in staffing and in legislative tools. Most provinces amended their planning laws, though Ontario, Prince Edward Island, and Saskatchewan introduced new legislation. The staff shortage was met by the import of planners from Britain and Europe, thus increasing the British influence on Canadian planning thought (Carver 1960: 3). The history of the evolution of the instruments (and the policies) of planning since that time reflects a continuous tug of war between what might be termed the firm zoning and the discretionary control philosophies of development control. The general trend has been from the former to the latter, though there is still considerable adherence, in theory less so than in practice, to the view that the certainty of zoning is superior to the uncertainty of any discretionary system. As we shall see, some provinces have achieved an extraordinary marriage of the two.

Discretionary Planning Controls in Canada

The key to the [Toronto Central Area] plan is on its last page, which tells developers and others "come and talk to us about amendments."

(Sewell 1983)[12]

The term "discretionary planning controls" is rather cumbersome, but it avoids the confusion attendant upon the simpler "development permit system." An official report on Ontario planning (Comay Report 1977) uses the latter to mean the former, but Manitoba uses "development permits" for permission given under a planning scheme or zoning bylaw. The City of Winnipeg Act, on the other hand, provides for "development permission." British Columbia has "development permits" for the implementation of its zoning bylaws, but Vancouver has "development permits" for the implementation of its discretionary control system. Rogers' standard legal text (1991: 5.25) defines "development control" as "the control of land use by permission rather than by regulation." It is not difficult to find further variations.

In essence, the system being referred to here is in accordance with Rogers' definition: it is one in which the owner's right to develop is controlled, not by a zoning bylaw, but by a planning authority's discretion. In fact, as is apparent from earlier discussion, it is misleading to think in terms of two totally different systems since, in practice, there is a continuum. At one extreme, theoretically, is a firm zoning provision which provides an owner with an uncontested right to develop as he wishes, subject only to the provisions of the zoning bylaw. At the other extreme (again theoretically) is a totally discretionary scheme which provides no guidance as to what might be allowed but simply gives the planning authority complete power to decide. In between, there are many possibilities, including discretionary elements in a zoning system, and a detailed land use plan complete with "guidelines" or "standards" in a discretionary system. Here, a brief account is given of some of the salient features of the planning system in Ontario, Alberta, Newfoundland, and Vancouver.

Distinct from the main provincial planning system (which in essence is a modified zoning scheme), Ontario has two areas in which a discretionary planning system operates: in the unincorporated areas of the northern part of the province, and in the Niagara Escarpment area. The former is a historical legacy: permits, issued by the Ministry of Natural Resources under the Public Lands Act, are required for building on Crown lands (which constitute most of the unorganized area).

Development control in the Niagara Escarpment is a very different matter. This is no legacy: it stems from legislation passed in 1973: the Niagara Escarpment Planning and Development Act. The necessity for a discretionary system for control was accepted on the grounds that traditional zoning instruments could not provide "the kind of control needed in a large, varied and environmentally sensitive area such as the Niagara Escarpment." In a designated "area of development control," all development is subject to approval (or otherwise) and to "such terms and conditions" as are considered desirable. Contravention of the Act is an offence for which a fine of up to $10,000 can be imposed. Any unapproved development is subject to an "order to demolish." In short, the provisions are essentially the same as those provided under the British system of discretionary control. The Niagara Escarpment is, of course, exceptional and it is unlikely that the development control system operating within the area will be applied elsewhere.

In Alberta, the evolution of land use planning has been, in significant ways, different from that in other provinces (Alberta Municipal Affairs 1987). For example, there has been a real attempt to devise an effective framework for regional planning. Even more important has been the "blending" of zoning and discretionary development controls. Laux (1971: 1) commented that the system in that province "combines the best features of zoning and [discretionary] development control while retaining the maximum flexibility for both the planner and the developer." The reference was to the pre-1977 situation, which provided for *both* zoning and discretionary development control. In that year the lengthy discussions on

the Alberta planning system,[13] finally culminated in a new Planning Act. The former dual system (which, as might be assumed, was not without difficulties) was replaced by a true hybrid: the land use bylaw (Macdonald 1984).

This neatly provides both for designated permitted uses of land, as in a zoning bylaw, and for discretionary uses. These broad powers are supplemented by a number of specific provisions. For example, ("without restricting the generality" of the powers), a council can provide in the land use bylaw for a wide range of matters commonly found in zoning bylaws (minimum and maximum area of lots, landscaping, parking provision, etc.), together with a discretion for a development officer "to decide upon an application for a development permit notwithstanding that the proposed development does not comply with the land use bylaw if, in the opinion of the development officer [sic], (a) the proposed development would not unduly interfere with the amenities of the neighborhood, or materially interfere with or affect the use, enjoyment or value of neighboring properties, and (b) the proposed development does not conflict with the use prescribed for that land or building in the land use bylaw."

In addition to this standard provision, municipalities can also designate "direct control districts" where "the council may regulate and control the use or development of land or buildings ... in such a manner as it considers necessary." The only precondition is that the municipality must have an adopted general municipal plan, presumably setting out the land use objectives and development criteria for the areas designated as direct control districts. Thus "a council may be more responsive to the needs of unique areas, such as inner city neighborhoods experiencing severe pressure for high density development, than would be possible if rigid land use classes were assigned" (Alberta Municipal Affairs 1980).

Direct control is clearly just a new name for discretionary control; but it is interesting to see, in the other areas, how discretionary control is fused with "permitted uses" (i.e. zoning control). On this, Laux writes:

> The philosophy behind the distinction between permitted and discretionary uses is simply that, where uses are shown as permitted within a particular district, they are regarded to be of that type as are clearly compatible with one another and, therefore, unlikely to adversely affect neighboring properties in the same district. On the other hand, discretionary uses are classed as such because, by their nature and although generally acceptable in a particular district, they may or may not be reasonably compatible with neighboring properties, depending upon the circumstances. Hence, it is necessary to confer upon the land use administrator a discretion as to whether or not to allow a particular application for such a use.
>
> (Laux 1979: 37)

To take a simple illustration, a day-care center is generally regarded as an appropriate use in a neighborhood where one-family dwellings are permitted uses. However, whether a particular day-care center on a particular lot in such a district is desirable will depend upon such factors as demand for such a facility, the amount of traffic in the neighborhood, the size and design of the structure proposed for the facility, the amount and location of available outdoor play area and the like. Consequently, it is considered appropriate to permit administrators to take into account such factors and either allow or reject the application, depending upon a reasonable assessment of all relevant variables.

Clearly, Alberta can be considered to be more inclined toward discretionary development controls than classic zoning practices.[14] But it is in Vancouver and Newfoundland that the former has, or is becoming, the standard system of land use control. The two areas are different in that Vancouver has steadily increased its use of discretionary controls, whereas Newfoundland simply modeled its system on that of Britain (its far-flung governing country until it entered Confederation in 1949).

In planning, as in many other areas, Newfoundland planning is different from the rest of Canada. The differences stem from its history, its geographic position, and its cultural and ethnic character. Even more than the other provinces, a quick perusal can give little of the real flavor of its planning system. Moreover, also unlike most other provinces, little study has been made of its government.

Perhaps the most eloquent illustration of the character of land use controls in Newfoundland is a quotation from the St. John's urban region plan:

> The general policies set out in this plan will be refined and amplified to individual local areas, as follows:
>
> 1   It is intended that detailed plans made in accordance with the requirements of the Urban and Regional Planning Act ... shall be required for areas where major development or redevelopment is to take place ...
> 2   ... within the local areas covered by detailed plans, such plans shall form the basis of Development Regulations necessary for implementation, while in remaining parts of the region development regulations shall be based directly on the policies contained in this plan.

A set of model regulations provides for the administration of a development permit system within which there is considerable discretionary power. For example, a municipality "may attach to a permit such conditions as it deems fit in order to ensure that the proposed development will be in accordance with the purposes of the approved plan and these regulations."

British Columbia has two separate planning systems, one in the city of Vancouver, operating under the Vancouver Charter, and the other in the remainder of the province, operating under the Municipal Act. The latter system has a relatively small degree of discretion and is essentially based on zoning (Thomas 1982). The matter is somewhat confused in that the term "development permits" is used for waivers to a zoning bylaw. The degree of flexibility this provides is real but tightly constrained to matters which are specified in the zoning bylaw and which are restricted by the provisions of the Municipal Act. Essentially they relate to matters of building structure and servicing. It is interesting to note that this system (which dates only from 1977) replaced a much more discretionary system of "land use contracts" (Corke 1983). The intention has been described by Mackenzie in these words:

> The most fundamental objective of this legislation is to return the municipality to a regulatory role as opposed to the present tendency of municipalities to conduct contractual regulations. It is intended that zoning be reinstated as the primary municipal control of land use. Certainty is to become the key word and the major factors of development are to be known at the outset.
>
> (Mackenzie 1978: 517)

There could be no greater contrast to Vancouver, where its development permits "can deal with virtually any aspect of a development including use and density requirements" (Ince 1984). Though the Vancouver Charter provides for zoning bylaws, the powers are of an enabling nature, and include "designating districts or zones in which there shall be no uniform regulations and in which any person wishing to carry out development must submit such plans and specifications as may be required by the Director of Planning and obtain the approval of council to the form of development." Another section of the charter provides for bylaws "prohibiting any person from undertaking any development without having

obtained a permit therefor."

There are some interesting features here. Firstly, there is a complete integration of zoning and development permit systems. Secondly, the discretionary character is made even more so by a power enabling the delegation to officials of "such powers of discretion relating to zoning matters which to council seem appropriate." These zoning matters include the relaxation of zoning bylaws "in any case where literal enforcement would result in unnecessary hardship."

The Role of Discretion

> The English system of discretionary development control would probably be as unacceptable for the overseas countries[15] as would their legally binding local plans be for England.
>
> (Davies *et al.* 1989)

The most striking, and fundamental, difference between planning systems in the three countries is the degree to which discretion is given to the planning agencies. In the United States this is relatively small. Discretion implies differential treatment of similar cases, and therefore runs foul of the equal protection clause of the Constitution. The Bill of Rights guarantees that individuals are to be free from arbitrary government decisions. This is a major constraint on planning and, of course, is the reason why in the United States so much emphasis is placed on zoning: a technique which is intended to determine the uses of land to which owners have a right.

By contrast, the British planning system provides for a great deal of discretion. This is further enlarged by the fact that the preparation of a local plan is carried out by the same local government that implements it. This is so much a part of the tradition of British planning that no one comments on it. The American situation is different, with great emphasis being placed on the separation of powers. (Typically the "plan" is prepared by the legislative body – the local government – but administered by a separate board.) The British system has the advantage of relating policy and administration, but to American eyes "this institutional framework blurs the distinction between policy making and policy applying, and so enlarges the role of the administrator who has to decide a specific case" (Mandelker 1962: 4).

There is little in the United States to compare with the "central control" which is exercised by the national government in Britain and the provincial governments in Canada. Plan making and implementation are essentially local issues, even though the federal government has become active in highways, water, and environmental matters, and a number of states have become involved in land use planning in recent years (discussed in Chapter 11). Moreover, there is a marked difference in the planning philosophy that underpins the different systems. Put simply (and therefore rather exaggeratedly), American "planning" is largely a matter of anticipating trends, while in Britain there is a conscious effort to bend them in publicly desirable directions.

Another striking characteristic of American land use planning is its domination by lawyers and the law. In this, it is different only in degree from other areas of American public policy. All government is assumed to be "an intrinsically dangerous and even an evil thing, to be tolerated only so long as its disadvantages are not outweighed by its defects" (Nicholas 1986: 11). By contrast, the law is a thing of great reverence.

The particularly strong presence of law in land use planning derives, of course, from the strong attachment to property – an attachment that is enshrined in, and protected by, the Constitution. Land use planning is thus inherently a matter of law. In Haar's words:

US officials[16] tend to assume, without question, that, in a free society, a system of land use control requires a flexible constitution, a judiciary to interpret it, lawyers to serve as advocates for developers and for governments to press their respective views of its content and application. As a result, city planners cannot avoid playing the lawyer's role. Hence, the United States is unique in requiring planners to demonstrate a deep knowledge of the often-contradictory case law as a prerequisite to designing land use programs that can deter or dissuade legal challenges.

(Haar 1984: 207)

The differences among the three countries are no accident: they stem from cultural differences (discussed in Chapter 13) and the consequent attitudes to land and property.

Underlying Attitudes to Land and Property

There has been much less sympathy in Britain than in North America for speculative land use and urban development, for the British aristocratic tradition stresses responsibilities far more than rights.

(Berry 1973)

The taking issue and the property rights argument simply do not operate on Canada.

(Lapping 1987)

The idea that one of the major ways to secure one's independence is through land ownership, and democracy through widespread ownership of land is deeply ingrained in the American political tradition, as it is in the American experience. All efforts to make land ownership collective in the United States have been decisively rejected, except in those limited cases where undeniable public goods are involved.

(Elazar 1988)

Perhaps the most tangible illustration of the American attitude to property is the constitutional safeguard. The Fifth Amendment provides that:

> No person shall be . . . deprived of life, liberty or property without due process of law; nor shall private property be taken without just compensation.

As already noted, there is no property right in the Canadian Charter of Rights. The Canadian "equivalent" of the Fifth Amendment provides that:

> Everyone has the right to life, liberty and security of the person and the right not to be deprived thereof except in accordance with the principles of fundamental justice.

Thus, in place of "property," the Canadian Charter has "security of the person." Court cases have sustained the interpretation that the Constitution does not protect property rights (Lapping 1987:123). Though this has precluded the vast amount of litigation on "the taking issue," it has not totally prevented it. In one case it was upheld that property rights *are* entrenched in the provisions of the Charter through extension of the notion of personal security (Mackenzie 1985: 13). A later case, however, refused to accept such "aggressive intellectual leaps" and concluded that no such rights existed. Mackenzie comments that the latter view is likely to prevail. Moreover:

> This being so, the existence of the Charter, while rapidly accelerating principles of fundamental fairness in other areas of Canadian law, could through the omission of property rights from the Charter severely curtail the development of principles of procedural fairness concerning property.

(Mackenzie 1985)

The point is, nevertheless, one of continuing debate. For instance, one legal scholar has argued that "the constitutional impetus to protect property rights is bound to expand" (Shumiatcher 1988: 207).

Britain has nothing equivalent, and though there have been attempts to introduce a new Bill of Rights, the focus has been on human rights.[17] There has been, however, a marked land preservation ethic, epitomized in the work of the Council for the Protection of Rural England (and its Scottish and Welsh counterparts) and, of longer standing, the husbandry of the landowning class. Added to this are the popular attitudes to the preservation of the countryside and the containment of urban sprawl (Hall *et al.* 1973). Without these attitudes, land use planning in its present form could not exist. In the United States, land has historically been a replaceable commodity that could and should be parceled out for individual control and development; and if one person saw fit to destroy the environment of his valley in pursuit of profit, well, why not? There was always another valley over the next hill. Thus the seller's concept of property rights in land came to include the right of the owner to earn a profit from his land, and indeed to change the very essence of the land, if necessary to obtain that profit (Large 1973). "Cheap land has as one of its consequences that of stimulating and universalizing acquisitive instincts and respect for property rights" (Philbrick 1938: 723). However, the time came when the ever-receding frontier ceased to be so; it was overtaken, and land became more valuable.

One might have expected the growth of a conservationist ethic, as is prevalent in western Europe. However, though this happened to a limited extent, particularly with environmentally valuable resources, the main effect was in the opposite direction: to increase the attractiveness of land as a source of profit. Speculation has never been frowned upon in the United States. In many countries, land is regarded as different from commodities: it is something to be preserved and husbanded. In the United States, the dominant ethic regards land as a commodity, no different from any other.

This attitude sits within the wider context of ambition to acquire wealth. Americans like to make money; but more important, they like to see other people making money. "Americans, when they see the rich, imagine themselves getting rich, rather then wanting to see rich people taken down a peg."[18] In Britain those who make money tend to be retiring on the subject, while those who observe them doing so are far from reluctant to make acerbic comment. This is particularly so in relation to land which, in Britain, has never been "commodified" as it has been in the United States. As Nicholas (1986: 13) writes in his *Nature of American Politics*, "few Americans feel entirely at ease with the slogan 'soak the rich', but the phrase 'deal me in' springs spontaneously and joyously to American lips."

There are, of course, many other issues which a fuller treatment would explore. Here we end this chapter by pointing to the major distinguishing features of land use planning in the three countries. Firstly, there is the matter of discretion: the British and many of the Canadian systems legally provide for a wide amount of this. The American system theoreti-cally denies it but, as has been stressed in earlier chapters, much discretion is achieved in spite of the law — by large lot zoning, for example, which forces developers to seek a rezoning and thus a discretionary action on the part of the local authority. Secondly, and allied to this, is the public policy nature of British and Canadian land use planning. While much of American planning is concerned with the resolution of local disputes (between developers and neighborhood groups for example), the British and Canadian systems are avowedly dedicated to serving the public purpose, even to the extent of overriding individual property interests. Thirdly, while the American system is characterized by an overwhelming role for lawyers and the courts, in the other two countries, these play only a minor role. Fourthly, the main agents of land use planning in the United States are autonomous local governments. There is little that is equivalent to central or provincial control. Fifthly, American zoning is blatantly discriminatory in its effects on low-income households and minorities, though the question as to how far British and Canadian practice differs is not directly addressed here. What is addressed is the extent to which reform might be possible to overcome the deficiencies in the United States.

Part VI
Conclusion

Chapter 16

Prospects for reform

Introduction

> "I moved here to escape the city.... I don't want the city following me here."

> "We made the town what it is and will fight to keep it what it is."
>
> (Quoted in Danielson 1976)

In this last chapter, the focus of attention is on the question of whether "the zoning system" might be reformed, or at least improved. The discussion starts with a review of some well-known proposals. The conclusion is quickly reached that most do not even address the most important deficiencies of zoning. Indeed, it has to be recognized that there are no easy solutions to hand. Any theoretical scheme, or any foreign model, which seems to meet the requirements fails to tackle the most important one – public acceptability. For the most part, Americans are not prepared to support a land use planning system that deprives them of control over development in their local area, or of a chance of making a profit from their land.

There is a long tradition of belief in the sanctity of the rights of property (which the law can be called upon to enforce) combined with a profligacy in its use (which the law is impotent to control). The "frontier" went many years ago, but it lives on in the national psyche.

However, there is a growing number of examples where restrictions have been accepted, perhaps reluctantly, but accepted nevertheless. These exceptional cases are not quite as rare as might be supposed, and they hold some promise for future reform on a wider scale. These innovatory planning systems are discussed later in this chapter, which concludes with the author's pessimistic view of the possibility of reform. First, however, it is necessary to discuss reforms which, though achieving some popularity, will not work, essentially because they fail to address the right questions.

Reforms That Will Not Work: Marketable Zoning

> With increasing frequency commentators have been urging greater reliance on the market mechanism to allocate resources in a variety of fields.
>
> (Ellickson 1973)

Economists have a favorite solution to the zoning problem: introduce some elements of a market. This panacea has little relevance outside polemical writing disguised as objective academic analysis. Thus, Ellickson argues that uniform standards could be enforced by a system of restrictive covenants, nuisance rules, and fines. This would provide a solution to the problem of "conflicts between neighboring owners." Both efficiency and equity would be better served by the use of use of nuisance law than by zoning. Perhaps so, but neighborhood conflicts are only one issue with which zoning deals. And is the exclusion of unwanted uses and people to be regarded as a conflict between neighbors? Surely not: these are matters of public policy that cannot be dealt with in a financial arrangement.

Of course, there are, as always, exceptions; and though they do not prove anything, they are worth examining. A particularly interesting one concerns the siting of a nuclear waste dump in New York State. Such dumps are a classic case of the unwanted NIMBY; but can a price be offered that would persuade a community to have one "in their back yard"? Apparently there is with the rural area of West Valley, a community of 2,100 people south of Buffalo. Though there is not unanimity (and the five-member Town Board is putting the matter to a referendum rather than taking the decision itself), there is considerable interest in the "price" which has been offered by a consortium of waste generators. This consists of "$4.2 million in benefits that would include a new town park, trust funds for the library and the fire department, road improvements, and a scholarship fund for local youths headed to college."[1] Additionally, the town would receive around $1.5 million more in taxes and fees, roughly double its current tax revenues.

This is by no means a unique example. The problem of finding sites for nuclear waste has increased since the federal government mandated every state to store its waste within its own boundaries, a move necessitated by the closure to out-of-state waste of the only existing disposal sites, in South Carolina, Nevada, and Washington. Different states are following different tactics, and no doubt other bargain offers will be made. There is even a publication, *The Radioactive Exchange*, which is keeping track of efforts.

This may be a telling illustration of how a market works: increased prices will bring about sufficient supply. But it remains to be seen whether it circumvents the difficulties of efficiency and equity that attends zoning, and indeed whether it works at all, other than in exceptional circumstances. And is there anyone asking whether the bargained site is desirable on wider social, economic, or environmental grounds? Surely this is more than a private matter between "neighbors."

A market mechanism operating in a different way is the neighborhood buyout, or "assemblages" as they are known in the real estate world (Haar and Wolf 1989: 1094; R.H. Nelson 1985: 5). There is a number of variations; in one, a group of homeowners, realizing that commercial development is creeping up on them get together and negotiate with the developer for the sale of the whole area. When this works, all homeowners receive higher prices than they could have otherwise expected, while at the same time the deal also makes negotiations simpler for the developer. He may well end up paying more, but he saves some time and, of course, "time is money." Presumably, on completion of this market transaction the zoning designation is obligingly changed.

Such examples of market-type land transactions also supposedly deal with the inadequacy of zoning in that, in Ellickson's words, they correct "the changes in wealth distribution" that zoning causes. There is a major question here as to who "owns" (or should own) the values attaching to particular pieces of land. Some hold that owners have the right to windfalls which follow zoning changes; some also accept that some loss might also have to be borne (as long as some value is left). By contrast, there is a body of argument which maintains that land values are largely the result of nature or of public action, and that therefore increases in land values should accrue to the public coffers.

This is too big an issue to deal with in passing. It has already been noted that Britain nationalized development rights after World War II, that British Columbia implemented a massive downzoning in the Lower Fraser Valley, and that American local governments are constantly downzoning particular pieces of land. The matter is thus not as simple as it might appear at first sight; but the essential issue is the extent to which land changes and their associated values should be a subject of market operation, or of public policy. We shall return to his later.

Another "cure for the zoning system" is R.H. Nelson's "marketable zoning":

> Formal public-planning institutions – and government generally – should play as small a role as possible in determining the specific uses of land. Their primary role should be to establish the necessary institutions and to provide a proper legal framework within which private decision making can operate. The extensive direct involvement of public agencies in current land-use determination is largely a result of the inadequacies of the legal framework that supports existing institutions.
>
> (R.H. Nelson 1977: xii)

The starting point for this argument is as follows:

> The development of zoning has represented a major change in property-right institutions in the United States. The essence of a property right is the authority it creates to control the use of property.... Zoning divides the control of the use of land, and thus the property rights to that land between the personal owner and the local government. As a consequence, even though it is not usually realized, zoning in effect creates collective rights that are held by local government.
>
> (R.H. Nelson 1977: 1)

On this view, the control of zoned property is shared between the owner (who receives no "compensation" for the public "acquisition" of a share of his property rights) and the relevant public authority.

For the owner, it is something more than a nuisance control, and something less than a sequestration of increases in land values (though other systems could be devised which would be either of these extremes or aimed at any point in between). Nelson continues that "the basic purpose of the collective property rights created by zoning is to provide an incentive for establishing and maintaining high-quality neighborhoods and communities" (R.H. Nelson 1977: 1). If this highly dubious statement is accepted, it is not a much further step to ask whether the task could not be done as well or better by the neighborhood residents themselves. The residents could then decide whether to preserve their neighborhood as is, or allow it (perhaps because of encroaching uses) to accommodate change. The residents might even obtain the required zoning change and thus enhance the price at which they could sell. At the extreme, a whole neighborhood could be sold for redevelopment on the lines indicated earlier. This is precisely in line with Nelson's arguments. To quote from a section entitled "basic principles":

If private tenures were to replace neighborhood zoning, the collective private-property rights created in place of zoning could be sold, as ordinary private-property rights are now. By combining all neighborhood personal rights with collective rights in a single package, all the rights needed for development of a whole neighborhood could be assembled and sold to the highest bidder.

(R.H. Nelson 1977: 195)

Of course, if zoning is regarded differently, as a matter of public policy, other conclusions would emerge. The basic fault with the premise of arguments of this nature (not to mention the lack of reality in the arguments themselves) is that zoning is but a tool though it has, of course, been widely and irresponsibly used for decades as if it were more than this. No discussion of the reform of zoning is possible until there is an appreciation of the role which this tool should be serving.

Before debating this, it is necessary to examine the unique case of the city of Houston, popularly known in planning circles as "the city without zoning."

Houston: The Fallacy of a City Without Zoning

For over a quarter-century, Houston has tried hard to satisfy both homeowners and commercial developers without adopting conventional land use zoning. [However] enforcement is haphazard, and the constitutionality of the entire operation is questionable.

(Mixon 1990)

Houston is constantly in the forefront of the zoning debate as being the city which proves that zoning, and all the problems that accompany it, is unnecessary: a superior system is one in which owners enter into private covenants which provide all the protection that is needed. This is a fallacy.

Houston has no citywide traditional zoning laws and, to date, the powerful real estate interests have obtained sufficient support to ensure the defeat of proposals to introduce it (Feagin 1988: 157). This power is rooted in the particular political history of the city: its absence of immigrant political machines and of grass-roots politics; and above all its dedication to using all levels of government to promote business goals. If ever there was an American city of which it could be said that "its business is business," that city is Houston.

In place of zoning, Houston uses (but only in part, as will be noted) a system of deed restrictions, or restrictive covenants, as the primary method of land use control. At one time, these restrictions covered the vast majority of subdivisions, but many have expired or fallen into disuse. Even those that are still valid may be unenforceable because of cumulative infringements.

The terms of the covenants vary greatly, and are often the subject of agreement between the developer and her mortgage lender, and they are recorded prior to the sale of any lots. Siegan gives further details:

If either FHA or VA approval is desired, its recommendations on covenants will have to be observed. The only other control over provisions of the covenants rests with the marketplace, whether they will be acceptable to potential purchasers. Once there has been a sale, the developer loses all power to make changes without the approval of each purchaser (unless otherwise provided, which is rare). Accordingly, unless there is a provision to the contrary, any change in the covenants requires the approval of all owners in the subdivision. Such consent might, of course, have to be purchased.

Frequently, covenants contain controls not normally found in zoning ordinances.

Provision governing architectural requirements, cost of construction, aesthetics, and maintenance of the lot and exterior of the house found in covenants are rarely found in zoning or other city ordinances and might be illegal if they were.

Almost all the covenants contain specific termination dates, and it is these provisions that could affect the future character and development of Houston. When restrictions expire, so do virtually all controls over land uses in these areas.

(Siegan 1972: 34)

Since 1965, Houston has had legislative authority to assist and spend municipal funds on the enforcement of private deed restrictions. This has given the city an important land planning technique.

However, this system of restrictive covenants is by no means the only land use control operating in the city. In addition, Houston has developed an abundance of planning controls and processes (Henderson 1987: 139). It has a building code which, among other provisions, specifies minimum distances between buildings, and designates setback requirements. It has a capital investment program, a comprehensive planning process, and subdivision regulations which apply to newly developing areas. It also has a major thoroughfare plan, which requires developers to dedicate right-of-way if the major streets adjacent to their property are not wide enough to meet city standards.

Nevertheless, Houston is still referred to as a model to be copied by localities which currently are burdened with zoning. The most elaborate exposition of this argument is still that provided by Siegan, whose study was published in 1972 (and was so popular that it was reprinted the following year). In this, he not only gives a very sympathetic analysis of "non-zoning" in Houston, but also argues for its general acceptance in place of zoning.

He sees some overwhelming advantages in doing this. In particular, homeowners would

have the protection of the market instead of the uncertainty of the zoning system, which can be, and is, changed to meet changing circumstances. This neatly turns the common argument (about the certainty afforded by zoning, and the uncertainty of the marketplace) on its head. He also argues that changes in values following the abolition of zoning would generally be gradual, except where unimproved properties have a scarcity value created by zoning. Vacant properties zoned for apartment and commercial classifications would tend to fall in price as the supply increased. On the other hand, vacant land previously zoned for single-family use might increase in value. Overall:

Some will benefit, others will not – as occurs in any comprehensive rezoning, and this is one of the risks zoning has created for property owners. One result is certain, however: the game of producing wealth through the beneficence of the city fathers thereafter will terminate.

(Siegan 1972: 234)

To support his argument, Siegan asks the reader to drive through any subdivisions which are unrestricted: "a great many areas contain only homes despite the expiration many years past or non-existence ever of restrictive covenants: the explanation is ... minimal pressures for diverse uses." Generally:

The experience of non-zoning has shown that commercial and industrial uses tend to locate and separate without compulsion much as they are supposed to do under the compulsion of zoning. Land values and other economic restraints are quite effective in excluding uses from areas where they would be harmful to values of commercial and industrial real estate, possibly more so than zoning. Adjoining owners may also enter into agreements to protect their mutual interests. Indeed, under zoning, owners of land and property zoned for multiple family, commercial, and industrial uses constantly face the hazard that subtly or directly the ordinance will be amended in ways that will erode values.

(Siegan 1972)

These, of course, are generalities, but Siegan proceeds to analyze in detail six areas where efforts were made to establish new covenants. His conclusion is that, "for most homeowners in these areas, the new covenants will continue to maintain a similar life style in perhaps a more circumscribed setting. This, then is a price that would have to be paid for the elimination of zoning. Fortunately, it is a small one; actually minuscule when compared to the benefits."

Siegan's arguments are not, of course, universally accepted.[2] A visiting English civil servant had some very different impressions of Houston:

The effects of no control are to be seen in the less well-to-do areas of the city, where the lack of density controls produces really bad conditions. Apartments are built to absurd densities on grossly inadequate sites. Most cheaper houses, notably in the Negro quarters, are badly overcrowded, with no setback from the pavement and practically no space between buildings. Oddly enough this was also evident in one of the most expensive new developments – Walnut Bend. Here was a site eight miles from the city center with homes in the $30,000 class [1962 prices], yet side yards were so inadequate that in one case the fancy roofs of two neighboring houses overlapped. But the worst effects are naturally in the cheapest developments. In the absence of subdivision requirements, about eighty percent of new residential developments provide normal site improvements (surface roads, sidewalks, sewers, and drainage) because the market demands them; but that leaves twenty percent ready-made slums.

(Delafons 1969: 91)

A broader indictment is made by Feagin in his study of *Free Enterprise City*:

By the early 1980s, if not before, the rank-and-file Houstonian was becoming increasingly concerned about the low-service city and its array of problems. Periodic flooding, subsidence, peeling paint and lung diseases, polluted water, late buses, water pressure problems, chronic air pollution, shortage of park land, slow police and fire department response times, defective traffic signals, variable garbage service, and absent zoning protection for neighborhoods made the ideological blandishments of the power elite about the benefits of the free market, laissez faire city seem increasingly hollow.

(Feagin 1988: 238)

These concerns have become widespread, and public disquiet has grown about the deterioration of unprotected neighborhoods. A measure of the degree of the change in attitude is given by the unanimous decision of the city council, in 1991, to create a comprehensive zoning ordinance. There are many elements in this change in traditionally held anti-zoning attitudes in addition to the visual evidence of the inadequacies of the system of restrictive covenants. Firstly, there has always been some support for zoning in Houston: what has happened is that the proportion of supporters in the city has grown. Secondly, it does not escape the notice of the accountants of Houston that within the city's enormous area of 580 square miles there are several separately incorporated cities, which have strict zoning laws. In these cities, residential lot prices are considerably higher than those in adjacent non-zoned parts of Houston where land owners lack "the security that comes from governmental regulation of their own and other areas within the cities" (Dillon 1991: 9).

There can be no underestimating the daunting nature of the task of introducing zoning into the city (or, to be more precise, those areas that do not have it). The necessary public hearings would take months, perhaps years, of official, consultant, and citizen time. Mixon, who has stressed this point, argues that "what is clearly needed is a different accommodation between public and private interests – one that lies somewhere between comprehensive zoning and the current chaos" (1940: 4).

For the purpose, he suggests "neighborhood zoning areas," determined in accordance with a comprehensive plan, as envisioned in the Standard State Zoning Enabling Act. At the same time, the comprehensive plan would take a citywide view of the location of "undesirable" NIMBYs, such as AIDS housing facilities, prisons, and group homes. (It might be noted in passing that few "comprehensive" plans deal with such matters.) Mixon comments that as a result of the comprehensive approach:

> every neighborhood stands as a potential
> site for these unwelcome neighbors.
> Neighborhood zoning could provide
> concerned neighborhoods with a mechanism
> for making rational locational decisions on
> such facilities.
>
> (Mixon 1990: 16)

It is not appropriate here to go further in examining Mixon's interesting proposal. Current indications are that Houston might lose its unique place as America's non-zoned city.

Withdrawing Zoning From Local Government?

> Land use control powers, as we know them,
> have outlived their usefulness, and should
> largely, if not totally, be withdrawn from
> municipalities.
>
> (Delogu 1984)

One suggestion that is made from time to time is that zoning powers should be removed from

local governments and administered by an agency under direct state control. A well-presented version of this is presented by Delogu in a 1984 paper. The article outlines a major "realignment of rights and duties between the public and private sectors, and between local governments and higher levels of government." The thesis may be difficult to realize politically, but it is conceptually simple, as the quotation at the head of this section indicates. The reasons for this are the same as those extensively set out in this book. Delogu marshals them with supporting references. He indicts local government for implementing a hidden agenda of reprehensible, antisocial, self-seeking policies: "fiscal zoning" aimed at fostering development of high ratable value, and keeping out land uses which involve high costs of servicing; maintaining the existing character of the area; excluding unwanted minorities and low-income groups; preventing undesirable developments ("from power plants to junkyards, multi-family housing to mobile homes, prison facilities to homes for unwed mothers"); and generally preventing change.

Though there is little that is new in Delogu's analysis,[3] he presents the evidence in a particularly useful and persuasive manner. He also goes further than many critics of zoning in setting out "the hidden defects." These he lists as:

1 The absence of a comprehensive planning framework (inevitable with "non-comprehensive jurisdictional entities");

2 The predominance of municipal self-interest and the lack of a mechanism to allocate undesirable but socially necessary land uses to optimal sites;

3 The inherent inability of local governments to address larger environmental questions;

4 The essentially negative character of local controls. Little of a positive nature (e.g. affirmative action) can be achieved.

The gross inadequacies of the situation are such "that one is tempted to argue that if we do nothing more than repudiate local land use controls in toto, leaving the whole land use development process to the market, we should hardly be worse off" (Delogu 1984: 133).

Some have maintained that this would be a positively good thing (Siegan 1972; Ellickson 1977). Karlin (1981: 566), for instance, argues that "it cannot be emphasized too strongly that the zoning, not simply the abuse, produces the distorting effects." Delogu does not go quite as far as to propose the abolition of zoning, though he admits that he finds it "appealing." His solution is to abolish *local* land use controls, to introduce at the state level "a broad and effective range of performance standards addressing all of the potential harms which development activities give rise to," and also to establish state or regional reviews of major development proposals.

In this system, the state plays the pivotal role; but there is also a positive role for the market. Siegan's arguments are quoted with approval:

> Economic forces tend to make for a separation of uses even without zoning. Business uses will tend to locate in certain areas, residential in others, and industrial in still others. Apartments will tend to concentrate in certain areas and not in others. There is also a tendency for further separation within a category: light industrial uses do not want to adjoin heavy industrial uses, and vice-versa. Different kinds of business uses require different locations. Expensive homes will separate from less expensive ones, townhouses, duplexes, etc. . . . When the economic forces do not guarantee that there will be a separation, and separation is vital to maximize values or promote tastes and desires, property owners will enter into agreements to provide such protection. The restrictive covenants covering home and industrial subdivisions are the most prominent example of this. Adjoining owners (such as those on a strip location) can also make agreements not to sell for a use that will be injurious to one or more.
>
> (Siegan 1972: 75)

To some extent therefore, things would continue as before, but without local government interference. A rejuvenated market would bring innovations in such matters as "restrictive covenants, easements, and ownership agreements to fashions limitation and to provide protections and enforceable rights that some developers and purchasers of property desired." The rest is left largely for further elaboration.

These proposals would apply to a wide range of matters ("every facet of development activity"): water, air, and noise pollution; building construction safety; fitness for purpose (warranty-type standards); parking, interior, and approach roads; soils, slope, and erosion; water supply; solid waste handling; and aesthetic considerations.

Implementation would be similar to the review processes already carried out by some states. Though these vary, they do have some similarities:

> These mechanisms all focus on major development proposals, although some states define this term more broadly than others. They all require some showing of compatibility with existing comprehensive plans, pollution control laws, and compliance with state or local subdivision and site infrastructure standards. They all provide some type of hearing and public participation process in which local governments and citizen groups may raise concerns and be heard with respect to the proposed project. They all have some capacity to condition an approval, thereby reshaping original development proposals that have some troubling aspects to them but which the state reviewing agency does not really want to turn down. Finally, they all

provide some form of administrative and/or judicial review of the final administrative decision.

(Delogu 1984: 303)

Further details are not spelled out but, as this quotation shows (and as earlier chapters have demonstrated), states have wide experience in land use and similar types of regulation. There should, therefore, be little problem of administration. It should also be remembered that the proposals envisage a reduction in the overall amount of controls. This is in harmony with much other analysis in this field, including the reports of several presidential inquiries such as the Douglas Report (1968), the Kaiser Report (1968), the President's Commission on Housing (1982), and the Advisory Commission on Regulatory Barriers to Affordable Housing (1991).

The President's Commission on Housing is quite clear what changes are required:

> To protect property rights and to increase the production of housing and lower its cost, all state and local legislatures should enact legislation providing that no zoning regulations denying or limiting the development of housing should be deemed valid unless their existence or adoption is necessary to achieve a vital and pressing governmental interest. In litigation, the governmental body seeking to maintain or impose the regulation should bear the burden for proving it complies with the foregoing standard.
>
> Vital and pressing governmental interests that zoning ordinances should serve include adequate sanitary sewer and water services; flood protection; topographical conditions that permit safe construction and accommodate septic tank effluence; protection of drinking water aquifers; avoidance of nuisance or obnoxious uses; off-street parking; prohibition of residential construction amidst industrial development; and avoidance of long term damage to the

vitality of historically established neighborhoods.

> The density of development should be left to the conditions of the market except when a lesser density is necessary to achieve a vital and pressing governmental interest.

(President's Commission on Housing 1982: 200)

There is a great deal of analysis and commentary in similar vein. Much of it is presented as if the points made are self-evidently true. Some of them undoubtedly are, but others are questionable, to say the least. A great deal is "half-true." For example, it seems clear that zoning and environmental regulations raise housing costs (Seidel 1978); but that does not mean that they are unjustifiable. Rather it is that the cost implications have not been thought through (this is usually a matter for a different government agency), and no measures have been taken to protect lower-income households from their impact (a matter for "housing policy").

Some regulatory schemes are horrendously complex, particularly when taken in combination, which is how the private developer typically experiences them. Any review of zoning, or related issues such as affordable housing, inevitably has a "shock list" of mind-boggling illustrations of bureaucratic monstrosities.[4] There is usually some way of "streamlining" (a favorite word of the Thatcher administration) the regulatory process, but the outcome can usually be predicted to be disappointing.

It would take us too far afield here to attempt to devise a workable reform – if, indeed, there is one. The reader can be referred to a large shelf of previous endeavors. Instead of adding to these, the discussion in these final pages is focused on the key points which need to be addressed in any approach to the problems. These include the conflict between the goals of flexibility and certainty; the tyranny of localism; the necessity for state controls; mandatory planning for local governments with approval of policies by the state; and, most difficult of all, public acceptability.

Certainty v Flexibility: Further Foreign Evidence

The contrast in England is that the control is much more flexible, more capable of accommodating change relatively quickly and easily in administrative terms. What is lost is the comparative certainty of the overseas systems, the greater predictability of outcome at least of the minor, routine proposals for development, or indeed of major developments in policy areas where there is a strong, legally enforceable consensus such as in much of the countryside in every country or, more narrowly for instance in the policy for preventing out-of-town regional shopping centres in the Netherlands.

(Davies *et al.* 1989)

Any planning system involves a compromise between certainty and flexibility. Previous chapters have shown that Britain explicitly operates a flexible system, but it might well be argued that the delegation of this degree of discretionary power to planning agencies would be unacceptable in the United States. Perhaps so; but all is not what it seems with the American system: appearances belie realities. While American zoning appears to provide certainty, the system is in fact manipulated to give a high degree of flexibility.

British development control contains so much room for discretion that there appears to be little certainty but, over the years, policies have become established and accepted to a point that the system displays a great deal of certainty. Curiously, both countries are, in varying ways seeking more certainty. The

American Supreme Court has awoken from a long neglect of land use planning to begin tentatively to lay down certain principles, such as that any charge or condition imposed upon developers should be rationally related to the development for which they are seeking approval (*Nollan v California Coastal Commission* 1987). The British, who see the shortcomings of their planning system much more clearly than its merits, are seeking help from studies of foreign practice (Wakeford 1990, Davies *et al.* 1989, Delafons 1990a and 1990b). Indeed, a valuable wealth of material is becoming available on comparative planning systems.

While it would take us too far afield to examine all this material, it is useful to look at the experience of two countries, the Netherlands and France, both of which have attempted to design planning systems having a high degree of certainty. The discussion is brief, and it skirts round the complexities of the Dutch system and the impenetrability[5] of French planning law.

The planning system in the Netherlands is based on the principle of "legal certainty" which is fundamental to all Dutch law and administration. Thus the *bestemmingsplan* (a local plan) has the force of law. If a development proposal satisfies the plan's requirements, the proposal must be approved. If it is not in compliance with the plan, it must be rejected. Like American zoning, Dutch planning combines the planning and control functions. The contrast is with Britain where, in exercising control, local authorities take into account the provisions of their plan *and* any other material considerations.

There is, however, a very big difference. A zoning ordinance is "timeless": it provides a determination of the uses that may be allowed when development pressures arise. Because of this disregard for time, and the changes that take place over time, the zoning ordinance frequently requires amendment before the development can take place. Indeed, this is often the exact intention: the zoning is designed to ensure that developers will seek a change, and thus provide the local government with a more flexible degree of control.

The *bestemmingsplan*, on the other hand, is a positive planning instrument, prepared within the framework of a regional plan, which is used not only to regulate development, but also to promote it. Municipalities frequently buy land, and sell it to developers after it has been serviced.[6] The active role played by municipalities is well illustrated by the following quotation, which deals with certain types of development, particularly those which involve municipal acquisition:

> The *bestemmingsplan* is accompanied by an economic feasibility study which is compulsory, under the [Physical Planning] Act, and has to be approved by the municipal council. The study contains a financial appraisal of the development, the *exploitatie opzet*, and a statement of the conditions under which the municipal executive will cooperate in the development of the land, covering arrangements for the donation of land and the allocation of costs of provisions made in the public interest. Together they form the basis for the agreements, the *exploitatie overeenkomst*, between the municipality and the developer, whether housing association or private sector organization, for the acquisition, disposal and development of the land. They are subject to approval by the provincial executive.
>
> (Davies *et al.* 1989: 375)[7]

The apparent certainty provided by the Dutch plans is thus strengthened by the active role played by the municipalities in implementation. This system is obviously foreign, in more ways than one, to the United States.

But this is to be too much persuaded by outlines of the law and by descriptions of what is supposed to happen. Is it really like this in practice? It does not take a great deal of inquiry to establish the fact that the Dutch system works only because ways have been found of surmounting its difficulties. The main one of these is that it is impossible to produce *bestemmingsplannen* at the speed required, and equally impossible to keep them up to date with changing circumstances. As Thomas *et al.* (1983: 228) comment, "it is significant that new *bestemmingsplannen* tend to be prepared when the need arises, rather than in accordance with the statutory requirements" which call for periodic revision.

One main way of dealing with the problem is to use a legal provision of the Physical Planning Act, which enables a municipality to act as if a new plan were in force. With this trick, all that is necessary is for the municipality to declare its intention (whether real or not) to prepare a new plan: it can then issue building permits in anticipation of this plan, even if the proposed development conflicts with the existing plan. And so flexibility wins out over certainty.

Similarly in France, with its *plans d'occupation des sols*. These (POS) plans are of various types, but their basic characteristics are:

1 The local plan (POS) enshrines development rights for developers, and adherence to its provision ensures an authorization.
2 An approved POS is a prerequisite for the devolution of planning to the local level, and in its absence all development outside the built up areas is effectively prohibited, leaving the remainder under control of national regulations.
3 The definition of development is precisely and all-embracingly drawn, and covers all acts of construction, subdivision, enclosure etc, and controls the external appearance of development.
4 Development control has a built-in enforcement system to ensure conformity of planning provisions.[8]

5 Notwithstanding decentralization, the state keeps significant control on planning practice through the law, which prescribes the format of the POS and the procedures of decision making, and continues to exercise technical control through the state services in most communes.

6 The existence of a comprehensive system of legal control over and recourse for the applicants, local authorities, affected individuals, and environmental groups provides an appeal system and legal recourse for all actors in the process.

(Davies *et al.* 1989: 233)

This sounds fine; but closer examination reveals "that formulating rules does not appear to create certainty either for decisionmakers or for developers" (Booth 1989: 413). There is a continuing attempt to elaborate rules to cover eventualities not previously foreseen. As a result, the rules become increasingly complex and difficult to apply. Inevitably, developers resort to negotiations in advance of submitting a formal application. And so, as in the United States, the formal system exists largely in law books; it is the informal system that makes this workable.

Booth concludes his study of French development control as follows:

The findings of the case studies suggest that the search for certainty through the legalization of plans is a delusion. The fixed limits of the regulations all too easily falter in the face of the particularities of development proposals and the committed interests of the actors involved. Bargaining is as much part of the French system as it appears to be of the US, British, or Dutch, and the time taken to process applications, quite apart from preapplication discussions, appears to be no shorter in a legalized system than in a fully discretionary one.

(Booth 1989: 413)

Davies and his colleagues, in their major comparative study of five European countries, though apparently finding Denmark, France, the Netherlands, and West Germany commendable for placing emphasis on legal certainty, have to admit that "all administrative systems require discretion to make them work, either in whether to apply the rules or how to interpret them." As a result, "certainty becomes an ideal to be strived for as far as possible, rather than something which in all instances characterizes the overseas systems" (Davies *et al.* 1989: 412).

Toward a Reform

No one is enthusiastic about zoning except the people.

(Babcock 1966)

If reform is to be achieved, the states must be involved far more extensively than they are now.

(Advisory Commission on Regulatory Barriers 1991)

Zoning has many critics and few defenders, except among the electorate. Its inadequacies have been abundantly, though not exhaustively, documented in this volume. Its major defect is that it is used to serve local interests to the detriment of wider interests; and so it always has been. Thus, lower-income housing, group homes, and major facilities ranging from power plants to hospitals, are targets of exclusion, while preference is given to development which harmonizes with the existing character of the area and produces "good ratables."

Such policies are not surprising. Indeed, it can be argued that they are not even reprehensible, since it is the raison d'etre of a political body to represent the inhabitants of its area, not those who might wish to move into it. This is, of course, a selfish attitude; but, given the normal political dynamics which operate at the local level, selfishness is to be expected.

Two things are at fault: the extreme localism which characterizes zoning, and the fact that zoning never became the servant of planning. The first point was majestically articulated by Madison:

The smaller the society, the fewer will be the distinct parties and interests composing it; the fewer probably will be the distinct parties and interests, the more infrequently will a majority be found of the same party; and the smaller the number of individuals composing a majority, and the smaller the compass within which they are placed, the more easily they will concert and execute their plans of oppression. Extend the sphere and you will take in a greater variety of parties and interests; you make it less probable that a majority of the whole will have a common motive to invade the rights of other citizens; or if such a common motive exists, it will be more difficult for all who feel it to discover their own strength and to act in unison with each other. Besides other impediments, it may be remarked that, where there is a consciousness of unjust or dishonorable purposes, communication is always checked by distrust in proportion to the number whose concurrence is necessary.

(Madison 1788: Essay 10)

Larger communities tend to embrace a wider range of needs (and abilities), a greater concern for the common good, and a broader conception of the role and purpose of public policy. In short, they are less selfish. Of course, exclusionary attitudes do not wither away; but they do not flourish as they do when isolated from other influences. Instead, they are kept in rein by "the sheer number whose concurrence is necessary."

Unfortunately, this is too simple a view of the matter, and increased size by itself is not necessarily enough. A large constituency may be nothing more than an assembly of small

bodies, each of which works the political process to achieve its particular aims. Where this is so, mutual accommodation of interests forces out broader thinking.

It follows that local governments, whether small or large, require a compelling framework of social responsibility within which to operate. This has two advantages. Firstly, it provides policy objectives which have to be followed. These move to the forefront of political attention: they develop, to use Donald Schon's phrase, into "ideas in good currency," (1971: 123) and they become a spur to action. Secondly, it gives elected members a defence against those who would have them work to a narrow interest: "we have no alternative," the fainthearted can plead, "we must do what we are required to."

Thus the rationale for policy can be a matter of high principle or of base political calculation: it makes no difference. What matters is that a public authority is operating in the broad public interest, not to the advantage of a powerful minority; and this means that it is including all needs in its civic policy.

Such an approach paves the way for devising a system in which zoning becomes the servant, rather than the master, of local land use planning. This was the original intent, though it was rapidly forgotten as zoning spread across the nation with remarkable rapidity. Zoning seemed clearly to provide immediate protection against the rapid changes which were taking place during the early years of this century; and it still does.

The first step is for states to expressly prohibit local restrictions on such matters as the number of bedrooms, floor areas, lot requirements, and limits on the total population of an area and the number of building permits allowed annually. Where necessary, restrictions on some of these matters could be imposed on a statewide basis.

The states should adopt a positive declaration of a commitment to rooting out exclusionary practices:

The whole idea of exclusion, the belief that individual towns may adopt land use policies that sharply limit growth, would be dealt a sharp blow if the legislature clearly stated that contrary view. A new, broadly worded statement of purposes should be added to the state's planning and zoning enabling legislation endorsing, among other things: the principles and policies of inclusion rather than exclusion; the right to travel[9] and to settle where one will; the right to use land in any reasonable manner which does not threaten the public's health, safety, and general welfare; and the right of all citizens to affordable housing.

(Delogu 1980: 63)

Secondly, state and regional goals need to be elaborated to provide a basis for the preparation of local plans which have some real meaning, rather than being simply a mapping of market trends or wishful thinking. Oregon is the best example here, as in other ways. This is not a state plan: it is a set of objectives and procedures that have been agreed to be desirable for the future of the people of Oregon. The nineteen goals (summarized in Chapter 11) provide a policy basis for the preparation of local plans (which are mandatory).

Goals, objectives, declarations, and such like are, of course, the stuff of political discourse, and they can be totally devoid of meaning. There are many plans like this: they adorn the libraries of planning schools, and sometimes they are useful to a planning historian – as long as his or her concern is with plans and not with planning.

Planning is a process, and like all processes must have an engine to drive it. To prepare plans, and to state goals, is one thing: to give them substance is another. Thus, without in any way belittling the symbolic value of declarations, it is also necessary to devise mechanisms to prompt, attract, bribe, and if necessary force, local governments to formulate and implement local plans which accord with the state goals.

The precise way in which this can be done will vary from state to state. Oregon is one model, but it may not work elsewhere. Each state has to devise ways of devising, debating, and agreeing goals, and of implementing them. Another system which is worthy of study, and is less well known in the United States, is that developed by the Canadian province of Alberta. Alberta's planning objectives include the preservation of local autonomy over land use and development by providing a broad regional framework and authority in which independently determined objectives can be expressed and realized.[10] The Planning Act has made it mandatory that any action or thing done by any municipality should conform to the regional plan. Consequently, every statutory plan, replotting scheme, or land use bylaw of a municipality must conform with the regional plan that affects the municipality. Urban municipalities with a population of 1,000 or more, and rural municipalities with a population of 10,000 or more must adopt a general municipal plan.

One thing is clear: whatever reform is attempted, a lot of talking is involved. No plan will be operable unless there is widespread support for it. Public participation is essential at all levels. The start may be a state commission: there are several examples in Chapter 11. The passage of legislation is the first major hurdle, because then people begin to realize that the matter is a serious one: something is going to happen; and their interests may well be affected. Compromises will probably be necessary, much to the disappointment to some of the protagonists, but little is better than nothing; and if the enacted legislation proves inadequate, there then exists the basis for pressing for something more effective.

The legislation will lay down duties and responsibilities. There has to be machinery to ensure that these are fulfilled. Oregon has a very useful innovation in its bipartisan Joint Legislative Committee on Land Use which acts "as a watchdog over the activities of state and local governments in carrying out the mandates" of the legislation (DeGrove 1984: 246).

Local governments will need financial help, and possibly technical assistance as well. As Baer has written:

> The problem in top-down planning lies in implementing the concept. Local governments, not the State, are the traditional means of implementing land-use development. They, not the State, are closest to the scene; and they, not the State, are continuously involved in day-to-day decisions over land use. Some cities actively resisted the State mandates by lobbying to change the law; others more passively "went limp" – doing nothing to comply. In both cases they were implicitly helped by state legislators who postured about the need for more housing but provided little funding to already hard pressed local governments. Plans without finance are but a nuisance.
>
> (Baer 1988: 270)

Technical assistance to local governments (particularly the smallest ones) may be necessary. This can be rendered by way of joint teams of local and state (or regional) officers. These also can be a good source for the development of understanding between different levels of government, but it is important that elected members are involved, or they may find themselves estranged from a system which they do not understand and which they may begin to despise as being remote and technocratic.

Another Oregon innovation which demands more study than it has so far received is its Land Conservation and Development Commission and its Land Use Board of Appeals. The first (the LDCD) is responsible for ensuring compliance with the state goals. It has the duty "to review and accept or reject comprehensive plans, and to develop standards to carry out the intent of the law through full implementation of the goals." DeGrove, from whom this quotation is taken, continues:

The LDCD assists the implementation of goals through grants and technical assistance, and through explanatory "policy papers" that guide the implementing jurisdictions. The 1981 Oregon legislature directed that all such commission policies be adopted by rule if they have statewide impact. The goals adopted by the commission have the force of law. The commission may conduct hearings and issue orders to require regulations, plans, and land use actions to conform to state goals. It is significant, especially in the national perspective, that those powers have been used; between 1973 and 1979, more than 133 petitions for review were heard by the LCDC under this provision.

> (DeGrove 1984: 256)

At some point, quasi-judicial and judicial procedures are required. The state circuit court is the first step, followed by the Land Use Board of Appeals (LUBA). The board consists of three hearing officers appointed by the governor.

> It reviews land use decisions of any local or state agency previously heard in the state circuit court, and also recommends orders to the LCDC regarding compliance with statewide planning goals. The board's decisions on land use and the LCDC decisions on goal compliance can be reviewed and enforced by the courts, starting with the court of appeal and ultimately reaching the Oregon supreme court. In the past, Oregon courts have generally supported the legislation, which has generated significant legal as well as administrative review.
>
> (DeGrove 1984: 257)

This is, of course, only one model. In some states, it would probably be unworkable or unacceptable (for instance, the role of LUBA). But the details are unimportant at this stage. The point is that the state has to have some machinery for enforcement (in addition to financial and other incentives); and at some point the normal courts have to become involved.

Given this type of approach, plaintiffs before the courts would have considerable difficulty in demonstrating why they should not comply with a thoroughly well thought-out plan and its procedures but, no doubt, issues would arise which need to go to the courts. A major problem is that the policies which are desirable (particularly the control of development, and even the prohibition of any development in certain areas) can also be used in a totally unacceptable way to exclude minorities or poor rateables. Indeed, the whole battery of zoning techniques, from density control to the phasing of development over time, can be used both for and against the wider public interest. Unfortunately, as the *Euclid* decision put it, "the line which separates the legitimate from the illegitimate assumption of power is not capable of precise definition." The matter thus becomes one for determination by either the courts or a higher level of government, or both.

Policies Reserved to the States

> Americans take federalism for granted, and are seldom aware of the role states play in their lives.
>
> (Judd and Robertson 1989)

Another technique which is well worthy of consideration is used in Florida, where some issues are reserved to the states: Areas of Critical State Concern (areas of special environmental, historical, or other regional or statewide importance which should be protected) and, to a lesser extent, Developments of Regional Impact. (These are discussed in Chapter 11.) In both cases, the legislation provides time for the state to get a measure of the problem and to

establish what special planning measures are required and how they shall be implemented.

Delogu (1981) presents an alternative to Florida's Developments of Regional Impact. The concern here is with the problems of siting major facilities such as electric generating facilities; solid waste and hazardous waste disposal areas and facilities; regional hospital and correctional facilities; regional airport and harbor facilities; regional industrial, commercial, and recreational facilities. Where these appear to promise increased property tax revenues, local governments may compete among themselves to attract the facility (sometimes offering generous bribes), but often forgetting to examine the other impacts on development in the wider region, such as traffic congestion. When the facility has little or no tax allurements, or is of such an unpleasant character as to outweigh tax considerations, strong opposition can be expected, with no impartial assessment of the best location for both the developer and the state. This is not to be unexpected: it is the political responsibility of a local government to represent the *local* interests. Wider interests should fall to a higher level of government which in turn is elected precisely to look after these wider matters. The American problem is that these higher levels of government do not operate effectively in this area.

Delogu's solution involves the establishment of state siting standards for major facilities and the identification of actual locations within the state which meet these standards. Secondly, a mechanism should be devised, again at the state level, for sharing the tax benefits and costs which developments of this type create. Finally, states should develop a system which provides for the evaluation of any local government exclusion of an otherwise legal type of development.

In this scheme of things, the administrative agency ("the reviewing body") would not be a super zoning mechanism. Rather it would be a means for allocating unpopular and difficult-to-locate activities among the many areas of the state. Its role would be, "not to facilitate the egalitarian distribution of undesirable but socially necessary land use activities"; instead its task would be "to look at physical realities to determine if a particular developer and activity have been totally and unreasonably excluded from an entire town. If this has occurred, the administrative body should authorize the developer to proceed with his project on the proposed site under whatever conditions the circumstances may warrant" (Delogu 1981: 28).

Making Plans Happen

Stop me before I plan again.

(Hedman 1977)

It is axiomatic that plans should be realistic, though many are not. To be "realistic" means that they are capable of being implemented. In practice this typically implies that development pressures are apparent and need to be accommodated (or resisted, and accommodated elsewhere) with some degree of urgency. Plans that lack such market signals are superfluous at best, and visionary at worst. Yet it is supposedly the business of planning to project the future (in the short term) and to translate future needs into land use requirements. Obviously, this is not a precise science, though there are established techniques available which assist with some dimensions of change. The important task is to ensure that sufficient land, together with the necessary services, will be available in appropriate places for the foreseeable development. (A time span of five years has proved to be about the limit attainable; but continuous monitoring and revisions are essential.) This means that there are two major elements in a comprehensive plan; these relate to land use and to capital investment.

An interesting example of a positive planning approach is given by a visiting British civil servant:

One of the most effective applications of zoning certainty that I observed was the temporary density bonus available on the west side of Midtown Manhattan. By making it clear that the bonus provision would be available only to developments started in the six years before a "sunset" applied, and sticking to its original intentions rather than wavering and extending the scheme, New York City refocussed growth from an area under pressure to a nearby location where there was rather more public transport capacity. The scheme worked because of its precision and the temporary certainty that it created. It was clear how the developers' sums would work out with and without the scheme.

(Wakeford 1990: 264)

In undeveloped areas, there is much to be said for large-scale land assembly on the part of a public agency, though experience shows that this can meet fierce public resistance on grounds of loss of local power and also restricted compensation. In the American Law Institute's *Model Land Development Guide* (1975: 221), Article 6 deals with this at some length under the heading of "land banking."

There is abundant experience from Europe that systems of advance purchase have great planning advantages and, where there is a large demand, some such scheme is far preferable to a development timetable that is determined by the time at which farmers decide to retire and sell off their land for development (or are simply enticed by the prevailing rate of land prices to sell at the point when they think that prices have reached their peak).

The British white paper which explained the proposed provisions of the postwar planning legislation stated quite categorically that: "the best way of putting plans into effect is often to make the land available for any development which the plan shows to be desirable" (GB 1947). More striking is the situation in the Netherlands, where history and geology have combined to create a strong favorable attitude to public land acquisition. It is worth explaining this:

> In the Netherlands, direct interference by the citizen with and his direct influence on matters of public interest are much smaller than on the other side of the Atlantic Ocean. . . . His influence works almost exclusively via the governmental representative bodies – municipalities, provincial councils and parliament – which he has elected himself. As political life in the Netherlands has a certain stability, and the average Netherlander has confidence in the representatives of his choice – and rightly so – he leaves a lot to the discretion of these representative bodies.
>
> Public acceptance of expropriation is so widespread that in carrying out town plans at the municipal level, expropriation is the rule rather than the exception, be it that a settlement by agreement often follows after the onset of the expropriation procedure. Perhaps it is also the result of the centuries long struggle against the water, namely in the sense that the Netherlander has a deep-rooted understanding of the fact that the free disposal of the often threatened land had to be restricted in the general welfare involved in the planning, construction, and maintenance of the system of dikes, sluices, and watercourses.
>
> (deCler 1967: 1150–3)

The historical and cultural background in the United States is, of course, quite different, and there is always the seductive vision of a large land value profit to prevent voluntary compliance with acquisition schemes. Nevertheless, vast acquisitions were undertaken in connection with the interstate highway program, and it might not be too much to hope that smaller versions of this could be negotiated for other types of needed development.

With developed areas, policies are usually more concerned with managing change in the existing urban fabric. This is an inherently much more difficult matter on which there are few guidelines (which is why developers prefer virgin sites). There are numerous plans for small historic areas, waterfront sites, major commercial and retail developments, and such like. But over the generality of urban areas, the forces of change are allowed to work themselves out, usually to the detriment of the city. There is little that might be called urban management (in the sense of attempting to channel the forces of change in desirable physical, economic, and social directions). The result is a fearful degree of urban decay. It is of little surprise that the increasingly distant suburbs are so appealing.

This takes us beyond the issues intended for discussion in this book. They are added here as a sad reminder that an efficient system of land use planning should be as concerned with existing developed areas as it is with new developments.

Profits From Land

> Ninety percent of land use is making a buck.
> (A Maryland farmer)[11]

It was noted in the last chapter that American local governments can downzone land (which amounts to reducing its value) within the compass of the law. It was implied that there was some sort of parallel between this and the scheme for the nationalization of development values in postwar Britain (which, contrary to often-held opinion, still operates). The parallel may seem strained, but there is nothing in law to prevent a local government from preparing a new comprehensive plan and zoning ordinance which (for example) concentrates development along transport corridors and downzones large areas (which cannot be serviced adequately) to agricultural use. Of course, there would have to be a proper planning process, and no arbitrary or "capricious" action would be tolerable. More problematic is the need for public support. Though essential, there would undoubtedly be fierce opposition from those whose land was reduced in value.

Some have argued that, whatever the law, it is unfair for planning to operate in such a manner.[12] However, the private market operates in exactly this way, and it is difficult to see why a loss created by a change in public policy should be compensable, while a similar change brought about by changed market conditions is not. Nevertheless, the point is one on which strong argument can be expected.

In this respect, American local governments have even more power than their British counterparts. If a British local government changes the "zoning" for a property in such a way as to reduce its value, it is liable to pay compensation (Hagman and Misczynski 1978: 289). Not so in the United States: the legal restraints on rezoning are not onerous, though, of course, there must an absence of caprice or malice. Yet this does not make the task of the American local government any easier. On the contrary, the fact that there is no obligation to compensate owners for losses arising from rezoning may make the local government unwilling to use its powers. To pay compensation, however, can result in rewarding landowners for increases in land values created by public policy and investment (as discussed in the previous chapter).

There is no easy answer to this conundrum. Nevertheless, it is clear that, over time, the public can come to accept that land is not to be viewed as a source of profit. There is, unfortunately, no indication that this might happen in the United States.

Underlying all this are some ill considered ideas of fairness. An illustration of this was given in the Introduction: the local saga of shopping malls on Delaware's Concord Pike. This is continuing (and presumably will do so until the Pike is totally "gridlocked"). At the time of writing, the latest episode concerns the zoning of yet another (236,500 square-foot) shopping complex on the Pike. This was approved by the county planning department: the county planning director is reported as saying that the department's endorsement of the project was partly motivated by the fact that two previous rezonings had been approved, and the department wanted "a degree of consistency" in its recommendations.[13] (At the same time it was reported that "the State Department of Transportation has projected that even without new development, Concord Pike [US 202] will 'fail' in less than two decades. That means the road will be carrying 140 percent of its capacity, causing severe congestion.")

This is only an illustration but it clearly shows the great importance attached to acting "fairly" between developers, and the virtual absence of concern for the public interest. Babcock and Feurer (1977: 319) have pointed out that this has a long history. The Standard State Zoning Enabling Act of 1924 provides that zoning shall take into account "the character of the district." This leads inexorably to a policy of "more of the same is better." The first intrusion is the excuse, legally, for the next. "The presence of one gas station at an intersection is the justification for a second. The first highrise to enter the three-story brownstone neighborhood is the camel's nose under the tent for a wall of highrises . . ."[14]

Conclusion

It would help if planners grasped the real nature of the American federalist system of land use controls. It is so loose, so deliberately disjointed and open ended, that it is barely a system in the sense that European elite civil service bureaucracies understand the term. The right to make particular regulatory decisions shifts unpredictably over time from one level of government to another. No principle of administrative rationality, constitutional entitlement, economic efficiency, or even ideological predisposition truly determines the governmental locus of decisions. It is more often a matter of the inevitably uncertain catch-as-catch-can pluralism of democratic power politics.

(Popper 1988)

It is abundantly clear that there are no easy answers to the problems examined in this book. Most people are satisfied with things the way they are, and resist changes which reduce their political power or prevent them from making a profit on land transactions. In areas of great pressure, action has been taken; and there is a model of state planning which could be copied elsewhere. The problem is that there is no incentive to act until severe development pressures arise. Then it may be too late, since every landowner will see a potential windfall profit which stronger planning controls might destroy. A favorite alternative is to zone very generously, thus avoiding offence to any landowner, but at the same time, destroying the community benefits which planning is intended to achieve.

So is there a way out? Wakeford maintains that there is, and that it is happening. Though it is inconceivable that development rights could be nationalized "overnight," as happened in Britain, nationalization is in effect taking place in a cumulative manner:

In reviewing their ordinances, cities and suburbs are generally reducing the amount of development that is permitted as-of-right. The motivation may be to secure more discretionary review powers, or to make sure that bonus schemes work without destroying the local amenity, or to fund trading systems. The fact is that it is happening.

(Wakeford 1990: 261)

But, if true, this process will take an enormous length of time and, in the meantime, the unintelligent scatter of unrelated land use which characterizes much American development will continue. Any reform must restrict the present license of local government.

Zoning is, for many local governments, a major responsibility. They are not likely to relinquish it readily; and the states can gain little electoral profit from making war with local government, particularly when effective action involves not only state control but also increased taxes for infrastructure needed to service development which local governments have held up. The courts can help, as in New Jersey, by putting the states into the role of an unwilling controller and tax gatherer, but this can only be a catalyst to action, not its foundation.[15] The federal government can be even more effective, since it can wield both a stick and a carrot: regulatory requirements and financial assistance. But the federal government is very distant from local government: the states are clearly much more appropriate as the body to promote, encourage, assist, and pressurize local governments. Perhaps the federal government could provide a system of financial incentives and penalties that would spur states into action with their constituent local governments?

It is, of course, clear from Chapter 12 that the federal government has limited authority and competence in relation to local land use planning. Nevertheless, there is evidence that it could play a role of some effectiveness. It has certainly operated with force in the environmental area, and its earlier promotion of regional planning indicated the possibilities for influencing local action. Planning grants and mandatory requirements (e.g. for "consistency with the comprehensive plan" of proposals involving federal grants, and for the planned provision of affordable housing) were of some effectiveness, though the potential was "largely unrealized" (Rubinowitz 1974: 173). At the time of writing (1991), however, there is little suggestion of a greater role for the federal government.

No scheme of things will change the general hostility of Americans to government in general and regulation in particular. Indeed, removal of controls from the local to the state level (even if reduced in total as suggested by Delogu) may increase the antagonism. It will be more remote, with the state board being housed in the state capital (among a comparatively large bureaucracy) rather than in the city hall. This will not only make regulators more inaccessible, but will make them more difficult to influence. Such points serve to highlight the larger issue: Americans do not like regulation, even when it is in their own interest (Glazer 1988: 189).

If this is thought to be unduly pessimistic, corroboration can be found in the planning endeavors of the most "progressive" areas. Oregon is commonly acknowledged as a leading state in planning policy and practice. Yet its own (1991a) review laments the fact that "growth has begun eroding the livability of the state's urban areas." A long list of reforms is proposed, including the creation of a state agency to assist local governments with infrastructure finance, thus ensuring that positive planning can take place. (This is the missing link in Florida's "concurrence" policy.) But Oregon is the exception that proves the rule.[16]

Take, for instance, San Francisco (hardly an unprogressive area). There, the Bay Area "Bay Vision Commission" has stressed the diversity of views which were represented on the commission and the difficulty of reaching agreement:

> We have noted that current forecasts predict an increase in the Bay Area's population from the current six million to well over seven million by the year 2000. Some of us have concluded that there is a point beyond which the Bay Area's population must not be allowed to grow if the natural resources of the Bay Area are to be protected adequately. Others of us believe that such a population

> limit is neither desirable nor possible to achieve. Still others believe that the issue is not population growth itself, but the need to manage development so that natural resources are not degraded as population increases. All of us agree, however, that the environmental impacts of an increasing population and an expanding economy will require a new, more comprehensive ability to plan and make regional decisions for the Bay Area.
>
> (San Francisco 1991: 4)

It is not clear which view will prevail, though it is likely that some form of policy will be adopted which will allow continued growth. Hopefully, it is proposed that a regional commission be set up combining, for a start, the functions of the Bay Area Air Quality Management District, the Metropolitan Transportation Commission, and the Association of Bay Area Governments (and later the Regional Water Quality Control Board and the Conservation and Development Commission), but this still leaves sixty-two other agencies! There is, moreover, disagreement on the constitution of the proposed regional commission, its role in equalizing tax burdens, and its power to control developments of regional importance. But "we strongly believe in maintaining the integrity of existing local governments and their autonomy over local decision-making" (San Francisco 1991: 38). As Joseph E. Bodovitz, the commission's project manager, has commented, "there is no ground swell of readiness to plug into a regional political system. Indeed, there is great antipathy" (Stanfield 1991: 2330).

There is a clear parallel with the 1990 report on Los Angeles and Southern California by "The 2000 Partnership." Though existing governmental agencies "cannot adequately plan for and manage growth on a regional level," no new planning authority is proposed; instead a new council would consolidate the current planning powers of existing agencies, and subregional councils could be formed on the basis of cooperation between local governments. It is suggested that the new regional council "would have the authority to make and implement policies when a city, county, or special district was determined to have failed to meet regional objectives within a specified time limit." This sounds as if the regional council would have some teeth, but these are quickly drawn: regional and subregional plans would be subject to the agreement of the constituent authorities who would have ample opportunity for "consultation" and "bargaining." At the most, the proposal amounts only to control over "limited areas of regional impact."

If such are representative of the more enlightened areas, what hope is there for the more typical areas which do not even go through the motions of planning? As R. Scott Fosler, of the Committee for Economic Development, has commented:

We're totally unprepared to deal with regional issues in terms of political and governmental institutions, let alone mental concepts. . . . There isn't any one institutional mechanism that either totally comprehends the others or is strong enough to provide the clear leadership role [that the central cities did in the past].

(Stanfield 1991: 2317)

Though, by default, some of the states have tried to step into the gap, the localities are not happy about this "and the states, in many cases, are not happy about getting involved either" (Stanfield 1991).

In ending on such a depressing note[17], it is useful to recall Haar's wise words: "The ultimate test a planning system must survive is not its legality, or even its wisdom, but rather its acceptability to the public at large."[18]

Appendix 1 Resuscitation of the common law?
Haar and Fessler's 'revolutionary' rediscovery of the common law tradition[1]

The United States is a land of constitutions, constitutional law, and constitutional interpretation. Americans have an abiding faith in the doctrine and workings of judicial review for defining or checking a multiplicity of statutes, and they place great reliance upon basic documents like written constitutions, fundamental charters, and organic acts for guidance in carrying on public affairs.

(Grant and Nixon 1982)[2]

The reader who has come this far will not need telling that the courts play a highly significant role in the U.S. land use planning process. Whether this role is also effective is another matter. Most writers on possible reforms give little or no attention to it and, instead, focus either on a more powerful function for the states or a "return" to the halcyon days of the free market, or at least to a system which imposes fewer fetters on the operation of market forces.

There is, however, one notable book in which the authors place their faith in a new active role for the courts, not as interpreters of the Constitution but as defenders of the rights provided by the common law. The book, *The Wrong Side of the Tracks*, by Haar and Fessler, (1986), stands out for the magisterial eloquence of its argument and the restrained passion in which the case – unknown to the layman and probably forgotten by the legal expert – is presented. Its validity and its capability of being translated into a forceful anti-exclusionary leverage on local governments is difficult to assess, certainly for this writer. But Haar and Fessler write with such authority that one cannot help but be persuaded that here is a new way out of the current American difficulties. Even if this should prove to be an over-optimistic reaction to the book, there is no doubt that it should be read by anyone who is interested in seeing reforms in the area of land use planning.

Haar and Fessler are extremely concerned

about the problem that forms the centerpiece of their thesis: the gross inequalities of conditions and opportunities which exist in American society. They argue compassionately, and indeed passionately, for "legal foundations for a constructive judicial role in redressing inequality in the provision of municipal services." Their thesis is that the common law provides a "body of rights, owned by the public, for equality of treatment." Exploring this thesis takes us into some surprising areas, ranging over a vast span of history. It takes the form of

> a journey designed to demonstrate the ancient origin – and the persistent appearance in human affairs – of the judicial aspiration that all persons in similar circumstances be accorded similar treatment at the hands of entities that render a public service. In order to accomplish that goal, the spokespersons of the common law advanced the requirements that there be equality and adequacy in the rendition, and reasonableness in the pricing, of public or communal services and facilities.

The book was born of the despair at cases of outrageous discrimination; in particular, the case of *Hawkins v Town of Shaw*. Hawkins was a black, living "on the wrong side of the tracks" in the small town of Shaw, Mississippi: "an impoverished site of hard-core neglect, where streets were unpaved and unlit, water mains and fire hydrants lacking, and drainage and sanitary sewers nonexistent." Ironically known as "The Promised Land," this predominantly black area contrasted dramatically with the other part of Shaw which was predominantly white. Hawkins filed a class-action suit on behalf of all the black residents of Shaw, charging the town government with unequal distribution of essential municipal services, a violation of his and others' constitutional rights to equal protection of the laws. The court acknowledged the facts, but dismissed the complaint for failure to state a claim on which relief could be granted. The court suggested

that since the issues were ones of "municipal administration," they were to be "resolved at the ballot box."

In a *Prologue*, Haar asks whether the law was truly so impotent; and he castigates legal positivism: the view "that law is more akin to a species of logic than to a broadly humanistic vision of society." This places the individual lawyer in a schizophrenic posture: "From law school on, [the lawyer] is taught to view cases with a disinterested legal eye, to eschew common sense . . ." The alternate human eye sees things differently, but "a decision that appears as an affront to the humanistic vision can also appear 'correct' and quite consonant with existing principles and precedents."

Haar's indignation led him, with Fessler, not only to participate in further legal proceedings in the *Hawkins* case, but also to undertake the monumental research work (extending over ten years) of which *The Wrong Side of the Tracks* is the outcome. "Happily" he reports "we found an alternative in the theories and accomplishments of an era when the dyslexia between the human and legal eyes was less pronounced." In an English case dating back to 1444 (yes: 1444!), the court found that "a conferral of monopoly power – to mills, to ferries, to markets – was implicitly conditioned on a *duty to serve* all members of the community alike." Haar continues enthusiastically

> As we traced this principle through history, fascinating characters and unexpected occasions pranced into view. We read, for example, of an elderly woman, in charge of a ferry in fifteenth-century England, who was clamped into jail for refusing to rise up in the middle of the night to transport a frenzied traveler across the river. The price of her monopoly was an enforceable demand for adequate service.

Another principle of the earliest English common law was that the distribution of scarce goods that are public in nature has to be equitable and fair. The relevance of this to the town of Shaw is important:

> The common law duty to serve does provide an alternative basis for the class action in *Hawkins* – indeed, for future actions challenging the unequal provision of public services. . . . Once the town of Shaw is viewed as a monopolistic provider of essential services to the public, it assumes the common law duty to serve. Viewed from the perspective of a social contract, the citizens of Shaw had awarded their municipal government a privilege far more profound than monopoly power or eminent domain: the precious privilege of governing. So the duty to serve could now allow the deprived citizens of Shaw to collect on their side of the bargain.

This common law *duty to serve* predates the equal protection clause of the U.S. Constitution by more than seven hundred years. Its persistence over such a long period of time "speaks to the organic nature of the common law; were it not consonant with ultimate moral visions and recurring economic aspirations, the doctrine would long ago have disappeared." Haar and Fessler trace its development from feudal times up to the present: monopolies were judged acceptable and competition suppressed "but only if the monopolist could demonstrate that his facilities were sufficient to serve the needs of each and every member of the claimed populace in an equal, adequate, and non-discriminatory manner, and at a reasonable price." This was applied (though never without difficulty) to grinding mills, ferries, bridges, markets, and later to the railroads and other new enterprises such as the canal, water, gas, electric, telephone, and telegraph companies. The epitome of the concern for equality was perhaps an Indiana Supreme Court case where the court ordered a company to allow a prospective customer to hook up to its gas mains, whether or not there existed sufficient supplies: the court insisted on equality for all, "even if it

should mean that none would receive sufficient service."

"Four vehicles on the path to equal services" are identified:

1   The imposition of a common right to access drawn from the doctrine of services as *a public calling*, essential to individual survival within the community;
2   The duty to serve all equally, inferred from and recognized as an essential part of *natural monopoly power*;
3   The duty to serve all parties alike, as a consequence of the *grant of the privileged power of eminent domain*; and finally,
4   The duty to serve all equally, *flowing from consent*, expressed or (more frequently) implied. ("If a company has chosen to engage in providing a service to the public, then it has voluntarily assented to the appropriate requirements of a public service.")

Underlying these four rationales is a leitmotif: that of "the judges' universal pride and confidence in their abilities to analyze what makes society run. Judges simply continued to trust in their special ability to adapt a concrete conflict to the underlying needs and presuppositions of society . . . . The common underlying need was for the conscience of the court to ascertain, to formulate, to develop the equal service doctrine as society evolved and as new technologies (which judges have nearly always been confident would produce only progress and enlightenment) were introduced into the mainstream of the economy."

Of course, not all judgments fall neatly into the overall pattern, and a judicial decision is not always followed by appropriate implementation. Nevertheless, "equality as the touchstone in the provision of public goods has sustained its worth despite changing social and historic circumstances, and is a major embodiment in one realm of the distinctly modern spirit." What then of the other side of the tracks?

On this, Haar stresses that the study demon-strates "the potential for alternative legal avenues of appeal and redress. Reinterpreting for this generation the common law duty to serve can translate lofty injunctions into concrete achievements and make equal access to public services a reality of life in American cities." Fessler notes that one of the lessons of the book is that contemporary liberals and con-servatives can "join in the quest of constructive judicial involvement in the alleviation of social ills ... the most important point emerging from our work is the joint advocacy of litigation strategies designed to provoke a corrective response from elected officials at the local or state level rather than through a federal judicial decree."

The next operational steps are not spelled out, but the book ends its eloquent and noble plea in characteristic language:

> The principle of equality of access to municipal services has been and, if society is to avert morally debasing outcomes, must continue to be a compelling and accessible alternative to federal constitutional law for advocates seeking to render irrelevant those tracks that delineate the conditions of life in America's villages, towns, and cities even today. If in the local areas of close, physical interdependence we can enhance the humanistic foundations of social collaboration, it will be because we have heeded Santayana's advice on the lessons of the past and continue to refine them into a powerful living tradition for the present and the future.

Those who are persuaded by this will search the current law cases to see if there are, indeed, any signs of litigation building on the rediscovered common law tradition. The reaction of legal reviewers has been that Haar and Fessler's anal-ysis does not go far enough (it fails to address a considerable literature on the theory of local government for example); that there are ques-tionable implications for the political role of local governments, and the likelihood of tension between them and the courts (cf. *Mount Laurel*); that no account is taken of the impli-cations for municipal finances and possible "white flight"; and that it is likely that the state courts will be less enamored of the principle of equality of service as are Haar and Fessler: after all, the common law doctrine of *inequality* is more genuinely consistent.[3] But no one doubts that this remarkable book has opened up a new line of inquiry. It remains to be seen whether anything positive comes of it.

Appendix 2 Extracts from the Fifth and Fourteenth Amendments of the United States Constitution

Amendment V

No person shall be held to answer for a capital, or otherwise infamous crime, unless on a presentment or indictment of a Grand Jury, except in cases arising in the land or naval forces, or in the militia, when in actual service in time of war or public danger; nor shall any person be subject for the same offense to be twice put in jeopardy of life or limb; nor shall be compelled in any criminal case to be a witness against himself, nor be deprived of life, liberty, or property, without due process of law; nor shall private property be taken for public use without just compensation.

Amendment XIV

1   All persons born or naturalized in the United States, and subject to the jurisdiction thereof, are citizens of the United States and of the State wherein they reside. No State shall make or enforce any law which shall abridge the privileges or immunities of the United States; nor shall any State deprive any person of life, liberty, or property, without due process of law; nor deny to any person within its jurisdiction the equal protection of the laws.

5   The Congress shall have power to enforce, by appropriate legislation, the provisions of this article.

Notes

Introduction

1. The reference is to the author's *Town and Country Planning in England and Wales*, Unwin Hyman, 10th edition, 1988, and *Urban and Regional Planning in Canada*, Transaction Books 1987.
2. The 1991 Act abolishes the need for central government "approval" of a structure plan but, if the plan fails to accord with government policy, action can be taken to ensure conformity.
3. *State v City of Rochester* 1978, quoted in Haar and Wolf 1989: 337. Weaver and Babcock (1979: 271) comment that the Oregon Supreme Court's decision is based "on the notion that 'correct' zoning decisions depend merely upon a disinterested assessment of objective facts." Moreover, there is "the unrealistic premise that precedural reform can guarantee that 'illegitimate' considerations will not find their way into the decisionmaking process." For further discussion see the section on zoning amendments in Chapter 4.
4. Babcock is quoted frequently in this book. The reader is particularly referred to Babcock 1966, and Weaver and Babcock 1979.
5. Crozier (1984: 88) argues that the United States is extreme in its preoccupation with the short term: "Americans trust that if the short term is handled alertly enough the long term will take care of itself." He also suggests that the problem is compounded by the fact that, unlike European countries, the United States has no organ of government that is concerned with the long run (such as the British higher civil service).

1 The institutional framework

1. Subdivision, as the term suggests, is the division of raw land into smaller parcels for the purpose of sale or/and development. The granting of subdivision approval is subject to regulations which establish requirements for streets, lot lines, etc. In short, subdivision is a land use control very similar to zoning: the difference is that while both deal with the physical development of a lot, zoning deals also with the *use* of the land. Throughout this book no distinction is made between the two types of land use control. For further discussion see Ducker (1988: 198), and for a detailed history see Yearwood 1971.

   A traditional definition of subdivision is "the division of land, lot, tract, or parcel into two or more lots, parcels, plats, of sites, or other divisions of land for the purpose of sale, lease, offer, of development, whether immediate or future" (Ordinance of Arvada, California, reproduced in Burrows 1989: 33).

Controls over subdivisions preceded comprehensive zoning, but it was not until the years after World War II that subdivision ordinances grew into development codes. Increasing conditions have been imposed through subdivision controls (exactions, dedications, impact fees). For further discussion see Freilich and Levi (1975), and Callies and Grant (1991: 229). The line between zoning and subdivision is currently of little importance for our purposes.

2. The British planning provisions are, of course, much more complicated than this might suggest (and they are subject to constant revision). There is further discussion in Chapter 15.

3. There are some notable exceptions which are discussed later; see particularly Chapter 5.

4. A famous statement of the nature of the police power was made by Justice Harlan in *House v Mayes* (quoted in Metzenbaum 1955: 82):

[The police power] "is not granted by or derived from the federal constitution, but exists independently of it, by reason of its never having been surrendered by the state to the federal government." The police power is "the power to regulate the relative rights and duties of all within its jurisdiction as to guard the public morals, the public safety, and the public health, as well as to promote the public convenience and the common good".

5. The Standard State Zoning Enabling Act is reproduced in several legal texts, e.g. Callies and Freilich (1986: 5–8), and Mandelker and Cunningham (1990: 168–90).

6. Bair (1984: 120, fn.1) notes that an early version of the Standard State Zoning Enabling Act included the following note: "A comprehensive plan: sound planning implies a comprehensive plan. The zoning should be applied to the whole municipality at once. Piecemeal zoning is dangerous, because it treats the same kind of property differently in the same community". See also Mandelker (1976b: 902 and 1988: 82).

7. The California legislation requires zoning ordinances to be consistent with a comprehensive plan. Unusually, the term "consistency" is defined. The definition includes the requirement that "the various land uses authorized by the ordinance are compatible with the objectives, policies, general land uses, and programs specified in such a plan" (California Code S.65860). The Florida legislation provides a fuller definition:

A development approved or undertaken by a local government shall be consistent with the comprehensive plan if the land uses, densities, or intensities, capacity or size, timing, and other aspects of the development are compatible with and further the objectives, policies, land uses, and densities or intensities in the comprehensive plan and if it meets all other criteria enumerated by the local government.

(Mandelker 1988: 85).

8. *Montgomery County v Woodward and Lothrop* 1977, and *Chapman v Montgomery County* 1970; both cited in *West Montgomery County Citizens Association v Maryland National Capitol Park and Planning Commission* 1987.

9. Special districts take many forms. Their number has increased dramatically in recent years in response to "tax revolts" and the like: "apparently, voters who stand to benefit directly from improvements made by a special district are more willing to pay for those improvements than for higher property taxes" (Porter, Lin, and Peiser 1987: 1).

Confusingly, the term "special district" is also used for a special zoning area. This is discussed in Chapter 4.

10. A discussion of the variety of administrative arrangements is to be found in So and Getzels (1988: 403). For a detailed legal guide to the provisions in one state (New Jersey), see Kienz (1990).

11. *Ginzburg v U.S.* 1966.

12. For a discussion of "The Economics of Public Use" see the paper with that title by Merrill (1988). A blistering attack on the judicial history of "public use" is to be found in Paul 1987.

13. The projected cost of the condemnation to Detroit was over $200 million; the land was conveyed to GM for slightly more than $8 million. It should be recorded that there were vigorous dissents: Judge Ryan wrote that "With

this case the Court has subordinated a constitutional right to private corporate interests. As demolition of existing structures on the future plant site goes forward, the best that can be hoped for, jurisprudentially, is that the precedential value of this case will be lost in the accumulating rubble". See Merrill (1988: 147), and also Wylie (1989).

14. "The fact that in a specific case the city commission could argue that it had, in fact, considered the listed criteria only and that there had therefore been no arbitrary action would not remedy the constitutional defect of the ordinance. It is the opportunity for arbitrary action, not the fact itself, that renders an ordinance void for vagueness" (Blaesser and Weinstein 1989: 37).

15. The *Euclid* case provides a good example. In accepting the constitutionality of the zoning ordinance, the court noted that "the great increase and concentration of population" had led to increasing problems "which require, and will continue to require, additional restrictions in respect of the use and occupation of private lands in urban communities. Regulation, the wisdom, necessity and validity of which, as applied to existing conditions, are so apparent that they are now uniformly sustained, a century ago, or even half a century ago, probably would have been rejected as arbitrary and oppressive ... while the meaning of constitutional guarantees never varies, the scope of their application must expand or contract to meet the new and different conditions which are constantly coming within the field of their operation."

More generally on the manner in which the Supreme Court changes with the times, see Goldman and Jahnige (1985, chapter 1, especially pp. 6–7), McCloskey (1960), Bickel (1962), and Mason (1968).

16. There are 111 federal courts and 18,252 state and local courts (Holland 1988: 11). In 1982, more than *25 million* civil and criminal cases were filed (Goldman and Jahnige 1985: 15).

17. See, for example, Freund (1904), Dunham (1958), and Sax (1964). There is a useful summary in Mandelker (1988: 23–29).

18. Cases may be "facial" or "as applied". In the *Metromedia* case the challenge was a facial one,

i.e. the ordinance was unconstitutional in itself. The alternative is a challenge, not to the ordinance itself, but as it is applied in particular circumstances. The distinction is well illustrated by the cases of *Euclid* and *Nectow*, discussed in the following chapter.

19. A contrasting case which raised the same issue was *Linmark Associates v Township of Willingboro* 1977. Here the Supreme Court unanimously concluded that a prohibition of "For Sale" and "Sold" signs violated the First Amendment. (The prohibition was aimed at stopping "white flight.")

20. There has been much criticism of the aversion of the courts to interference in land use planning cases. Sager (1978: 1421) comments (in relation to the Supreme Court) that "the court's willingness to evaluate land use policies under the due process clause seems restricted to actual fits of municipal madness; mean and self-serving acts of exclusion are apparently to be received as jeweled exercises of the police power." In another paper (Sager 1979) he argues that state courts – some of which were then becoming more active in "curbing municipal abuses of the zoning power" – should be more "vigorous" in judging zoning cases in the context of state (rather than federal) constitutions. State courts should "approach decisions of the Supreme Court in the land use area as skeptics rather than as the devout".

The most famous case based explicitly on a state constitution – to preclude federal review – is *Mount Laurel*: see Chapter 5.

21. See particularly *Fasano v Board of County Commissioners* 1973, and *Neuberger v City of Portland* 1979. The *Fasano* case is discussed in the Introduction.

22. Blaesser and Weinstein (1989: 41). There is much controversy on this issue. Of particular interest is C.M. Rose (1983: 846) who argues that "piecemeal local land decisions should not be classed as either 'legislative' or 'judicial'; these rubrics are drawn from a separation-of-powers doctrine more appropriate to larger governmental units. Piecemeal changes are quintessentially local matters, and any jurisdictional test of the reasonableness of piecemeal changes must identify and build upon factors that lend legitimacy and institutional competence to local decisionmaking."

2 Historical background

1. Stoebuck actually wonders "how it [the just compensation clause] got into our constitutions at all" (quoted in Bosselman, Callies, and Banta 1973: 100).
2. Bosselman, Callies, and Banta (1973: 105) note that "after the adoption of the federal Bill of Rights, the various states gradually amended their own constitutions until by mid-century the great majority of state constitutions contained taking clauses." See also Grant (1931: 70).
3. The case is quoted in Bosselman, Callies, and Banta (1973: 106). The authors comment that "this case is perhaps the earliest example of judicial approval of a governmental regulation of the use of land designed to promote environmental enjoyment."
4. The New York ordinance is perhaps the best known, but it was not the only one that was passed around this time. See M. Scott (1969: 152); and, for a succinct overview, Gerckens (1988: 34). See also Fogelson's account of Los Angeles (1967: 247).
5. There is here a striking difference from the stand taken in cases such as Hadacheck.
6. "The committee was heavily composed of men who had come out of the urban reform movement. Three of them (Edward Bassett, Lawrence Veiller, and Nelson P. Lewis) had been intimately involved in the New York effort" (Toll 1969: 201).
7. See also Arnold (1972) and Hawley (1974).
8. Boyer later (318) gives a reference to an article entitled "Does Your City Keep its Gas Range in the Parlor and its Piano in the Kitchen?" (Swan 1920).
9. The promotion of home ownership was considered desirable both as a means of dealing with some aspects of the housing problem and as a guard against social unrest. The logic was unflawed: "Many ... men although willing to acquire homes, were afraid to do so lest they later ruined their investment if an apartment, stable, laundry or public garage were built next door. ... Big industries and businessmen therefore have good reason to work for the establishment of protected residential zones, as a definite encouragement to home ownership and to more stable labor conditions" (Cheney 1920a: 32).

The number of nonfarm homeowning households increased by 3.5 million during the 1920s (Goldfield and Brownell 1990: 291).

10. The inside story of the *Euclid* case is a fascinating one which has become almost a specialized field of historical research. The most ambitious work is Haar and Kayden (1989a); a brief overview is given in M. Scott (1969: 237). See also McCormack (1946) and Flack (1986).
11. "The Euclid ordinance is almost an exact duplicate of the New York City Zoning Ordinance, except as to local names and locations" (Metzenbaum, *Brief on Behalf of Appellants*, quoted in Toll 1969: 231).
12. R.H. Nelson (1977: 26) suggests that the Supreme Court became much more circumspect about interfering with legislative enactments partly because of the adverse reaction to the Court's intervention in overturning the New Deal legislation. "The result was that the courts declined except in all but the most extreme cases of abuse to consider the merits of challenges to community zoning ordinances." Some state courts, however, did lay down an extensive body of law upholding zoning provisions which went far beyond the initial ideas of districting. This will be readily apparent from the discussion in the following chapter.

3 The objectives and nature of zoning

1. *Borough of Cresskill v Borough of Dumont* 1953. See Mandelker (1988: 135), where the higher court case is cited: *Borough of Cresskill v Borough of Dumont* 1954. The latter case is excerpted in Wright and Gitelman (1982: 805).
2. The Standard State Zoning Enabling Act includes the following provisions:
    (i) *Grant of Power.* For the purpose of promoting health, safety, morals, or the general welfare of the community, the legislative body of cities and incorporated villages is hereby empowered to regulate and restrict the height, number of stories, and size of buildings and other structures, the percentage of lot that may be occupied, the size of yards, courts, and other open spaces,

the density of population, and the location and use of buildings, structures, and land for trade, industry, residence, or other purposes. (ii) *Districts*. For any or all of said purposes the local legislative body may divide the municipality into districts of such number, shape, and area as may be deemed best suited to carry out the purposes of this act; and within such districts it may regulate and restrict the erection, construction, reconstruction, alteration, repair, or use of buildings, structures or land. All such regulations shall be uniform for each class or kind of building throughout each district, but the regulations in one district may differ from those in other districts.

3.   See also the *Ladue* case referred to in Chapter 1.

4.   Justice Marshall, in his dissent, stated:

I think that the First Amendment provides some limitation on zoning laws. It is inconceivable to me that we would allow the exercise of the zoning power to burden First Amendment freedoms, as by ordinances that restrict occupancy to individuals adhering to particular religious, political, or scientific beliefs. Zoning officials properly concern themselves with the uses of land – with, for example, the number and kind of dwellings to be constructed in a certain neighborhood or the number of persons who can reside in those dwellings. But zoning authorities cannot validly consider who those persons are, what they believe, or how they choose to live, whether they are Negro or white, Catholic or Jew, Republican or Democrat, married or unmarried.

5.   For a fuller discussion see Dear and Wolch (1987). A comprehensive study by Lester D. Steinman concluded that: "nationwide, recent legislative and judicial developments generally manifest acceptance of the group home as a permitted use in residential zones. However, persistent community opposition to the location of such uses, particularly in single family neighborhoods, will continue to fuel litigation" (Steinman 1988). See also Dear and Taylor (1982), U.S. General Accounting Office 1983 and Gordon and Gordon (1990).

6.   A significant part of the "expert evidence" was the testimony of the Committee on the Hygiene

of Housing of the American Public Health Association. This discussed the relationship between the adequacy of living space and marital and emotional health. (See Haar 1953: 1056.)

7.   Haar (1953), Nolan and Horack (1954), Haar (1954), and Williams and Wacks (1969).

8.   Mobile homes in the United States are often far from mobile, as the discussion in this chapter shows. In many other countries the term has a meaning nearer to what it suggests, i.e. a home that is "mobile". In Britain, "the term applies to any structure which is designed or adapted for human habitation and can be moved from one place to another by towing or other motor transportation". See Dyer (1991: 4).

9.   See the condemnation of apartments by Justice Sutherland in the *Euclid* case: Chapter 2 above.

10. Unfortunately, the court did not indicate what rights the owners of the Girsh property had as a result of its decision. Nether Providence subsequently zoned several pieces of land – but not the Girsh property – for apartments, claiming that it had thereby complied with the court's decision. The Girsh property owners disagreed, and after two years won a clarifying order from the court which directed the township to grant the permits required for the development of their site. In the meantime, the township had begun procedures to condemn the Girsh property for a public park (Krasnowiecki 1972: 1080). This is a typical example of the way in which the drama of land use disputes is played out.

11. For a more extended discussion of mobile homes and manufactured housing see Bair (1968), Flynn (1983), and Jaffe (1983).

12. See, for example, a report by the State of California (1986a).

13. See also the book-length treatment of the subject (entitled *Wheel Estate*) by Wallis (1991).

4 Zoning with a difference

1.   The adjective "special" is necessary to distinguish this category from an unadorned "exception." "The latter usually involves a non-discretionary determination in that the ordinance simply

provides that certain types of structures or uses are excluded from the application of all or part of the ordinance" (Wright and Wright 1985: 151).

2. "Conditional rezoning", otherwise known as "contract rezoning", is also discussed later in this chapter.

3. New Jersey Municipal Land Use Law 40: 55D-67.

4. Williams comments:

> There is a curious dualism about the use of the phrase "downzoning", which reflects (perhaps unconstitutionally) the users' views as to what is really important. Under one view (apparently particularly in the Northeast) "downzoning" means making the regulations less restrictive – that is, reducing the protection available to neighbors. Under the opposite view (apparently particularly in the West) "downzoning" means reducing the developer's opportunities for profitable intensive land use.
>
> (Williams 1985–90)

Rathkopf and Rathkopf note the confusion and comments:

> Confusion is sometimes engendered by the labeling of residential and sparser intensities as "higher" land uses, and less desirable activities as "lower" land uses – an ordering which is sometimes reflected in the hierarchy of ordinance designations themselves. In point of fact, this conceptualization of higher and lower uses is the opposite of the ordering implied by upzoning and downzoning.
>
> (Rathkopf and Rathkopf, 1988: 27.05)

See also Bartke and Lamb (1976).

5. Terminology is confused. Some courts use the terms interchangeably; others distinguish between them, using "conditional" to refer to a case in which conditions are accepted by an owner but with no commitments on the part of the local government, i.e. a unilateral agreement (see Shapiro 1968: 269); by contrast "contract" implies a bargain between the owner and the local government (i.e. a bilateral agreement). However, the distinction is tenuous: "what appear to be unilateral conditions have often been hammered out over a period of time involving numerous discussion sessions" (Wright and Wright 1985: 144). Moreover, "unless developers believe that local legislative and

administration officials will fulfill their promises to rezone and grant the necessary permits, they will have no inducement to enter into such an agreement" (Rathkopf and Rathkopf 1988: 27.05). Meshenberg (1976: 11) notes that "the distinction between conditional zoning and contract zoning is fuzzy and seems to revolve around which is emphasized more, the conditions or the contract." A more recent term is "concomitant agreement zoning": see Brown (1986) and also Wegner (1987: 977) who uses the generic term "contingent zoning". For our purposes, the term "conditional" will suffice. (There is further discussion of this matter in Chapter 7.)

6. New Jersey Municipal Land Use Law S.40: 55D-41.

7. On PUDs generally see Tomioka and Tomioka (1984).

8. The distinction between cluster zoning and PUDs is neither clear nor generally accepted, but a PUD is supposedly distinctive because of the legal framework which it provides. "Consequently, unlike clustering, PUD represents a substantive alternative to the use of traditional zoning regulations, and can be adapted to the development of commercial and industrial uses as well" (Moore and Siskin 1985: 5). However, a Minnesota State Planning Agency report (1979) notes that "the term PUD has as many definitions as there are people defining the term".

9. According to Yannacone, Rahenkamp, and Cerchione the "basic elements" of impact zoning include:

1. Determination of the capacity of the ecological, societal, and economic systems of the community and its region to accommodate existing and future growth;
2. Identification and analysis of natural, societal, and economic constraints upon development;
3. Formulation and enumeration of community goals for future growth and development;
4. Legal analysis of the extent of local land use regulatory authority.

> (Yannacone, Rahenkamp, and Cerchione 1976: 441)

One Massachusetts town has incorporated a requirement for an "impact statement" in its zoning ordinance:

In order to evaluate the impact of the proposed development on town services and the welfare of the community, there shall be submitted an Impact Statement which describes the impact of the proposed development on (1) all applicable town services, including but not limited to schools, sewer system, protection; (2) the projected generation of traffic on the roads of and in the vicinity of the proposed development; (3) the subterranean water table, including the effect of proposed septic systems; and (4) the ecology of the vicinity of the proposed development. The Impact Statement shall also indicate the means by which town or private services required by the proposed development will be provided, such as by private contract, extension of municipal services by a warrant approved at town meeting, recorded covenant, or by contract with homeowner's association.

(Haar and Wolf 1989: 592)

10. See Kendig (1980), and Porter, Phillips, and Lassar (1988). Performance zoning for industrial sites became popular in the 1960s (Mandelker and Cunningham 1990: 275). An example of the use of "performance standards" in zoning is given in Haar and Wolf:

All industrial operations in any district shall, as a condition of permitted use, emit no obnoxious, hazardous or annoying odors, but shall, by the installation and operation of suitable deodorizing equipment, totally eliminate or diminish such odors to the highest possible degree.

(Haar and Wolf 1989: 277)

11. For further discussion of negotiations and agreements see Chapter 7.

12. The Standard State Zoning Enabling Act made no provision for floating zones, and thus there is no general rule as to which zoning agency shall approve them. Hagman (1980: 569) notes that: "The power to change zoning, regarded as a legislative matter in most states, is delegated to administrative bodies in a few eastern states."

13. See *Rodgers v Village of Tarrytown* 1951, and *Miss Porter's School Inc. v Town Plan and Zoning Commission* 1964.

14. See, for example, Callies and Freilich (1986: 283). A useful discussion of the case, and the implications for "referendability" is to be found in Callies, Neuffer, and Caliboso (1991), from which the quotation is taken.

15. There is a succinct discussion of initiative and referendum in Callies, Neuffer, and Caliboso (1991). For an extended account see Caves (1991).

16. R.H. Nelson points out that though neighborhood zoning "generally protects well-off people from the entry of the less well-off", sometimes it can work the other way round:

Consider SoHo, a New York neighborhood with old manufacturing lofts, that has attracted many artists. Because a sizable number of well-to-do New Yorkers like to live in neighborhoods with artists, the artists of SoHo were threatened by an influx of nonartists, with possible eviction from the neighborhood because of increased rent levels, a situation that had already occurred in nearby Greenwich Village. Following the example of many suburban neighborhoods, the SoHo artists decided to meet the threat by establishing restriction on entry. In 1971, they persuaded the New York City government to establish an Artists Certification Committee that would certify applicants as professional artists and then to make such certification a zoning requirement for SoHo residency.

(R.H. Nelson 1977: 15)

5 Exclusionary zoning and affordable housing

1. Excluded from the discussion is the (now rare) provision of public housing, and other housing policies such as housing vouchers – though the latter can form a very useful element in a coordinated policy for the provision of affordable housing.

2. The quotation is from the *lower court* case: *Ambler Realty Co. v Village of Euclid* 1924.

3. The Housing Act of 1949 proclaimed the goal of "a decent home and suitable living environment for every American family", but it did not (as, for instance, did the comparable British legislation) impose a duty on anybody to do anything.

4. See, for example, Haar (1953 and 1954), N. Williams (1955), Douglas Report (1968), Kaiser Report (1968), N. Williams and Wacks (1969), N. Williams and Norman (1971), Aloi and Goldberg (1971), Davidoff and Davidoff (1971), and Brooks (1972).

5. Reilly and Schulman (1969), Danielson (1976: 306), Osborn (1977), Zimmerman (1983: 99), and Heiman (1988).

6. Sherer (1969), Vaughn (1974), Danielson (1976: 300), and Krefetz (1979).

7. The cases were *Fisher v Bedminster Township* 1952, *Lionshead Lake Inc. v Wayne Township* 1952, and *Vickers v Township of Gloucester* 1962.

8. For a useful discussion see Listokin (1976: 16), and Burchell *et al.* (1983: 4).

9. This 1971 case should not be confused with the identically named 1977 case held after the *Mount Laurel* decision. It is the later case which has received attention in the legal literature. This account of the earlier case is based on Burchell *et al.* (1983: 7).

10. Several state courts had held that municipalities had an obligation to consider regional housing needs in their zoning ordinances. A list is given in Kayden and Zax (1983: 146).

11. The high points in the saga are the three New Jersey State Supreme Court decisions: *Southern Burlington County NAACP v Township of Mount Laurel* 1975 ("Mount Laurel I"); *Southern Burlington County NAACP v Township of Mount Laurel* 1983 ("Mount Laurel II"); and *Hills Development Company v Township of Bernards* 1986 ("Mount Laurel III"). A useful collection of articles on *Mount Laurel I* is to be found in Rose and Rothman (1977).

12. A 1976 survey showed the prevalence of large lot zoning in New Jersey. See Seidel (1978).

13. The complaint was written entirely in terms of racial exclusion. The court, however, ignored this, and chose to deal with it as a case against discrimination on economic grounds: see N. Williams (1984a: 832).

14. For a short and informative account of Governor Cahill's (and earlier) initiatives at the state level in New Jersey, see Listokin (1976: 87). Danielson (1976: 298) quotes Governor Cahill as warning the legislature in 1972 that "unless we act together to help open the way for needed housing, the courts will do it for us and will

continue to move strongly in the direction of bypassing home rule by judicial process".

15. It is not easy to obtain relevant figures of court cases, and reliable statistics of dwellings provided as a result of the *Mount Laurel* litigation are unavailable. The figures in the text are from McDougall (1987: 623, fn 4). Chall (1985: 19, fn 2) reports that, by 1985, 135 lawsuits had been brought under *Mount Laurel II*. This figure was taken from a press release of June 10, 1985, issued by the New Jersey Administrative Office of the Courts. Chall notes that there were "other related Supreme Court rulings", but he does not discuss them in his paper.

16. Kent-Smith (1987: 939, fn 72).

17. Information from New Jersey Council on Affordable Housing, April 2, 1991.

18. The *Hills* case "has had a devastating effect on preexisting settlements, including Mount Laurel Township, which is appealing its quota of 950 units, achieved through twelve years of litigation". However, the court has refused transfer of one case (*Haveis v Far Hills* 1986) to the Council on the grounds of municipal "opportunism and obstructionism" (Kent-Smith 1987: 947 fn 136).

19. See, for example, Citizens Commission on Civil Rights (1983), (U.S.) Commission on Civil Rights (1983), Lake (1981), R.A. Smith, (1985), Metcalfe (1988), Aoki (1989), Schwemm (1989 and 1990), and U.S. HUD (1991). For an account of the British experience see MacEwen (1990).

20. One commentator has stated that the decision is "well outside the mainstream of modern legal development" (Baade 1974: 441, fn 12).

21. For a succinct discussion of the legal issues raised by the inclusionary zoning concept see Mallach (1984: 28–37). A fuller discussion, with a summary of the comments of a symposium panel, is to be found in chapter 4 of Merriam, Brower and Tegeler (1985).

22. California Government Code, section 65915 (a); see also California (1990a).

23. In *United States v Certain Lands in Louisville* 1935. (N. Williams 1985–90: 67.05, fn 45) describes this decision as "appalling".

24. Rosenberg's analysis is devoted to civil rights; abortion and women's rights; and the environment, reapportionment, and criminal

law. It is, however, clear that his conclusions apply equally in the field of land use regulation.

25. Rosenberg (1991: 343). But what, the reader might ask, of the successes of the court in the civil rights movements? After careful analysis, Rosenberg concludes that the court was a follower, not a leader:

> The combination of all these factors – growing civil rights pressure from the 1930s, economic changes, the Cold War, population shifts, electoral concerns, the increase in mass communication – created the pressure that led to civil rights. The court reflected that pressure; it did not create it.
>
> (Rosenberg 1991: 169)

26. See also the conference papers on "The Fair Housing Act After Twenty Years" in *Yale Law and Policy Review* (1988, 6: 331–92).

6 Development charges

1. Some writers use the term exactions to refer to land dedications or other in-kind contributions, while development fees are monetary payments (Snyder and Stegman 1986: 26, fn 21). In A.C. Nelson (1988b) the authors use the term development impact fees to refer to payments for facilities located outside the development site on which they are levied, but others use the term in a more generic way. Other terms include infrastructure fees, system development charges, capital facility fees, building and occupancy fees. Sometimes a term is used in a specialized sense; other times the terms seem to be interchangeable. It is perhaps significant that Burrows' *Survey of Zoning Definitions* (1989) excludes reference to any of these.

    For a brave attempt to bring some order into this confused field see Alterman and Kayden (1988). The clearest statement is to be found in Callies and Grant (1991: 222).

2. For an interesting short history see Smith (1987). Callies and Grant (1991) provide a succinct and clear account together with an insightful comparison with British parallels.

3. U.S. Department of Commerce (1928) *A Standard City Planning Enabling Act*.

4. There is an enormous, and continually growing, literature on impact fees. Currently, the best single source (both of information and of references) is A.C. Nelson (1988b); but see also Blaesser and Kentopp (1990) for a detailed review of the development of the law relating to impact fees.

5. For a discussion of the range of attitudes taken by the courts, see Snyder and Stegman (1986, chapter 5, especially pp. 56–7), and Callies and Freilich (1986: 372–5).

6. See, for example, *Jordan v Village of Menomonee Falls* 1965, *Contractors and Builders Association of Pinellas County v City of Dunedin* 1976, *Banberry Development Corporation v South Jordan City* 1981, *Home Builders and Contractors Association of Palm Beach v Board of County Commissioners* 1983, *Nollan v California Coastal Commission* 1987.

    The 1987 Texas statute *Impact Fees for Capital Improvements or Facility Expansions* provides a detailed plan for the operation of the impact fee system in Texas. Impact fees can be levied only for specified purposes.

7. See for example, Nicholas (1988). A useful survey by a British observer, published after this typescript was complete, is Delafons (1990a).

8. For a revealing account of some of the problems as experienced in Florida, see Bosselman and Stroud (1985).

9. Others include *Keystone Bituminous Coal Association v DeBenedictis* 1987, and *First English Evangelical Lutheran Church of Glendale v County of Los Angeles* 1987.

10. Among the large number of commentaries and analyses of *Nollan*, see Epstein (1987), Freilich and Morgan (1987), Best (1988), Freilich and Morgan (1988), R.A. Williams (1988), Curtin (1989), *Harvard Law Review* (1989), Kelly (1989), Lawrence (1989), Roddewig and Duerksen (1989), and Crew (1990). For a discussion of "The First Applications of the *Nollan* Nexus Test" see Lemon, Feinland, and Deihl (1990).

11. Ellickson (1977: 399, fn 34) gives a list of examples from both court decisions and the academic literature.

12. The complex problem of calculating the impact of growth controls (restrictions and charges) and zoning regulations on house prices is not

addressed directly here. There is a large literature on the subject, though many studies are limited to a local area and ignore the impacts over the larger housing market area. See, for instance, Sagalyn and Sternlieb (1973), Danielson (1976), Muth and Wetzler (1976), Davies (1977), Franklin and Muller (1977), Greenspan (1978), Seidel (1978), White (1978), Frieden (1979), Case and Gale (1981), Elliott (1981), Nicholas (1981), Peiser (1981), Dowall (1984), Schwartz, Hansen, and Green (1981 and 1984), Segal and Srinivasan (1985), Weitz (1985), Lillydahl and Singell (1987), Davidson and Usagawa (1988), Brueckner (1990), and Pollakowski and Wachter (1990).

13. The program was revised in 1985 when it was renamed the Office-Affordable Housing Production Program (OAHPP). Details are given in Goetz (1989: 75, endnote 1). At the date of Goetz's paper, there had been few new office approvals in San Francisco since 1985. The new program thus could not be evaluated.

It is not self-evident that offices should be picked out as the sole type of development to be subject to the linkage policy. It "seems to be more a consequence of political expediency than social or economic rationale. It reflects the nature, and the perception by the public, of new office developments as large, discrete structures, conveniently visible, politically unpopular, and relatively few in number" (Dawson and Walker 1990: 160).

14. The program was applicable to projects of over 50,000 square feet until an ordinance of March 1990 lowered it to 25,000 square feet (Taub 1990: 684).

15. Hausrath (1988: 215). The reasoning underlying these calculations, and a more extended discussion of them is to be found in Hausrath (1988). An updating note is in Taub (1990: 683).

16. See, for example, Ontario Ministry of Treasury and Economics, *Financing Growth-Related Capital Needs*, 1988, and the Development Charges Act 1989. A useful legal text which includes the latter is Macaulay and Doumani (1991).

7 Planning by agreement

1. The account does not deal with the fascinating and complex story of downtown commercial and redevelopment agreements. On this see, for example, Schultz (1977), National League of Cities (e.g. NLC 1979), Fossler and Berger (1982), New York (1982b), Fainstein *et al.* (1986), Mollenkopf (1983), Moore (1983), Farr (1984), MIT Center (1984), Levitt and Kirlin (1985), Levitt (1987), Lassar (1990); and also Frieden and Sagalyn (1989), which, in addition to being an excellent analysis, has a wonderful bibliography.

2. For a review of cases see Hagman (1979). See also Siemon, Larsen and Porter, (1982). An illustrative case is *Avco Community Developers Inc. v South Coast Regional Commission* 1976; see Callies and Freilich (1986: 178).

3. For a comparison of the California, Hawaii, Florida, and Nevada development agreement schemes see Wegner (1987: 995). There is a discussion of the Florida scheme in Rhodes (1988), and in Taub and Rhodes (1989). On the Hawaii scheme see Callies (1989b).

4. The certainty is not absolute. For instance, a local government cannot freeze the laws of a higher jurisdiction. In addition a local government can modify an agreement if the property owner does not comply in good faith with the terms of the agreement.

5. The "subsequent reimbursement" provision presumably "contemplates reimbursement from subsequent developers on a pro rata basis" (Goldwich 1989: 251, fn 13).

6. For a detailed discussion of one particular agreement (Colorado Place) see Fitzgerald and Peiser (1989).

7. Whyte (1988: 233). Unfortunately, Whyte does not reveal the basis for this intriguing comparison.

8. For an excerpt from New York's code relating to the urban plaza bonus see Lassar (1989: 131). See also New York (1989b) and, more generally, New York (1990).

9. The lesson may have been clear, but this does not mean that it was learned – as the Columbus Center project, discussed below, shows. See also New York (1988a) and (1989a).

10. This raises the interesting question as to whether

there should be a right of appeal against a bonusing decision. During the course of a discussion of this, Bermingham (1988: 19) comments that "I do not believe that anyone could force a municipality to enter into the bonusing 'game' but once into the game of its own volition, the municipality will have to deal fairly and equitably with the land owners who are players. The absence of any effective appeal right seems to me to turn the bonusing scheme from a planning exercise into a sort of bidding war".

11. *Municipal Art Society of New York v City of New York* 1987. For a succinct account see Lassar (1989: 34), and Babcock (1990: 23). (Babcock provides brief outlines of other similar projects in Richmond, Virginia, Albuquerque, New Mexico, and San Diego, California.) A detailed bibliography of the Columbus Circle controversy is to be found in Frieden and Sagalyn (1989: 345).

    The Municipal Art Society of New York is "a well-established, nonprofit organization interested in city design and historic preservation issues, and a frequent opponent of large real estate development projects" (Kayden 1990: 103).

12. For a discussion of the legal issues involved see Kayden (1990: 101). The city eventually altered its plans for the area, and the Municipal Art Society dropped its suit. However, other suits were brought alleging violations of air quality standards and of procedures used for developer selection (Babcock 1990: 29).

13. Curiously, Columbus Circle figures in an earlier chapter in the eventful history of planning in New York: The Coliseum (which occupies the site) was developed as a slum clearance project – "a classic case of abuse of the legislation", according to Weiss:

    Columbus Circle at 57th and Broadway was a valuable commercial site in the early 1950s when Moses induced the New York City Planning Commission and the Department of Slum Clearance and Urban Redevelopment [DSCUR] to approve it as a Slum Clearance Project. This designation was based on the argument that a small number of aging tenements at the far end of the project's carefully drawn boundaries, constituting less than one percent of the total property value

of the project area, were "substandard" and "insanitary".

    The redevelopment plan for the two-block area called for the construction of a commercial exhibition hall (the New York Coliseum) occupying fifty-three percent of the site, and a luxury high-rise housing development occupying the other forty-seven percent. Since the "predominantly residential" rule [imposed by the legislation] was defined by the DSCUR as being at least fifty percent of the total square footage of the project area, Moses needed to tip the balance by three percent. He announced that the tenants of the new apartment building could park their cars in the Coliseum's underground garage if their own parking lot was full. The New York City Planning Commission and DSCUR then designated 18,000 square feet of the Coliseum's underground garage as "residential" which made the entire project "predominantly residential" in its reuse.

(Weiss 1985: 267)

14. Low-income housing is discussed separately in Chapter 5.

15. There has been an accompanying increase in publications. See particularly Cibulskis and Ritzdorf (1989) and their bibliography. See also Caplan (1989), and the publications of the "Children and Families in Cities Project" of the National League of Cities (e.g. NLC 1988); also Cohen (1989).

16. One notable exception has been the International Business Machines Corporation. The *New York Times* reported on December 12, 1990, that the company had announced its decision to spend $3 million in 1991 to build child-care centers near its offices and plants around the country to serve 530 preschool children. The director of IBM's Work/Life program (*sic*) commented that "the number of male employees at IBM had declined by twenty percent over the last thirty years, while the number of female employees had tripled, so child care had become a vital program to keep women in the work force".

17. The *New York Times* of November 6, 1990 reports on several cases where "entrepreneurial mothers are tangling with puzzled local officials

over zoning and day care: the outcome can be community resentment and, at worst, the shutdown of scarce day care centers that leaves hard pressed parents scrambling for alternatives". One example of community resistance included a complaint that playground equipment and grass in common areas would wear out, leading to an increase in maintenance fees. In this particular case, opponents used scare tactics such as staking out the house in which a small day-care center was operating and photographing automobile license plates as children were dropped off or picked up.

18. These states are merely illustrative. Many others have taken similar action. The source is Cibulskis and Ritzdorf (1989: 12).

8 Aesthetics

1. Lord St John of Fawsley, in the Introduction to Judy Hillman's *Planning for Beauty* (1990).
2. See, for example, Dewey (1934), Berger and Luckmann (1966), and the wide-ranging discussion in Costonis (1982). See also Dukeminier (1955), Crumplar (1974), and S.F. Williams (1977).
3. Strictly speaking, the aphorism is "where there's muck there's brass".
4. In Maryland, Texas, and New Jersey, the courts for many years refused to permit use zoning which prohibited commercial establishments from residential areas. It was held that such exclusions were primarily an aesthetic matter. See N. Williams (1985–90: 11.05 and 84.04).
5. For an analysis of the relevant decisions in each state see Bufford (1980).
6. The same logic is also applied to pornography.
7. For an account of Lady Bird Johnson's role in the highway beautification movement see Gould (1988).
8. Draft federal guidelines were opposed and watered down to such an extent that they became completely ineffective: for instance, the guideline for the size limit of billboards was 650 square feet – a size larger than all but 2 percent of all existing signs. The guidelines were eventually abandoned and replaced by a model agreement drawn up by the Federal Highway Administration in cooperation with the Outdoor Advertising Association of America, which states could use, if they so wished, during the negotiation process. In this the size limit was increased to 1,200 square feet (Floyd 1979b: 117).

9. The first billboard removed under the Highway Beautification Act was in Shreveport, Maine in May 1971 (ten months after the "final compliance date" which was originally set for the removal of *all* nonconforming signs). The compensation paid amounted to $6,000. The property tax valuation was $820 (Floyd 1979b: 122).

10. In 1990, the Bush administration rescinded a policy which had allowed tree cutting in front of billboards for the sole purpose of improved billboard visibility. This action was in line with the President's avowed goal of "planting millions of trees across the country" (*Scenic America: Sign Control News*, Summer 1991, 3).

11. *State ex rel Saveland Park Holding Corporation v Wieland* 1955.

12. See Chapter 1 above. A later case, *Members of City Council v Taxpayers for Vincent* (1984), clarified some of the problems of the Metromedia case. A succinct account of the *Metromedia* and *Vincent* cases is to be found in Mandelker (1988: 424–7).

13. It has also been argued that a "guidebook" does not help much: "For the competent designer a handbook on design is unnecessary, and for the incompetent it is almost useless as a medium of instruction" (Holford 1953: 71). In similar vein, Barnett (1982: 10) has commented that "cities are not designed by making pictures of the way they should look twenty years from now. They are created by a decision-making process that goes on continuously, day after day".

14. See Chapter 4, p. 59.

15. The Lake Forest ordinance embraces *both* the "excessive difference" and the "excessive similarity" rules:

> 'The City Council hereby finds that excessive similarity, excessive dissimilarity or inappropriateness in exterior design and appearance of buildings ... in relation to the prevailing appearance of property on the vicinity thereof adversely affects the desirability of immediate and neighboring

areas and impairs the benefits of occupancy of existing property in such areas, impairs the stability and taxable value of land and building in such areas . . . and destroys a proper balance in relationship between the taxable value of real property in such areas and the cost of the municipal services provided therefor.'

> (Lake Forest Building Code, quoted in Poole 1987: 305)

16. N. Williams notes that the "economics-plus-aesthetics" rationale is illustrated by the fact that "After all, conventions do not go to New Orleans to enjoy the salubrious climate, and they *do* go to Vermont to enjoy the mountain views in the background." He adds:

> A particularly striking example of this economics-plus-aesthetics rationale, in the realm of political action, took place in Vermont. In the late 1960s, when the Legislature was considering the bill to prohibit (*and* to remove) billboards in Vermont, the local motel and hotel association strongly supported it and helped its passage: they understood that no one was going to come up and stay in their motels to look at a row of billboards.

> (N. Williams 1990: 4)

9 Historic preservation

1. The acquisition of the national battlefield at Gettysburg gave rise to the first significant legal case concerning preservation and the important Supreme Court ruling that the acquisition was for a valid public purpose. (*United States v Gettysburg Electric Railway Company* 1896).

2. The 1906 Act was declared to be unconstitutionally broad in *United States v Diaz* 1974. It was replaced and supplemented by the 1979 Archeological Resource Protection Act (Dworsky *et al.* 1983: 238).

3. New Orleans was actually the first to establish a preservation commission in 1921, but its legal basis was inadequate. A new and strengthened commission (still in existence) was set up after the state of Louisiana amended the state constitutional provisions in 1936 (Duerksen 1983: 6).

4. The ACHP is composed of the heads of federal agencies whose departmental activities regularly affect historic properties; experts in historic preservation; a governor; a mayor; private citizens appointed by the President; and representatives of the National Trust and the Conference of State Historic Preservation Officers. For a useful discussion of the role of the Council see Storey (1987).

5. The Register is separate from the earlier list of National Historic Landmarks. A national landmark, as its name suggests, is a property of *national* significance such as Mount Vernon and the Alamo. There are some 1,600 National Historic Landmarks, and over 50,000 Historic Places. Many states have developed state registers similar to the national one; while these (and the statutory provisions relating to them) vary in their effectiveness, they do serve to increase awareness and understanding of historic preservation.

6. Before 1980 owners had no right of objection to listing. The rationale for this was:
   (a) the Register is a list of properties that meets an objective evaluation, which applies criteria and professional standards regardless of a current owner's opinion of the property;
   (b) the owner's opinion has no bearing on whether a property is historic; and
   (c) inclusion in the Register does not directly restrict a private owner's use of his property in any manner.

   > (Duerksen 1983: 206)

   However, as a result of political pressures, the right to object was introduced in the 1980 amending Act. As a compromise the concept of "eligibility" for inclusion in the Register was introduced.

7. These are published in the Federal Register. A convenient reprint is published by the ACHP (1986). For a short but illuminating discussion of the section 106 process see Storey (1987: 37).

8. Section 1653 (f); commonly referred to (by a previous numbering) as section 4 (f).

9. *Citizens to Preserve Overton Park v Volpe* 1971. In Duerksen's words, "the court clearly perceived that preservation values are not comparably quantifiable, and that to hold otherwise would destroy the legislative protection afforded by section 4(f)" (Dworsky *et al.* 1983: 241).

10. The Railroad Revitalization and Regulatory Reform Act 1976 and the Amtrack Improvement Act 1973 both provide for grant aid and directives for the reuse of old railway stations.

11. "To the fullest extent possible, agencies shall prepare draft environment impact statements concurrently with and integrated with environmental impact analyses and related surveys and studies required by . . . the National Historic Preservation Act of 1966 . . . and other environmental review laws and executive orders." 40 C.F.R. 1502.25 (1981). The complex interconnections between the many environmental Acts is nicely illustrated in Hagman and Juergensmeyer 1986: 380.

12. This subject is a highly complex one. Details are given in Roddewig 1983. An annual analysis of tax incentives is published by the NTHP: see Chittenden 1988, and Blumenthal and Siler 1990.

   Since it is impossible to keep up to date with tax changes (whether of a "reform" nature or not), the reader will have to supplement any discussion of the important tax provisions for historic preservation with contemporary commentaries. A good source of these is the NTHP.

13. A good and succinct description of the *Penn Central* case is to be found in Haar and Kayden (1989b: 154–68). See also the interesting reference in the highly enjoyable Costonis (1989). There is a full discussion in N. Williams (1985–90: 71B).

14. Goldberger (1990) instances the commission's 1981 designation of the Upper East Side Historic District. This was criticized for being more an act of city planning than of landmark designation. Goldberger is of the opinion that "it now turns out to have been one of the most important, not to say prescient, land use decisions made in New York in the last generation. Without historic district status, the great blocks of the Upper East Side would have been devastated in the construction boom in the 1980s." Presumably the ends justifies the means? For a different view of this issue see J.B. Rose (1984).

15. For a complete list of preservation commissions, see Thurber and Moyer (1984).

16. "Municipalities can combine different types of zoning by use of "an overlay zone, a zone with special requirements (such as review procedures, height limits, or aesthetic review requirements) that covers more than one zoning district and does not change the underlying use and density standard" (Cook 1980: 21). See also Blackwell (1990).

17. Since this was written, a fuller account of Roanoke's historic preservation policies has been published in Collins, Waters, and Dotson (1991: 103).

18. "Resolution of July 11, 1969, at Second International Conference, Oxford, England," International Council of Monuments and Sites, Palais de Chaillot, Place du Trocadero, Paris; quoted in Fitch (1990: 78).

19. Collins, Waters, and Dotson (1991). For a useful short discussion on tourism and historic preservation see Fitch (1990: 77–81).

20. I am grateful to Carolyn Torma for making this point. This paragraph leans heavily on her work. See also Torma (1987) and chapter 12 of Murtagh (1988).

21. There is a large and burgeoning literature: see, for example, Quimby (1978), Schlereth (1980), Wells (1982 and 1986), and Upton and Vlach (1986).

22. In this connection, note the reorganization (during the Carter administration) of the federal preservation program within the Department of the Interior as the Heritage Conservation and Recreation Service (Stipe 1987: 274).

10 Growth management and local government

1. It may be wise at this point to note that, though other countries have attempted much stronger and more sophisticated techniques of "growth management", the policy has been beset with difficulties and doubts. See, for instance, Bourne (1975) and Sundquist (1975).

2. Three volumes were published in 1975 (R.W. Scott), and two further volumes later (Schnidman, Silverman, and Young 1978, and Schnidman and Silverman 1980). The list is on pages 24–31 of the first volume. In another study, Gleeson *et al.* (1975: 8) identify fifty-five "specific techniques or system elements".

3. The points scheme is set out in Callies and Freilich (1986: 826). For a useful short discussion of the criteria used in the plan, see ECO Northwest (1986).

4. The *Ramapo* decision did not consider whether this implicit quota was constitutional. This question was to the fore in *Petaluma* since this plan had an explicit quota (Mandelker 1988: 398). For further discussion of *Ramapo* see R.W. Scott (1975 Vol. 2), Bosselman (1973), and Franklin and Levin (1973).

5. The cap was to be effected by a 50 percent downzoning. For a discussion of the Boca Raton case see chapter 19 of Godschalk *et al.* (1979). This volume also contains another eleven case studies of growth management. A shorter discussion of thirteen cases (including Boca Raton) is to be found in Gleeson *et al.* (1975).

6. There can be dramatic local exceptions to this generalization. For instance, from 1972 to 1975 the growth control program of Fairfax County, Virginia, triggered 251 state court suits (Ellickson 1977: 388).

7. *Robinson v City of Boulder*, Supreme Court of Colorado, 1976. A municipality cannot be compelled to extend a utility service beyond its boundaries, but if it does so the courts may require the municipality "to serve impartially all those reasonably within reach of its supply system" (Callies and Freilich 1986: 330 – which contains a discussion of utility extension as a method of subdivision control).

8. An example of the development rights typically purchased by the city is given in Godschalk *et al.* (1979: 259). For further discussion see below.

9. Raup (1975), Vining (1979), Plaut (1980), Simon (1981), Crosson (1982), Fischel (1982), Raup (1982), Baden (1984), Delogu (1986).

10. Schnidman, Smiley, and Woodbury (1990: 14) give a table of state and local programs for the protection of agricultural land. See also Holloway and Guy (1990), who discuss a wider range of issues relating to the loss of farmland (including soil erosion and conservation).

11. Hanna (1982), Popp (1988: 524) and Thompson (1982). An interesting comparison of U.S. and Canadian right-to-farm legislation is to be found in Penfold *et al.* (1989).

12. See Schnidman, Smiley, and Woodbury (1990: 314), and Thompson (1990: 55). A useful

bibliography is given by Tripp and Dudek (1990: 515) in a paper dealing with other applications of TDR such as waste reduction credit exchanges, and air and water pollution rights. For a more general discussion of the latter see Dales (1968). An interesting account of the experience in safeguarding agricultural land in Marin County, California is given in Hart (1991).

13. In 1987 the Maryland Court ruled against the Montgomery County scheme, but solely on the grounds that it had not been incorporated in the zoning ordinance: there was no disapproval of the use of TDRs (*West Montgomery County Citizens Association v Maryland National Capitol Park and Planning Commission*, 1987). The position was regularized in the same year when the county council enacted seven new TDR zoning districts, and approved sectional map amendments designating the TDR zones on properties as receiving areas (Banach and Canavan 1989).

14. For an obviously sympathetic account of the Montgomery County program, see the paper by the County Planning Director: Tustian (1984).

15. There is, of course, the irony that while some public policies are geared to keeping land in agricultural use, others (particularly federal agricultural subsidies) are aimed at limiting output. For an absorbing account of *How the US Got Into Agriculture and Why it Can't Get Out*, see Rapp (1988). The U.S. is by no means alone in this: see Sanderson (1990).

16. But see the eloquent *Vermont Landscape* (Williams, Kellogg, and Lavigne 1987).

17. James and Windsor (1976), Ellickson (1977), M.J. White (1978), and Frieden (1979).

18. Hammer, Siler, George Associates (1975: 4); quoted in Babcock (1980: 267).

11 Growth management and the states

1. Anyone embarking upon a summary of the role of the states in land use planning is heavily indebted to Healy and Rosenberg (1979) and DeGrove (1984). Parts of this chapter lean heavily on their work.

It should be noted that this chapter deals only with the role of the states in growth management. For a review of state actions in relation to a wider range of issues see Zimmerman (1983).

2. The heading comes from the title of David Callies' 1984 book *Regulating Paradise: Land Use Controls in Hawaii*.

3. DeGrove (1984: 58). See also the references quoted in Myers (1976), Mandelker (1976a), DeGrove (1984), and Callies (1984).

4. See also Goodin (1984) and Keith (1985).

5. The heading is the title of a chapter by David G. Heeter (1976). Heeter provides a detailed account of the politics of the Vermont legislation, and other measures which were taken at the same time.

6. However, a central planning office was created in 1963 to coordinate state agency policies and the various planning activities at state, regional, and local levels. In 1968, legislation was passed banning billboards throughout the state. For this earlier history, see Byers and Wilson (1983).

7. Daniels and Lapping (1984: 502). For a discussion of the enforcement provisions of Act 250 (both as originally passed and as amended in 1985) see Bashaw (1985).

8. A detailed account is given in Heeter (1976).

9. The heading is taken from Turner (1990a: 39).

10. A prelude to this was the earlier decision to abandon the program of draining the 3 million acres of the Everglades!

11. The initial recommendation for designation can come from any group or citizen. Designation is made by the governor and cabinet, following review by the legislature.

12. The Act was an "omnibus" one since it incorporated several pieces of legislation: revisions to the Local Government Comprehensive Plan, legislation relating to the coast, and revisions to the DRI legislation.

13. At this date Florida had no personal income tax, and exempted all inheritance and estate income from taxation above the federal level. The report commented that "in reality, ours has become a tax haven state".

14. For another account of Florida's "outmoded tax structure" and the constitutional limits to reform, see Zingale and Davies (1986):

Among the 50 states and the District of Columbia, Florida ranks 48th in state and local taxes as percent of income; 44th in per capita state taxation; 40th in per capita general expenditures of state and local governments; 51st in per capita expenditures for public welfare; and 45th in per capita expenditures of state and local governments for capital outlay.

(Taylor, C.A., unpublished paper quoted in Audirac, Shermyen, and Smith 1990: 478, fn 1)

15. For a review of the position in early 1990 see Turner (1990b). Governor Martinez lost the 1990 election to Lawton Childs. One of Childs' first actions was to set up a review of the state's growth management efforts. The report (Florida 1991b) was issued in February 1991. It is replete with criticisms of the lack of coordination between state agencies, and the inadequate concern for a coordinated capital plan and budget. It is stressed that there are "funding shortfalls caused not only by infrastructure deficits, but also by inadequate revenue mechanisms to fund future growth at the local level". There is, however, little indication as to how the necessary funding is to be obtained. In short, the report does not appear to have added much to the account given above. But, in this politically charged situation, unexpected initiatives might appear. On the other hand, they might not.

16. The initiative establishing coastal planning and regulation in the state was Proposition 20.

Writing in 1989, Curtin and Jacobson (1989: 491) noted that "since 1971 at least seventy-six local growth management measures have been adopted by popular vote through the initiative process in California cities and counties". Between 1971 and 1988, at least 152 local land use related initiatives and referenda were voted upon (Callies and Freilich 1988: 116). For a general discussion of initiatives and referenda related to growth management and land use, see Berwanger (1987), Glickfield, Graymer, and Morrison (1987), Fountaine (1988), and Nitikman (1988). On Proposition 13, see Raymond (1988).

17. For a brief account of this earlier history see Squire and Scott (1984). Duddleson (1978: 3) has commented: "During the ten years over

which California's present system of coastal development controls was envisioned, built, tested, enacted into law, and put into operation, more effective participation by more people took place than has occurred in any other land use or environmental planning and regulatory program in the United States."

18. The regional commissions were abolished in 1981, since when there has been just the one state commission.

19. Other states, such as Michigan and North Carolina, have combined regulatory functions with those of restoration and access development. Fischer (1985: 318) comments that merger has "significantly increased both programs' political acceptability, even popularity".

The California Coastal Management Program approved under the federal Coastal Zone Management Act of 1972 consists of the enabling statutes of the Bay Conservation and Development Commission, the California Coastal Commission, and the State Coastal Conservancy.

20. For a detailed history of land use planning in Oregon in the postwar period up to the beginning of the 1980s see Leonard (1983).

21. See Chapter 5.

22. See the quotation at the head of this chapter. For a discussion of growth management from a *national* point of view, see Chinitz (1990).

23. Debate on new planning legislation is under way in many states. See, for example, Holman (1990) for a discussion of North Carolina; Maryland (1991). A useful update is provided by *Developments: The Newsletter of the National Growth Management Leadership Project* (published by 1000 Friends of Oregon).

24. Of particular interest is a study undertaken by DuPage County (1991). This showed that there was "a significant statistical relationship between development (both residential and nonresidential) and increases in personal property taxes". Another conclusion was that "nonresidential development (which includes commercial, office, and industrial land uses) is a major contributor to property tax increases in DuPage County". Nonresidential development was found to have two effects:
1 It brought into the county additional workers,

shoppers, and others. In turn this led to demands for increased government services.
2 The change in the environment to a more urban character gave rise to a demand for more urban services and amenities.

25. The legislation provides:
If the commission determines that a certified local coastal program is not being carried out in conformity with any policy of this division it shall submit to the affected local government recommendations of corrective actions that should be taken. Such recommendations may include recommended amendments to the affected local government's local coastal program.

Recommendations submitted pursuant to this section shall be reviewed by the affected local government and, if the recommended action is not taken, the local government shall, within one year of such submission, forward to the commission a report setting forth its reasons for not taking the recommended action. The commission shall review such report and, where appropriate, report to the legislature and recommend legislative action necessary to assure effective implementation of the relevant policy or policies of this division.
(California Coastal Act of 1976, as amended to January 1990, section 30519.5)

12 The Federal Government and Urban Policy

1. Donna Shalala, "A Pilgrim's Progress: Moving Toward a National Urban Policy", address at Trinity College, Hartford, Connecticut, April 20, 1978; quoted in Dilger (1982: 139).

2. *Statistical Abstract of the United States 1990*, Table 340 shows that of the total 2.271 billion acres, the federal government owns 727,000.

3. *New York Times*, July 13, 1975; quoted in Fitch (1990: 36).

4. A further factor was the tax benefits of the Tax Code. These constitute "massive tax benefits for housing" (Aaron 1972: 53). The largest of these is the exemption from taxation of the net

imputed rental income from the house, and the deductibility of mortgage interest and property tax payments. Whether or not there was (and still is) any public policy justification for this positive discrimination in favor of home ownership is a debatable matter. Compare Aaron with Slitor (1985). See also Hartman (1975).

5. The literature is large. Of particular relevance to this book is Dilger (1982). Bernard and Rice note that "the South and the West were the big winners" in the deliberate efforts of the armed forces "to relocate their personnel and training facilities around the country and to spread out defense contracts in order to make bombing and even invasion more difficult for the enemy". Morris (1980) argues vociferously that "the entire Sun Belt is incredibly dependent on defense dollars for its economic growth and survival". See also Weinstein, Gross, and Rees (1985), who argue that the importance of federal spending to the economic development of the Sunbelt has been "much overstated"; and Malecki (1982), who examines the impact of federal research and development expenditure on metropolitan economies.

It should be noted that there is no agreed definition of the belts – whether they be sun, snow, or frost! See, for example, Bernard and Rice (1983: 2) and Abbott (1981: 7).

(Since this was written, a major new study by Markusen *et al.* [1991] has been published. As with all good studies, this shows how complex the issue at debate is. One particularly interesting point made is that "the rise of the gunbelt . . . has worked as a kind of underground regional policy"; the Pentagon has "quietly pursued the most geographically targeted program of any government agency"; "American taxpayers have financed one of the most impressive population redistributions in history." See also Morrison [1991].)

6. The nearest that the federal government has ever come to devising a "national urban policy" was with Roosevelt's New Deal (Mollenkopf 1983: 55). There is, however, a much earlier national *land* policy: in the nineteenth century this was to sell off public lands to private owners as quickly as possible. See, for example, Carstensen (1962).

7. There is not space here to analyze the biennial reports issued in accordance with the 1970 Act and the amended provisions of the Housing and Community Development Act of 1977. These are characterized by their breadth, wooliness, and indeterminacy. However, initially they represented a first attempt to bring together the disparate issues which need to be addressed in a "national urban policy"; but they became increasingly perfunctory. See, for example, Ahlbrandt (1984) and Reiner (1979).

8. Quoted in Arnold (1973: 37).

9. The 1937 Act was "an act to provide financial assistance to the states and political subdivisions thereof for the elimination of unsafe and unsanitary housing conditions, for the eradication of slums, for the provision of decent, safe, and sanitary dwellings for families of low income and for the reduction of unemployment and the stimulation of business activity, to create a United States Housing Authority, and for other purposes."

Goldfield and Brownell put the matter in more earthy terms:

The Roosevelt administration "sold" Congress on USHA by promoting it both as an employment package and as a way to protect threatened center city property values. In turn, cities countered the charges of creeping socialism by touting the housing program as a boon for free enterprise. As Atlanta's housing chief Charles F. Palmer (himself a real estate developer) explained to his colleagues, "wiping out the slum area would enhance the value of our central business properties."

(Goldfield and Brownell 1990: 331)

10. In his 1949 inaugural address, Truman presented twenty-one major policy items. The Housing Act was the only one to pass. Among the defeated proposals was aid to education, fair employment practices, national health insurance, and middle-income cooperative housing (Mollenkopf 1983: 77).

Among the slogans used by the opposition to public housing were: "There are too many jokers in the public housing deal"; "Can you afford to pay somebody else's rent?" "Government operated politically controlled housing has no place in the American plan of living"; and "Public housing means the end of racial segregation in Savannah!" (Davies 1966: 127). Davies gives a detailed account in his *Housing Reform During the Truman Administration*.

11. Judd (1988: 263). Judd also lists the forces of opposition at both national and local levels (Judd 1988: 267). Weiss (1985: 54) argues that "Urban renewal owes its origins to the downtown merchants, banks, large corporations, newspaper publishers, realtors, and other institutions with substantial business and property interests in the central part of the city. . . . The state laws passed in the 1940s and 1950s and the federal law passed in 1949 fulfilled the goal that this powerful coalition had set for itself." His thesis is that there is a "remarkable continuity" between the objectives of the program's original proponents in the early 1930s and "the ultimate results."

12. The grant for "write-down", technically known as the net project cost, was met two-thirds from the federal government, and one-third by the local authority. Even with this subsidy, new suburban sites were typically more attractive to lenders, builders, and home buyers.

13. For an atypically active city (Newark, New Jersey), see Kaplan (1963). Anderson (1964) gives a scathing critique which was attacked for its errors, misleading use of statistics, and distortions (Groberg 1965, Kristof 1965, Smith 1965, Keith 1973: 155). Wilson (1966) provides a more balanced account. Another useful collection of papers is Bellush and Hausknecht (1967). The classic case study of the "selection" of sites for public housing is Myerson and Banfield (1955).

14. A summary of achievements and weaknesses is given in the Douglas Report (1968: 165).

15. See, for example, Jacobs (1961), Anderson (1964), Abrams (1965), Greer (1966), Wilson (1966), Wolfinger (1974), as well as Barnekov, Boyle, and Rich (1989: 47).

16. Gelfand notes that the quirk in the law which had the result of virtually excluding public housing from renewal areas:

> Public housing could be built on renewal sites, but the cities would have to pay more of the costs than if it was built in other areas. Besides having to pick up the normal local share of public housing subsidies, municipalities also had to contribute one-

third of the write-down expense for renewal land. Rather than contract for the "double subsidy", most cities put their low-rent housing elsewhere or, as sometimes happened, built none at all. Thus Title I, which public housers had hoped would solve many of the land price problems of their program, only made public housing construction more difficult.
> (Gelfand 1975: 209 and 423)

17. Mollenkopf (1983: 91–3), Judd (1988: 316–21), Goldfield and Brownell (1990: 363–7), and also Piven and Cloward (1971: 270).

18. Frieden and Kaplan (1977: 11).

19. One suggestion was that there should be only a single demonstration city – Detroit (Frieden and Kaplan 1977: 47).

20. Rodwin (1970: 257). Banfield (1973: 125) records that Johnson was informed (immediately after taking office in November 1963) that President Kennedy had approved a suggestion that a program should be developed to alleviate poverty. Johnson was enthusiastic, but he wanted something big and bold. "A program for just a few cities could never be propelled through Congress and, in any case, it would be regarded as another example of 'tokenism' by black leaders whose growing anger was a matter of concern to him."

The term "demonstration" was dropped because of the association of the term with the riots of the previous summer. Johnson commented that the new model cities would "spell the difference between despair and the good life" (Banfield 1973: 140).

For an analysis of model cities grants, see Arnold (1979: 165).

21. Other references on model cities are Kaplan, Gans, and Kahn (1970), Banfield (1973), Marris and Rein (1973), and Howitt (1984: ch. 5).

22. Bureau of the Budget Staff Memorandum; quoted in Gelfand (1975: 198).

23. For an account spanning the period from Roosevelt to Nixon, see Graham (1976). A particularly important document is the 1964 report of the Advisory Commission on Intergovernmental Relations, *Impact of Federal Urban Development Programs on Local Government Organization and Planning*. This was "the first attempt to promote a systematic survey

of requirements for local government organization and planning in all federal programs affecting physical development in urban areas, and to assess the extent and nature of inter-agency coordination of these programs within the federal government".

Among later works, special mention must be made of Markusen (1987), which also has a good bibliography.

24. A useful account is given in Benko and Hand (1985). See also McDowell (1985). (Both are in So, Hand, and McDowell 1985, which has several chapters on various aspects of this subject.)

25. Quoted in Judd (1988: 333).

26. For a brief discussion see Judd (1988: 332) and Mollenkopf (1983: 122). Much fuller accounts are given in Reagan (1972) (and the revised edition: Reagan and Sanzone 1981), Dommel (1974), Haider (1974) and Nathan, Manvell, and Calkins (1975). The attitudes of city mayors changed dramatically when it became apparent that Nixon was not merely reorganizing grant-aid, but also cutting it. On this see particularly Dommel (1974).

27. Originally it was intended that the eligible localities should be metropolitan cities, defined as every central city of a metropolitan area and suburban cities with a population of more than 50,000 population. This totaled 445 jurisdictions: 313 central cities and 132 suburban cities (Dommel and Rubinowitz 1982: 35). However, political pressures quickly led to substantial increases.

28. Judd (1988: 348), Goldfield and Brownell (1990: 391), and Pierce and Guskind (1985: 13). For an interesting succinct discussion of some of the regulatory difficulties, see Kettl (1981: 121); a fuller account is given in Kettl (1980). For an analysis of the CDBG as applied to small cities see Jennings et al. (1986).

29. U.S. HUD (1978b).

30. "Not everyone was content to leave the task of formulating a national urban policy to the president. Among those who wished to offer their own proposals was Congressman Henry Reuss, who conducted hearings in 1977 on urban policy options. Reuss had been a member of the National Committee on Urban Growth Policy. After the hearings, Reuss issued a proposal for a national policy based on the notion that older cities, with their existing infrastructure and compact patterns of development, were great 'conservators of land, energy, and other resources'. An elaborate public and private effort to create employment was the key to saving the cities (Reuss 1977). The proposal received little attention, for it was promptly upstaged by the Carter plan" (Eisinger 1985: 22, endnote 31). Among the many other alternatives circulating about this time was that of John Kain of Harvard University (Kain 1978).

31. Conlan (1988: 96) adds: "Renewed program fragmentation and regulation were not confined to block grants however. Between 1975 and 1980, ninety-two new categorical programs were created, bringing the total to a record 534" (ACIR 1984: 2).

32. The commission issued a main report and nine panel reports on such topics as energy, science and technology, the quality of life, and the relationship between America and the World Community. The report which is of particular relevance here (and from which the quotations are taken) is Urban America (President's Commission on a National Agenda for the Eighties 1980).

33. This is not the place to discuss the merits of the Urban America report. The interested reader is referred to Byrne, Martinez, and Rich (1985), on which the following section leans heavily.

34. Barnekov, Boyle, and Rich (1989: 100) comment that, though Reagan's privatization policies represented "a radical departure from previous national urban policies", a broader historical perspective shows that "they were an extension of the long and continuous tradition of policy efforts to promote the private city".

35. Though this policy approach is indelibly (and rightly) associated with Reagan's name, there was a considerable (even if moderated) support for it. Apart from the National Agenda, there were the reports of the Committee on National Urban Policy of the National Research Council (Hanson 1982a, 1982b, and 1983) and the Urban Institute (for example Palmer and Sawhill 1982 and 1984, and Peterson and Lewis 1986), and a host of others (e.g. Baer 1976 and Savas 1983).

36. Ahlbrandt tells a fascinating story of the drafting

of this report. The rejection of a draft written by E.S. Savas may have prompted the article he wrote for *Urban Affairs Quarterly* (Savas 1983).

37. There are another two in Northern Ireland. A compendium of information on the British zones is published by the Department of the Environment; see GB (1990).

38. See Tym (1984) (and earlier reports for 1982 and 1983) and Hall (1991: 190). A list of references to British experience is given in Cullingworth (1988: 109, endnote 53).

39. For a discussion of state enterprise zones see Butler (1981); Mier and Gelzer (1982); Walton (1982); Rubin and Zorn (1985); Green and Brintnall (1987 and 1988); Brintnall and Green (1988); Wilder and Rubin (1988); Green (1991).

40. Including the *Report of the Ad Hoc National Committee on Urban Growth Policy* which recommended the construction of a hundred new towns of 100,000 population each, and ten cities of at least a million each (Canty 1969). For a reference to some of the other numerous reports on, or relevant to, "national urban growth policy" see Wingo (1972).

41. This proposal was to be found in the American Law Institute (1968) Code, and was incorporated in the Coastal Zone Management Act.

42. For a collection of essays on urban impact analysis, see Glickman (1980). See also "A Changing Agenda for Urban Planning" which is the last chapter of Meltzer (1984).

13 Cross-cultural perspectives

1. Part of this section is based on the last chapter of the author's *Urban and Regional Planning in Canada* (1987).

2. This is a question which I hope to address at a later time. Among the important sources are Delafons (1969); Clawson and Hall (1973); Haar (1984); and Wakeford (1990).

3. See Lipset (1963b), Sharpe (1973), and Arnold and Barnes (1979).

4. There is widespread confusion in the use of the terms "England" and Britain". The following

explanation shows how justified this is: Great Britain (which is so called to distinguish it from "Little Britain" or Brittany) consists of England, Wales (annexed in the 13th century), and Scotland (joined in union by a common king in 1603). In 1922, following the division of Ireland into the Irish Free State (now the Republic of Ireland) and Northern Ireland, the latter joined Britain to constitute the United Kingdom (in full: the United Kingdom of Great Britain and Northern Ireland). The Channel Islands and the Isle of Man have a special constitutional status. All together these form the British Isles. For further discussion see de Smith and Brazier (1990: chapter 3).

5. For a brief account, see Tarn (1980). A fuller treatment is given in Ashworth (1954). For a detailed account of one city (Birmingham) see Briggs (1952).

6. A short account is given in Perks and Smith (1985: 1879). See also Perks and Robinson (1979); Rose (1980); Simpson (1981); Stelter and Artibise (1982); Hodge (1985).

7. Over a third of a million houses were demolished in the decade before World War II. By 1938, demolitions were running at the rate of 90,000 a year. Had it not been for the war, over a million of Britain's older housing stock might have been cleared by 1951. Instead, the program came to an abrupt halt, not to be resumed on a significant scale until the mid-1950s.

8. Though the betterment system was dismantled, development rights in land remained nationalized, and still are.

9. CMHC also promoted planning education and financed the establishment of the Community Planning Association of Canada (now the Canadian Institute of Planners). In 1949 there were forty-five practicing planners in Canada; by 1984 the membership of the CIP was around 3,500 (Perks and Smith in *The Canadian Encyclopedia* 1985: 1882).

10. For an account of British urban policy, see Stewart (1987), Barnekov, Boyle, and Rich (1989: chapters 6–7), Parkinson (1989), and the references quoted therein.

11. See, for example, Savas (1982), Ascher (1987), and Barnekov, Boyle, and Rich (1989).

12. For a discussion of the contrast between Thatcher's centralization and Reagan's

decentralization see Wolman (1988). The essential point is that centralization and decentralization are not goals in themselves but different means to the same end, i.e. the reduction in government expenditure. The different means are the result of the dissimilar institutional contexts in the two countries.

Thatcher policies were a strange mixture of centralization and decentralization ideas. Thus she wanted to enable local electorates to be better able to control local affairs (meaning local expenditure). This was exactly in line with Reagan's philosophy which included a belief that "the federal government has distorted local priorities, lessened administrative accountability, and suffocated the growth of institutions which are closer to the citizenry than is Washington" (Cohen 1983: 304).

13. A useful account of the court systems in Canada is to be found in Waltman and Holland (1988).

14. The only rights which cannot be overridden are the "democratic rights," "mobility rights," "minority rights," "minority language educational rights," rights associated with the equal status of French and English in some parts of Canada, and an interpretation clause providing that rights in the Charter are guaranteed equally to male and female persons (Bayefsky 1987: 817).

15. For a very different view on "judicial supremacy" see Rosenberg (1991).

16. On the Loyalists, it has been commented that: "Modern Canada has inherited much from the Loyalists, including a certain conservatism, a preference for 'evolution' rather than 'revolution' in matters of government, and tendencies towards a pluralistic and heterogeneous society" (Wilson, B.G., "Loyalists" in The Canadian Encyclopedia 1985: 1042).

Northrop Frye (1980) has quipped that "historically a Canadian is an American who rejects the revolution."

17. The wording is: "It shall be lawful for the Queen, by and with the Advice and Consent of the Senate and the House of Commons, to make Laws for the Peace, Order and good Government of Canada, in relation to all Matters not coming within the Classes of Subjects by this Act assigned exclusively to the Legislatures of the Provinces . . ."

18. For an essay which conveys some Quebecois feelings, see Dion (1989). An enlightening account of "French-English Duality and Canadian Federalism" is given in Smiley (1987). Other references include Gagnon (1984), Latouche (1986), Linteau, Durocher and Ricard (1991), and, more generally, Whittington and Williams (1984).

19. Bill of Rights 1689 and Act of Settlement 1701 (limit the powers of the monarchy); Parliament Acts 1911 and 1949 (define the powers of the House of Lords); Representation of the People Acts 1948 and 1949 (set out the provisions of the electoral system).

20. New York Times Review of Books, June 9, 1991. The four books were Roland Flamini, Sovereign: Elizabeth II and the Windsor Dynasty, Delacorte Press, 1991; Charles Highham and Roy Moseley, Elizabeth and Philip: The Untold Story of the Queen of England and Her Prince, Doubleday, 1991; Anne Edwards, Wallis: The Novel, Morrow, 1991; and Celia Brayfield, The Princess, Bantam, 1991.

21. Bagehot's The English Constitution was first published in 1867. The edition used here is that of 1963 (published as a Cornell paperback), which contains an excellent introduction by R.H.S. Crossman (who, in addition to his academic credentials, was for a time the English Minister of Housing and Local Government).

22. For a short discussion of the Ontario Municipal Board, see Cullingworth (1987: 435).

23. Stare decisis is the rule by which the courts regard precedent as a compelling guide to decision in similar cases. Literally, it means "to stand by that which was decided".

24. An up-to-date treatise on the Canadian legal system is Gall (1990). Makuch (1983) provides a clear and useful account of the significant land use cases.

25. See, for example, Hoffecker (1983: 140 ff) who gives a graphic account of the history of the construction of the I-95 through the city of Wilmington, Delaware. She comments that "Although the 1956 Highway Act represented a series of compromises among various interest groups, one set of interests had not been included in the deliberations: that of the cities."

14   Local government systems

1. See Molotch (1976), Maurer and Christenson (1982), Logan and Molotoch (1987), and Logan and Zhou (1989).
2. More generally on this issue, see Self (1982).
3. Mills (1987: 54) has commented that "the successful campaign to abolish English metropolitan government in favor of local jurisdictions and joint boards suggests that some degree of American-style metropolitan fragmentation has been deemed worthy of emulation."
4. Nothing could be more different than the position in the big American cities. Particularly striking is the newly reformed New York City Council, which has the unusually large number of thirty-five members, each of whom receives $50,000 annual salary (twice the 1988 median wage). In addition, members have access to large sums of money for expenses and for reelection campaigns which, in relative terms, hardly exist in Britain.
5. The House of Commons has 650 members, while the House of Lords has 1,202 (mercifully most of whom are inactive). By contrast, the House of Representatives has 435 members, and the Senate 100. Canada, nicely selecting terminology from both countries, has 282 members in the House of Commons, and 104 in the Senate.
6. In the author's home area, the Delaware State Legislature has proposed that the New Castle County Council should be increased from seven to nine members. It is thought that this would enable a better service to be given to constituents. It is being opposed, however, on the ground that "every decision will take longer because we'll have nine speeches instead of seven" (*Wilmington News Journal*, May 27, 1991). A comparable British authority (e.g. the county of South Glamorgan) has sixty-two members!
7. The administrative counties exclude separate local governments within their boundaries. Figures on council membership and political affiliation are from *Whitaker's Almanac* (1990: 569), and relate to May 1989.
8. Thus Vogel in his study of *National Styles of Regulation*, though accurately portraying the operation of central government, comes astray in his account of local government when he writes:

    > An important purpose of the town and country planning system is to allow community residents affected by a proposed development to decide whether or not they wish to grant approval to it. While the central government overrides the preferences of local authorities on occasion, it does so relatively infrequently, and usually only when a planning decision has national implications.
    > (Vogel 1986: 270)

    Here, the British system is seen through American eyes, and "community residents" are confused with "local authority."
9. An exception was the widespread public concern over the procedures for designating the routes of major highways. On this see McAuslan (1980: 55).
10. Desmond Heap, then Comptroller and City Solicitor to the City of London, quoted in Elkin (1974: 100). Heap achieved international fame by selling London Bridge to Arizona. Rumor has it that Arizona thought that it was buying the much more impressive Tower Bridge!
11. See also Heidenheimer, Heclo, and Adams (1990: 302), and Peterson and Kantor (1977: 198). Though not subject to detailed analysis in the present book, it needs to be stressed that the concept of "the public interest" is an elusive one. For a thoughtful and stimulating discussion see McAuslan (1980).
12. I am indebted to Sharpe (1973) for these ideas and references.
13. There is a general power which enables a local authority to arrange for the discharge of any of its functions by a committee, a subcommittee, an officer of the authority, or by any other local authority (Local Government Act 1972, s.101). According to Grant:

    > A common pattern is for decisions on householder and similar minor applications to be delegated completely, and other categories of application to be delegated subject to a power of veto by the committee. A committee may also wish to approve an application in principle but reserve some detail to be approved by the officers before permission is issued.
    > (Grant 1982: 256)

Any decision made by an officer in accordance with such procedures is in law a decision of the local authority.

14. Parts of this section of the chapter are taken from the author's *Urban and Regional Planning in Canada* (1987).

15. Toronto City (1983). In a 1992 study Frisken writes:

> Metropolitan Toronto, with 240 square miles, is now the central city of a provincially-defined Greater Toronto Area, more than ten times that size. Thus its government is no longer in a position to undertake regional planning or provide major urban services to the whole of the Toronto-centered metropolis. The only government able to perform those functions is the provincial government, which under the Canadian Constitution has full responsibility for municipal affairs.
>
> (Frisken 1992)

16. To add a quantitative dimension to this: "Fred Kohler, who served as mayor, sheriff, and police chief in the 1920s left an estate of $400,000, although he never earned more than $15,000 a year" (Krumholz and Forester 1991: 13).

17. National Advisory Commission on Criminal Justice Standards and Goals (1973: 206); quoted in U.S. Department of Justice (1979: 41).

18. Quoted in Sharpe (1973: 150); see the sources quoted by Sharpe in his footnote 68.

19. Redcliffe-Maud Report (1974), and Salmon Report (1976). See also Doig (1984). For a collection of scandals at the parliamentary level, see Doig's 1990 book appropriately titled "Westminster Babylon."

15 Land use planning systems

1. On this point, the reader will find it useful and interesting to consult Stevens (1985), and also Nathanson (1977), who is quoted on page 442 of Stevens' article.

2. This provision was repealed in 1970: there was no way in which such broad and vague responsibilities could be legally enforced, or even recognized if they had been achieved. However, it did represent a statement of the political responsibility of the central government, and though no longer enshrined in legislation, this responsibility remains unchanged. (In passing, it is interesting to note that U.S. legislation is replete with such lofty phraseology: but it is distinctly un-British.)

3. For present purposes, we can ignore the generally slight differences to be found in Scottish planning law. Northern Ireland has a number of significant features which would have been discussed here had the coverage been the UK instead of Britain.

Parts of this discussion are taken from or are based on the author's book of British planning (Cullingworth 1988).

British planning law is subject to continual amendment. The account refers to the situation in 1990. It therefore ignores the Planning and Compensation Act of 1991.

4. Legislation passed in 1991 (the Planning and Compensation Act) provides that future structure plans will no longer require the approval of the secretary of state. However, he retains powers of "intervention", including in the last resort the power to "call in" a plan and direct a local authority to modify it. At the time of writing, only a "consultation draft" was available on these new provisions. Readers interested in the outcome of this (and many other innovations in planning legislation and policy) are advised to consult the *Journal of Planning and Environment Law*.

5. Town and Country Planning (Structure and Local Plans) Regulations 1982 (S.I. 1982, No. 555).

6. A brief account is given in the author's *Town and Country Planning in Britain*, 10th edition, 1988, chapter 5 (on which this section is based), and much more exhaustively in the Cabinet Office History of *Land Values, Compensation, and Betterment*, 1980.

7. This, of course, is not self-evident, though it has seldom been questioned in Britain; but see Munby (1954) and Leung (1979).

8. For an extensive discussion (on an international scale) see Hagman and Misczynski (1978).

9. Much of this section is based on Cullingworth (1987: chapter 4).

10. See, for example, Artibise and Stelter (1981) and Stelter and Artibise (1977, 1982, and 1986).

11. But see Perks and Jamieson (1991).

12. As reported in the Toronto *Globe and Mail*, (January 28, 1991).

13. These were largely focused on *Towards a New Planning Act for Alberta*, Alberta Municipal Affairs (1974).

14. For an updated and detailed exposition of Alberta planning law see Laux (1990).

15. The "overseas countries" referred to are Denmark, France, the Federal Republic of Germany, and the Netherlands.

16. Haar here is using the term "officials" in its American sense, which covers both professional staff and elected members.

17. See de Smith and Brazier (1990: chapter 22). On the issue of the "ideology of private property" in Britain see McAuslan (1980).

18. *The Economist* (October 12, 1991, p. 27). The anonymous writer suggests that this is one of the two explanations why "economic class populism does not take off in the United States." The other explanation is anticommunism.

16 Prospects for reform

1. Sam Howe Verhovek, "Town Heatedly Debates Merits of a Nuclear Waste Dump", (*New York Times* June 28, 1991).

2. Mixon (1990) adds a footnote which is irresistible: "In an amusing sidelight, some of the most ardent anti-zoning advocates that the author has met live in tightly zoned West University Place. They adamantly proclaim that zoning has nothing to do with their decision. It is worth noting that Bernard Siegan, who wrote the definitive scholarly defense of non-zoning in Houston, made his home in tightly zoned La Jolla, California." Actions may speak louder than words!

3. See, for example, N. Williams (1955), Babcock (1966), Douglas Report (1968), Kaiser Report (1968), Ellickson (1973, 1977, and 1981), Krasnowiecki (1980), Kmiec (1981), and President's Commission on Housing (1982).

4. The latest is the "Nimby" report, which cites a builder in New Jersey who had been trying *for*

sixteen years to obtain the permits necessary for a housing development. "The company has spent more than $95 million so far for carrying costs and regulatory costs, but has not yet started construction. The total amount of the outlay could add $68,000 to the price of each home once the development is finished" (Advisory Commission on Regulatory Barriers to Affordable Housing 1991: 2–13). Without in any way denying the likelihood that such intolerable circumstances arise, it is unfortunate that the evidence is typically hearsay, and no analysis is given as to how and why the situation developed as it did. The examples are quoted as being self-evidently true and condemnatory of "bad government." For a contrasting thorough analysis see Seidel (1978).

5. The term is John Punter's, in Davies *et al.* (1989: 234).

6. It is interesting to note that one of the problems to which municipal land acquisition in the Netherlands gives rise is the burden of loan costs. A study by Hamnett (1978: 41) showed that "the need to recover this through resale puts the municipality in a weak position in negotiations with potential developers."

7. See also Hamnett (1978), and Thomas *et al.* (1983).

8. When a development is complete, a *certificat de conformite* is issued. This attests that the development is in conformity with the permit which authorized it. No building can be sold before a certificate of conformity is issued.

9. Several notable cases have dealt with "the right to travel." In *Belle Terre*, appellants argued that an ordinance which prohibited three or more unrelated persons from living in a single household violated the appellants' "right to travel." The California Supreme Court (in both the majority and the dissenting opinions) held that "an ordinance which has the effect of limiting migration to a community does not necessarily abridge a fundamental right to travel" (Callies and Freilich 1986: 845, n. 22).

10. Alberta Municipal Affairs (1987: 5). A detailed account of the Alberta planning system is given in Laux (1990).

11. Quoted in Healy (1977: 2).

12. See, for example, Delafons (1969: 110), who complains that an obstacle to progress in

American planning is "the fact that while more and more precise and restrictive land-use controls have been brought under the umbrella of police power, there is still no obligation to compensate owners for the losses incurred as a result of those controls. Such a situation cuts both ways; it may facilitate control since its exercise is not a charge on public funds, but it may also stultify control by imposing too heavy a penalty on private property. There comes a point in the exercise of control where the community is reluctant to require that the cost of public benefit be borne entirely by private loss."

13. *Wilmington News Journal*, June 14, 1991.

14. "In land use law, the suggestion that regulation may have as a legitimate purpose the protection of existing uses from competition is greeted by the charge that such is improper. Some cases maintain that the use of zoning regulations to control competition is ultra vires or unconstitutional, while others suggest that control of competition cannot be the dominant purpose of zoning. These responses are bemusing in light of the monopolistic consequences of the traditional zoning treatment of nonconforming uses, and the common and accepted litany in zoning cases that apartments may be excluded from single-family districts because, among other consequences, apartments will damage existing single-family property values" (Babcock and Feurer 1977: 326; citations omitted).

15. Nevertheless, recent innovations in New England hold out some promise in relation to the provision of affordable housing. A Connecticut state law of 1990 requires local governments to justify the rejection of proposals that include affordable housing. If rejection cannot be justified on health or public safety grounds (as opposed to aesthetics or "local character") the court can order the housing to be provided. In Massachusetts, a zoning appeal process can overrule local zoning decisions. (G. Johnson, "Connecticut Developers Challenge the Power of Local Zoning", *New York Times*, November 24, 1991.) It is early to assess the effectiveness of these measures.

   More generally, see Nenno (1986) and Stegman and Holden (1987).

16. Stanfield (1991: 2330). More encouraging evidence from Oregon was published after this section was written: a joint report of 1000 Friends of Oregon and the Home Builders Association of Metropolitan Portland, entitled *Managing Growth to Promote Affordable Housing* (Oregon 1991b). This documents the success of Oregon's "Metropolitan Housing Rule" which requires the local governments in the Portland Metropolitan area to:

   1 provide adequate land zoned for needed housing types;
   2 ensure that land within the Metropolitan Portland urban growth boundary may accommodate the region's projected population growth;
   3 provide greater certainty to the development process; and
   4 reduce housing costs.

   The report finds that the implementation of this Rule in the early 1980s removed zoning constraints to the development of lower cost housing: "By mandating that certain *minimum* densities be allowed, the Metropolitan Housing Rule removed a regulatory barrier to development and encouraged the creation of smaller (higher density) less costly lots". The report is available from 1000 Friends of Oregon, 300 Willamette Building, 534 SW Third Avenue, Portland, Oregon 97204.

17. For a similarly depressing conclusion (on regional governance) see Morrill (1989).

18. Haar (1984: 209). It should also be added that planning policy cannot be divorced from other areas of public policy. For instance, underpinning the planning and housing policies of countries such as Britain and Canada are highly developed income-maintenance and comprehensive health care systems. These cover virtually the whole population, and underpin specific urban policies. Moreover, they are financed nationally (or, as in Canada, at least on a provincial basis); as a result there is an automatic transfer of resources from richer to poorer areas. Thus, poorer areas are relieved of some of the crushing burden which falls on U.S. cities, which have the much more limited underpinning by way of social security.

Appendix 1 Resuscitation of the common law

1. Part of this discussion originally appeared in a book review in *Urban Law and Policy* 8: 304–8 (1987). The full reference of the book is *The Wrong Side of the Tracks: A Revolutionary Rediscovery of the Common Law Tradition of Fairness in the Struggle Against Inequality*, by C.M. Haar and D.W. Fessler, Simon and Schuster 1986.

2. D.R. Grant and H.C. Nixon, *State and Local Government in America*, Allyn and Bacon, Fourth Edition, 1982.

3. Eisenstadt (1987), Ely (1987), Gillette (1987), Goode (1986), Kalsheur (1987), and Scurlock (1987).

Bibliography

Aaron, H.J. (1972) *Shelter and Subsidies: Who Benefits from Federal Housing Policies?* Washington: Brookings Institution.

—— **(1978)** *Politics and the Professors: The Great Society in Perspective*, Washington: Brookings Institution.

Abbott, C. (1981) *The New Urban America: Growth and Politics of Sunbelt Cities*, Chapel Hill: University of North Carolina Press.

—— **(1991)** "Urban Design in Portland, Oregon, as Policy and Process, 1960–1989," *Planning Perspectives* 6: 1–18.

Abraham, H.J. (1980) *The Judicial Process*, (4th edn.), Oxford: Oxford University Press.

Abrams, C. (1965) *The City is the Frontier*, New York: Harper and Row.

ACIR [Advisory Commission on Intergovernmental Relations] (1964) *Impact of Federal Urban Development Programs on Local Government Organization and Planning*, Washington: U.S. GPO.

—— **(1968)** *Urban and Rural America: Policies for Future Growth*, Washington: U.S. GPO.

—— **(1970)** *Eleventh Annual Report*, Washington: U.S. GPO.

—— **(1974)** *A Look to the North: Canadian Regional Experience*, Washington: U.S. GPO.

—— **(1981)** *Studies in Comparative Federalism: Canada*, Washington: U.S. GPO.

—— **(1984)** *Catalog of Federal Grants-in-Aid*, Washington: U.S. GPO.

Ackerman, B.A. (1977) *Private Property and the Constitution*, New Haven, CT: Yale University Press.

Ackerman, B.A. (ed.) (1975) *Economic Foundations of Property Law*, Boston: Little, Brown.

Adams, T. (1922) "Modern City Planning: Its Meaning and Methods," *National Municipal Review* 11, 6 (June 1922).

—— **(1937)** "Town and Country Planning in Old and New England," *Planners Journal* 3: 91–8.

Advisory Commission or Committee: *see also* President.

Advisory Commission on Intergovernmental Relations: *see* ACIR.

Advisory Commission on Regulatory Barriers to Affordable Housing (1991) *"Not in My Back Yard": Removing Barriers to Affordable Housing*, Washington: U.S. HUD.

Advisory Committee on City Planning and Zoning (1928) *A City Planning Primer*, Washington: U.S. Department of Commerce, U.S. GPO.

Advisory Council on Historic Preservation (1979) *The Contribution of Historic Preservation to Urban Revitalization,*

Washington: The Council.

—— **(1983)** *Federal Tax Law and Historic Preservation: A Report to the President and Congress, 1983*, Washington: The Council.

—— **(1984)** *National Historic Preservation Act As Amended* (2nd edn.), Washington: The Council.

—— **(1985)** *An Overview of Federal Historic Preservation Law: 1966 to 1985*, Washington: The Council.

—— **(1986)** *36 CFR Part 800: Protection of Historic Properties. Regulations of the Advisory Council on Historic Preservation Governing the Section 106 Review Process*, Washington: The Council.

—— **(1989)** *Report to the President and Congress of the United States 1989*, Washington: The Council.

Agnew, J.A. (1987) *Place and Politics: The Geographical Mediation of State and Society*, London: Allen and Unwin.

Agnew, J.A. and Duncan, J.S. (eds.) (1989) *The Power of Place: Bringing Together Geographical and Sociological Imaginations*, London: Unwin Hyman.

Agnew, J.A., Mercer, J., and Sopher, D. (eds.) (1984) *The City in Cultural Context*, London: Allen and Unwin.

Ahlbrandt, R.S. (1984) "Ideology and the Reagan Administration's First National Urban Policy Report," *Journal of the American Planning Association* 50: 479–84.

Ahmuty, S.J. (1977) "*Arlington Heights*: Closing Federal Courts to Exclusionary Litigation," *Albany Law Review* 41: 789–811.

Alberta Municipal Affairs (1974) *Towards a New Planning Act for Alberta*, Edmonton: Alberta Department of Municipal Affairs.

—— **(1980)** *Planning in Alberta: A Guide and Directory*, Edmonton: Alberta Department of Municipal Affairs.

—— **(1987)** *A Guide to Land Use Planning in Alberta*, Edmonton: Alberta Department of Municipal Affairs.

Alder, J. (1990) "Planning Agreements and Planning Powers," *Journal of Planning and Environment Law* 1990: 880–9.

Almond, G.A. (1956) "Comparative Political Systems," *Journal of Politics* 18: 391–409.

Almond, G.A. and Verba, S. (1963) *The Civic*

Culture, Princeton: Princeton University Press.

—— **(1965)** *The Civic Culture: Political Attitudes and Democracy in Five Nations*, Boston: Little, Brown (abridgement of original 1963 edition), Princeton: Princeton University Press.

Aloi, F.A. and Goldberg, A.A. (1971) "Racial and Economic Exclusionary Zoning: The Beginning of the End," *Urban Law Annual* 4: 9–62.

Aloi, F.A., Goldberg, A.A., and White, J.M. (1969) "Racial and Economic Segregation by Zoning: Death Knell for Home Rule?," *University of Toledo Law Review* 1: 65–108.

Alterman, R. (1988a) "Evaluating Linkage and Beyond: Letting the Windfall Recapture Genie Out of the Exactions Bottle," *Journal of Urban and Contemporary Law* 34: 3–49.

—— **(1988b)** *Private Supply of Public Services: Evaluation of Real Estate Exactions, Linkage, and Alternative Land Policies*, New York: New York University Press.

—— **(1989)** *Evaluating Linkage and Beyond: The New Method for Supply of Affordable Housing and Its Impacts*, Cambridge, MA: Lincoln Institute of Land Policy.

Alterman, R. and Kayden, J.S. (1988) "Developer Provisions of Public Benefits: Toward a Consensus Vocabulary," in Alterman (1988b).

Altshuler, A.A. (1965) *The City Planning Process: A Political Analysis*, Ithaca: Cornell University Press.

—— **(1977)** "Review of *The Costs of Sprawl*," *Journal of the American Institute of Planners* 43: 207–9.

Ambrose, P.J. (1986) *Whatever Happened to Planning?*, London: Methuen.

Amdursky, R.S. (1969) "A Public-Private Partnership for Urban Progress," *Journal of Urban Law* 46: 199–215.

American City (1913) "Efficiency in City Planning," *American City* 8: 139–43.

American Farmland Trust (1986) *Density-Related Public Costs*, Washington: American Farmland Trust.

—— **(1988)** *Protecting Farmland Through Purchase of Development Rights: The Farmers' Perspective*, Washington: American Farmland Trust.

—— (1990) *Agriculture and the Environment: A Study of Farmer Practices and Perceptions,* Washington: American Farmland Trust.

American Law Institute (1968) *A Model Land Development Code,* (Tentative Draft), Philadelphia, PA: American Law Institute.

—— (1975) *A Model Land Development Code: Complete Text and Reporter's Commentary,* Philadelphia, PA: American Law Institute.

American Planning Association (1989) *The Best of Planning,* Washington: American Planning Association.

American Public Health Association (1950) (Committee on the Hygiene of Housing) *Planning the Home for Occupancy,* Washington: American Public Health Association.

American Society of Planning Officials (1971) *Planning 1971: The Making of National Urban Growth Policy,* Washington: American Society of Planning Officials.

—— (1972) *Land Use Controls Annual 1971,* Washington: American Society of Planning Officials.

Ames, D.L., Callahan, M.H., Herman, B.L., and Siders, R.J. (1989) *Delaware Comprehensive Historic Preservation Plan,* Center for Historic Architecture and Engineering, College of Urban Affairs and Public Policy, University of Delaware.

Anderson, J. (1990) "The 'New Right', Enterprise Zones, and Urban Development Corporations," *International Journal of Urban and Regional Research* 14: 468–89.

Anderson, M. (1964) *The Federal Bulldozer: A Critical Analysis of Urban Renewal, 1949–1962,* Cambridge, MA: MIT Press.

Anderson, R.M. (1962) "The Board of Zoning Appeals – Villain or Victim?," *Syracuse Law Review* 13: 353–88.

Andrew, C.I. and Merriam, D.H. (1988) "Defensible Linkage," *Journal of the American Planning Association* 54: 199–209.

Anton, T.J. (1975) *Governing Greater Stockholm: A Study of Policy Development and System Change,* Berkeley: University of California Press.

—— (1989) *American Federalism and Public Policy; How the System Works,* New York: Random House.

Aoki, K. (1989) "Recent Developments: Fair Housing Amendments Act of 1988," *Harvard Civil Rights – Civil Liberties Law Review* 24: 249–63.

Appelbaum, R.P, Bigelow, J., Kramer, H.P., Molotch, H.L., and Relis, P.M. (1976) *The Effects of Urban Growth: A Population Impact Analysis,* New York: Praeger.

Arnold, J.L. (1971) *The New Deal in the Suburbs: A History of the Greenbelt Town Program 1935–1954,* Columbus: Ohio State University.

—— (1973) "City Planning in America," in Mohl and Richardson (1973).

Arnold, P.E. (1972) "Herbert Hoover and the Continuity of American Public Policy," *Public Policy* 20: 525–44.

Arnold, R.D. (1979) *Congress and the Bureaucracy: A Theory of Influence,* New Haven, CT: Yale University Press.

Arnold, S.J. and Barnes, J.G. (1979) "Canadian and American National Character as a Basis for Market Segmentation," *Research in Marketing* 2: 1–35.

Artibise, A.F.J. (1989) "Canada as an Urban Nation," in Graubard (1989).

—— (ed.) (1990) *Interdisciplinary Approaches to Canadian Society: A Guide to the Literature* (Association for Canadian Studies), Montreal: McGill-Queens University Press.

Artibise, A.F.J. and Stelter, G.A. (1981) *Canada's Urban Past: A Bibliography to 1980 and a Guide to Canadian Urban Studies,* Vancouver: University of British Columbia Press.

Artibise, A.F.J. and Stelter, G.A. (eds.) (1979) *The Usable Urban Past: Planning and Politics in the Modern Canadian City,* New York: Macmillan.

Ascher, K. (1987) *The Politics of Privatization,* New York: St Martins Press.

Ashley, T. (1975) "Congress and New Towns," *Public Administration Review* 35: 239-46.

Ashton, W. and Lightbody, J. (1990) "Reforming Alberta's Municipalities: Possibilities and Parameters," *Canadian Public Administration* 33: 506–25.

Ashworth, W. (1954) *The Genesis of Modern British Town Planning,* London: Routledge and Kegan Paul.

Asimow, M. (1983) "Delegated Legislation: United States and United Kingdom," *Oxford Journal of Legal Studies* 3: 253–76.

Audirac, I. and Zifou, M. (1989) *Urban Development Issues: What is Controversial in Urban Sprawl? An Annotated Bibliography of Often-Overlooked Sources*, Council of Planning Librarians, Bibliography 247.

Audirac, I., Shermyen, A.H., and Smith, M.T. (1990) "Ideal Urban Form and Visions of the Good Life: Florida's Growth Management Dilemma," *Journal of the American Planning Association* 56: 470–82.

Ayres, B.D. (1991) "Senators who Bring Home the Bacon," *New York Times*, September 6, 1991.

Baade, J.A. (1974) "Required Low-Income Housing in Residential Developments: Constitutional Challenges to a Community Imposed Quota," *Arizona Law Review* 16: 439–64.

Babcock, R.F. (1947) "The Illinois Supreme Court and Zoning: A Study in Uncertainty," *University of Chicago Law Review* 15: 87–105.

—— **(1959)** "The Unhappy State of Zoning Administration in Illinois," *University of Chicago Law Review* 26: 509–41.

—— **(1960)** "The Chaos of Zoning Administration: One Solution," *Zoning Digest* 12: 1.

—— **(1966)** *The Zoning Game: Municipal Practices and Policies*, Madison: University of Wisconsin Press.

—— **(1973)** *Exclusionary Zoning: Land Use Regulation and Housing in the 1970s*, New York: Praeger.

—— **(1980)** "The Spatial Impact of Land Use and Environmental Controls," in Solomon (1980).

—— **(1982)** "Houston: Unzoned, Unfettered, and Mostly Unrepentant," *Planning*, March 1982: 21–3.

—— **(1983)** "The Egregious Invalidity of the Exclusive Single-Family Zone," *Urban Law and Policy* 6: 185–97.

—— **(1990)** "The City as Entrepreneur: Fiscal Wisdom or Regulatory Folly?" in Lassar (1990).

—— **(ed.) (1987)** "Exactions: A Controversial New Source for Municipal Funds," *Law and Contemporary Problems* 50: 1–194.

Babcock, R.F. and Bosselman, F.P. (1973) *Exclusionary Zoning: Land Use and Housing in the 1970s*, New York: Praeger.

Babcock, R.F. and Feurer, D.A. (1977) "Land as a Commodity Affected with a Public Interest," *Washington Law Review* 52: 289–334.

Babcock, R.F. and Larsen, W.U. (1990) *Special Districts: The Ultimate in Neighborhood Zoning*, Cambridge, MA: Lincoln Institute of Land Policy.

Babcock, R.F. and Siemon, C.L. (1985) *The Zoning Game Revisited*, Boston: Oelgeschlager, Gunn and Hain.

Baden, J. (ed.) (1984) *The Vanishing Farmland Crisis: Critical Views of the Movement to Preserve Agricultural Land*, University of Kansas (Political Economy Research Center, Bozeman, Montana).

Baer, W.C. (1976) "On the Death of Cities," *The Public Interest* 45: 3–19 (Fall 1976).

—— **(1988)** "California's Housing Element: A Backdoor Approach to Metropolitan Governance and Regional Planning," *Town Planning Review* 59: 263–76.

Bagehot, W. (1867) *The English Constitution*, (reprinted 1963, with an introduction by R.H.S. Crossman), London: Collins/Fontana.

Bahl, R. (ed.) (1978) *The Fiscal Outlook for Cities: Implications of a National Urban Policy*, Syracuse: Syracuse University Press.

Bair, F.H. (1968) "Mobile Homes – A New Challenge," in Everett and Johnston (1968).

—— **(1984)** *The Zoning Board Manual*, Chicago: Planners Press.

Baker, N.F. (1925) "The Constitutionality of Zoning Laws," *Illinois Law Review* 20: 213–48.

—— **(1926)** "Zoning Legislation," *Cornell Law Quarterly* 11: 164–78.

Baldassare, M. (1981) *The Growth Dilemma: Residents' Views and Local Population Change in the United States*, Berkeley: University of California Press.

—— **(1986)** *Trouble in Paradise: The Suburban Transformation of America*, New York: Columbia University Press.

Baldassare, M. and Protash, W. (1982) "Growth Controls, Population Growth, and

Community Satisfaction," *American Sociological Review* 47: 339–46.

Banach, M. and Canavan, D. (1987) "Montgomery County Agricultural Preservation Program," in Brower and Carol (1987).

—— **(1989)** "Montgomery County, Maryland: A Transfer of Development Rights Success Story" in Hiemstra and Bushwick (1989).

Banfield, E.C. (1961) "The Political Implications of Metropolitan Growth," *Daedalus* 90: 61–78.

—— **(1973)** "Making a New Federal Program: Model Cities, 1964–68," in Sindler, A.P. (1973) *Policy and Politics in America*, Boston: Little, Brown.

—— **(1975)** "Corruption as a Feature of Governmental Organization," *Journal of Law and Economics* 18: 587–615.

Banfield, E.C. and Wilson, J.Q. (1963) *City Politics*, Cambridge, MA: Harvard University Press and MIT Press.

Banham, R., Barker, P., Hall, P., and Price, C. (1969) "Non-Plan: An Experiment in Freedom," *New Society* 26: 435–43.

Barewin, H.J. (1990) "Rescuing Manufactured Housing From the Perils of Municipal Zoning Laws," *Washington University Journal of Urban and Contemporary Law* 37: 189–213.

Barker, E. (1927) *National Character and the Factors in its Formation*, New York: Harper.

Barker, M. (ed.) (1984) *Rebuilding America's Infrastructure: An Agenda for the 1980s*, Durham, NC: Duke University Press.

Barlow Report (1940) *Report of the Royal Commission on the Distribution of the Industrial Population*, Cmd 6153, London: HMSO.

Barnekov, T.K. and Rich, D (1989) "Privatism and the Limits of Local Economic Policy," *Urban Affairs Quarterly* 25: 212–37.

Barnekov, T.K., Boyle, R., and Rich, D. (1989) *Privatism and Urban Policy in Britain and the United States*, Oxford: Oxford University Press.

Barnekov, T.K., Hart, D., and Benfer, W. (1990) *US Experience in Evaluating Urban Regeneration*, London: HMSO.

Barnekov, T.K., Rich, D., and Warren, R. (1981) "The New Privatism, Federalism, and the Future of Urban Governance: National

Urban Policy in the 1980s," *Journal of Urban Affairs* 3: 1–14.

Barnett, J. (1982) *An Introduction to Urban Design*, New York: Harper and Row.

Barro, S.M. (1978) *The Urban Impacts of Federal Policies: Vol.3, Fiscal Conditions*, Santa Monica, CA: Rand Corporation.

Barrows, R.L. (1982) *The Roles of Federal, State and Local Government in Land Use Planning*, Washington: National Planning Association.

Bartke, R.W., and Lamb, J.S. (1976) "Upzoning, Public Policy and Fairness: A Study and Proposal," *William and Mary Law Review* 17: 701–36.

Bashaw, J.R. (1985) "Vermont Legislation: Formal Enforcement of Act 250," *Vermont Law Review* 10: 469–77.

Bassett, E.M. (1924) "Constitutionality of Zoning in the Light of Recent Court Decisions," *National Municipal Review* 13: 492–7.

—— **(1940)** *Zoning: The Laws, Administration, and Court Decisions During the First Twenty Years*, (revised edn.) New York: Russell Sage Foundation.

Baumbach, R.O. and Borah, W.E. (1981) *The Second Battle of New Orleans: A History of the Vieux Carre Riverfront Expressway Controversy*, Birmingham: University of Alabama Press.

Baumol, W.J. (1981) "Technological Change and the New Urban Equilibrium," in Burchell and Listokin (1981).

Bayefsky, A.F. (1987) "The Judicial Function under the Canadian Charter of Rights and Freedoms," *McGill Law Journal* 32: 791–833.

Beatley, T. (1988) "Ethical Issues in the Use of Impact Fees to Finance Community Growth," in Nelson (1988b).

Beaton, W.P. (1991) "The Impact of Regional Land Use Controls on Property Values: The Case of the New Jersey Pinelands," *Land Economics* 67: 172–94.

Beatson, J. (1981) "A British View of *Vermont Yankee*," *TulaneLaw Review* 55: 435–64.

Bell, D. and Tepperman, L. (1979) *The Roots of Disunity: A Look at Canadian Political Culture*, Toronto: McClelland and Stewart.

Bellush, J. and Hausknecht, M. (eds.) (1967) *Urban Renewal: People, Politics and Planning*, Garden City, NY: Anchor Books.

Bendick, M. and Rasmussen, D.W. (1986) "Enterprise Zones and Inner-City Economic Revitalization," in Peterson and Lewis (1986).

Benko, R.G. and Hand, I. (1985) "State Planning Today," in So, Hand, and McDowell (1985).

Benveniste, G. (1990) *Mastering the Politics of Planning: Crafting Credible Plans and Policies That Make a Difference*, San Francisco: Jossey-Bass.

Berger, P.L. and Luckmann, T. (1966) *The Social Construction of Reality*, New York: Doubleday (Anchor Books 1967).

Bermingham, T.W. (1988) *Section 36: Density and Height Bonuses*, Canadian Bar Association, Toronto.

Bernard, R.M. and Rice, B.R. (eds.) (1983) *Sunbelt Cities: Politics and Growth Since World War II*, Austin: University of Texas Press.

Berry, B.J.L. (1973) *The Human Consequences of Urbanization: Divergent Paths in the Urban Experience of the Twentieth Century*, London: Macmillan.

Berton, P. (1982) *Why We Act Like Canadians: A Personal Exploration of Our National Character*, Toronto: McClelland and Stewart.

Berwanger, C. (1987) "Land Use Planning by Initiative," *Los Angeles Lawyer* 9: 43–9.

Best, R.K. (1988) "*Nollan* Sets New Test and Standards for Exactions," in Gordon (1988).

Bettman, A. (1927) "The Decision of the Supreme Court of the United States in the Euclid Village Zoning Case," *University of Cincinnati Law Review* 1: 184–92.

Bickel, A. (1962) *The Least Dangerous Branch: The Supreme Court at the Bar of Politics*, Indianapolis: Bobbs-Merrill.

Biddle, S.G. (1989) "Justice Stevens' Pro-Land Use Planning Views and Recent Regulatory Takings Cases," *Urban Lawyer* 21: 579–88.

Biggs, J.H. (1990) "No Drip, No Flush, No Growth: How Cities Can Control Growth Beyond Their Boundaries by Refusing to Extend Utility Services," *Urban Lawyer* 22: 285–305.

Birch, A.H. (1990) *The British System of Government (8th edn.)*, London: Unwin Hyman.

Birch, E.L. (1984) "The Planner and the Preservationist: An Uneasy Alliance," *Journal of the American Planning Association* 50: 194–207.

Bishir, C.W. (1989) "Yuppies, Bubbas, and the Politics of Culture," in Carter and Herman (1989).

Black, E. (1977) *Southern Governors and Civil Rights: Racial Segregation as a Campaign Issue in the Second Reconstruction*, Cambridge, MA: Harvard University Press.

Blackwell, R.J. (1990) "Overlay Zoning, Performance Standards, and Environmental Protection After *Nollan*," in Dennison (1990) (also in *Boston College Environmental Affairs Law Review* 16: 615–59).

Blaesser, B.W. and Kentopp, C.M. (1990) "Impact Fees: The 'Second Generation'," *Washington University Journal of Urban and Contemporary Law* 38: 55–114.

Blaeseer, B.W. and Weinstein, A.C. (1989) *Land Use and the Constitution: Principles for Planning Practice*, Chicago: Planners Press.

Blake, P. (1964) *God's Own Junkyard: The Planned Deterioration of America's Landscape*, New York: Holt, Reinhart and Winston.

Blucher, W. (1956) "Is Zoning Wagging the Dog?," *Planning 1955*, Washington: American Society of Planning Officials.

—— **(1960)** "Let Us Throw Out the Baby to Get Rid of the Bathtub," *Zoning Digest* 12: 145–8.

Blumenthal, S. and Siler, B. (1990) *Tax Incentives for Rehabilitating Historic Buildings: Fiscal Year 1989 Analysis*, Washington: National Trust for Historic Preservation.

Bollinger, S.J. (1983) "Federal Enterprise Zones: The Administration's View," *Economic Development Commentary* 7, 2: 2–3.

Boorstin, D.J. (1953) *The Genius of American Politics*, Chicago: University of Chicago Press.

Booth, P. (1989) "How Effective is Zoning in the Control of Development?," *Environment and Planning B: Planning and Design* 16: 401–15.

Bosselman, F. (1973) "Can the Town of Ramapo Pass a Law to Bind the Rights of the Whole World?," *Florida State University Law Review* 1: 234–65.

—— **(1985)** "Downtown Linkages: Legal Issues," in D.R. Porter (1985).

Bosselman, F. and Callies, D. (1972) *The Quiet Revolution in Land Use Control*, Council on

Environmental Quality, Washington: U.S. GPO.

Bosselman, F. and Stroud, N.E. (1985) "Pariah to Paragon: Developer Exactions in Florida 1975–1985," *Stetson Law Review* 14: 527–63.

Bosselman, F., Callies D., and Banta J. (1973) *The Taking Issue: A Study of the Constitutional Limits of Governmental Authority to Regulate the Use of Privately-Owned Land Without Paying Compensation to the Owners*, Washington: Council on Environmental Quality, U.S. GPO.

Bosselman, F., Raymond, G.M., and Persico, R.A. (1976) "Some Observations on the American Law Institute's Model Land Development Code," *Urban Lawyer* 8: 474–90.

Boston, City of (1988) *Design Guidelines for Neighborhood Housing*, Boston: City of Boston Public Facilities Department.

Bothwell, R., Drumond, I., and English, J. (1989) *Canada Since 1945*, (revised edn.), Toronto: University of Toronto Press.

Bourne, L.S. (1975) *Urban Systems: Strategies for Regulation – A Comparison of Policies in Britain, Sweden, Australia, and Canada*, Oxford: Clarendon Press.

Boyer, M.C. (1983) *Dreaming the Rational City: The Myth of American City Planning*, Cambridge, MA: MIT Press.

Bozung, L.J. (1982) "A Positive Response to Growth Control Plans: The Orange County Inclusionary Housing Program," *Pepperdine Law Review* 9: 819–51.

——— **(1983)** "Transfer Development Rights: Compensation for Owners of Restricted Property," *Zoning and Planning Law Report* 6: 129–35.

Bozung, L.J. and Alessi, D.J. (1988) "Recent Developments in Environmental Preservation and the Rights of Property Owners," *Urban Lawyer* 20: 969–1069.

Bozung, L.J. and McRoberts, M.R. (1987) "Land Use, Planning and Zoning in 1987," *Urban Lawyer* 19: 899–986.

Bradbury, K.L., Downs, A., and Small, K.A. (1982) *Urban Decline and the Future of American Cities*, Washington: Brookings Institution.

Brebner, J.B. (1960) *Canada: A Modern History*,

Ann Arbor: University of Michigan Press.

Brennan, W.J. (1977) "State Constitutions and the Protection of Individual Rights," *Harvard Law Review* 90: 489–504.

Briggs, A. (1952) *History of Birmingham*, Oxford: Oxford University Press.

Brindley, T., Rydin, Y., and Stoker, G. (1989) *Remaking Planning*, London: Unwin Hyman.

Brintnall, M. (1989) "Future Directions for Federal Urban Policy," *Journal of Urban Affairs* 11: 1–19.

Brintnall, M. and Green, R.E. (1988) "Comparative State Enterprise Zone Programs," *Economic Development Quarterly* 2: 50–68.

Brock, W.R. (1987) "Americanism," in Welland (1987).

Brogan, H. (1986) *The Pelican History of the United States of America*, Harmondsworth: Penguin (originally published by Longman, 1985).

Brolin, B.C. (1980) *Architecture in Context: Fitting New Buildings With Old*, New York: Van Nostrand Reinhold.

Brooks, A.V.N. (1989) "The Office File Box – Emanations from the Battlefield" [The *Euclid* case], in Haar and Kayden (1989a).

Brooks, M. (1972) *Lower Income Housing: The Planners' Response*, Washington: American Society of Planning Officials, Planning Advisory Service Report 282.

Brooks, R. and Lavigne, P. (1985) "Aesthetic Theory and Landscape Protection: The Many Meanings of Beauty and Their Implications for the Design, Control, and Protection of Vermont's Landscape," *Journal of Environmental Law* 4: 129–72.

Brough, M.B. (1985) *A Unified Development Ordinance*, Chicago: Planners Press.

Brower, D.J. and Carol, D.S. (1984) *Coastal Zone Management as Land Planning*, Washington: National Planning Association.

Brower, D.J. and Carol, D.S. (eds.) (1987) *Managing Land Use Conflicts: Case Studies in Special Area Management*, Durham, NC: Duke University Press.

Brower, D.J., Godschalk, D.R., and Porter, D.R. (1989) *Understanding Growth Management: Critical Issues and a Research Agenda*, Washington: Urban Land Institute.

Brown, J.G. (1986) "Concomitant Agreement Zoning: An Economic Analysis," in Deutsch (1986).

Brown, J.M. and Sellman, M.A. (1987) "Manufactured Housing: The Invalidity of the 'Mobility' Standard," *Urban Lawyer* 19: 367–99.

Brown, H.J., Phillips, S., and Roberts, N.A. (1981) "Land Markets at the Urban Fringe: New Insights for Policy Makers," *Journal of the American Planning Association* 47: 131–44.

Brown, P.G. (1975) *The American Law Institute Model Land Development Code, The Taking Issue and Private Property Rights*, Washington: Urban Institute.

Brownell, B.A. (1980) "Urban Planning, the Planning Profession,and the Motor Vehicle in Early Twentieth-Century America," in Cherry (1980).

Brownstone, M. and Plunkett, T.J. (1983) *Metropolitan Winnipeg: Politics and Reform of Local Government*, Berkeley: University of California Press.

Bruce, A.A. (1926–27) "Racial Zoning by Private Contract in the Light of the Constitutions and the Rule Against Restraints on Alienation," *Illinois Law Review* 21: 704–17.

Brueckner, J.K. (1990) "Growth Controls and Land Values," *Land Economics* 66: 237–48.

Brumback, B.C. and Marvin, M.J. (eds.) (1989) *Implementation of the 1985 Growth Management Act: From Planning to Land Development Regulations*, Miami: Florida Atlantic University and Florida International University Joint Center for Environmental and Urban Problems.

Bryden, D.P. (1977) "The Impact of Variances: A Study of Statewide Zoning," *Minnesota Law Review* 61: 769–840.

Bryden, R.M. (1967) "Zoning: Rigid, Flexible, or Fluid?," *Journal of Urban Law* 44: 287–326.

Bryson, J.M. and Einsweiler, R.C. (1988) *Strategic Planning: Threats and Opportunities for Planners*, Chicago: Planners Press.

Bucknall, B. (1988) *Of Deals and Distrust: The Perplexing Perils of Municipal Zoning*, unpublished paper (Toronto).

Bufford, S. (1980) "Beyond the Eye of the Beholder: A New Majority of Jurisdictions Authorize Aesthetic Regulation," *University of Missouri-Kansas City Law Review* 48: 125–66.

Bunce, H.L. and Glickman, N.J. (1980) "The Spatial Dimensions of the Community Development Block Grant Program: Targeting and Urban Impacts," in Glickman (1980).

Bunting, T. and Filion, P. (eds.) (1991) *Canadian Cities in Transition*, Oxford: Oxford University Press.

Burchell, R.W. and Listokin, D. (eds.) (1981) *Cities Under Stress: The Fiscal Crises of Urban America*, New Brunswick, NJ: Center for Urban Policy Research, Rutgers University.

Burchell, R.W., Beaton, W.P., Listokin, D., Sternlieb, G., Lake, R.W., and Florida, R.L. (1983) *Mount Laurel II: Challenge and Delivery of Low-Cost Housing*, New Brunswick, NJ: Center for Urban Policy Research, Rutgers University.

Burns, J.M. (1989) *The Crossroads of Freedom* (vol III of *The American Experiment*), New York: Knopf.

Burrows, L.B. (1978) *Growth Management: Issues, Techniques, and Policy Implications*, New Brunswick, NJ: Center for Urban Policy Research, Rutgers University.

Burrows, T. (ed.) (1989) *A Survey of Zoning Definitions*, Planning Advisory Service Report 421, Washington: American Planning Association

Butler, D. and Butler, G. (1986) *British Political Facts 1900–1988* (6th edn.), London: Macmillan.

Butler, J.A. and Getzels, J. (1985) *Home Occupancy Ordinances*, Planning Advisory Service Report 391, Washington: American Planning Association.

Butler, S.M. (1981) *Enterprise Zones: Greenlining the Inner Cities*, New York: Universe Books.

—— **(1991)** "The Conceptual Evolution of Enterprise Zones," in R.E. Green, (1991).

Butowsky, H.A. (1986) *The U.S. Constitution: A National Historic Landmark Theme Study*, Washington: National Park Service, U.S. GPO.

Byers, N.G. and Wilson, L.U. (1983) *Managing Rural Growth: The Vermont Development Review Process*, Montpelier: State of Vermont Environmental Board.

Byrne, J., Martinez, C., and Rich, D. (1985) "The Post-Industrial Imperative: Energy, Cities, and the Featureless Plain," in Byrne J. and Rich, D. *Energy and Cities*, New Brunswick: Transaction Books.

Byrne, T. (1990) *Local Government in Britain* (5th edn.), Harmondsworth: Penguin.

Cahan, R. (1979) "Rescuing Our Architectural Heritage," *Barrister* 6: 47–52.

Cairns, A.C. (1988) *Constitution, Government, and Society in Canada: Selected Essays by Alan C. Cairns* (edited by Williams, D.E.), Toronto: McClelland and Stewart.

California, State of (1976) *California Coastal Act of 1976* (as of January 1990).

—— **(1986a)** *Mobile Homes in California*, Sacramento: State of California, Department of Housing and Community Development.

—— **(1986b)** *California Affordable Housing Legislation: A Study of Local Implementation* (2 vols.), Sacramento: State of California, Department of Housing and Community Development.

—— **(1989)** *Planning, Zoning, and Development Laws 1989*, Sacramento: Governor's Office, Office of Planning and Research.

—— **(1990a)** *State Density Bonus Law*, Sacramento: State of California, Department of Housing and Community Development.

—— **(1990b)** *Manufactured Housing for Families: Innovative Land Use and Design*, Sacramento: State of California, Department of Housing and Community Development.

California Coastal Commission (1989) *Briefing on Transfer of Development Credit Program in Malibu/Santa Monica Mountains*, San Francisco: California Coastal Commission.

California Coastal Conservancy (1986) *The California State Coastal Conservancy*, San Francisco: California State Coastal Conservancy.

—— **(1988)** *Annual Report 1988*, San Francisco: California State Coastal Conservancy.

California: The 2000 Partnership (1990) *Managing Growth in Southern California*, San Francisco: California: The 2000 Partnership.

Callies, D.L. (1980) "The Quiet Revolution Revisited," *Journal of the American Planning Association* 46: 135–44.

—— **(1984)** *Regulating Paradise: Land Use Controls in Hawaii*, Honolulu: University of Hawaii Press.

—— **(1985)** "Developers' Agreements and Planning Gain," *Urban Lawyer* 17: 599–612.

—— **(1988)** "Property Rights: Are There Any Left?," *Urban Lawyer* 20: 597–629.

—— **(1989a)** "Review Essay: Impact Fees, Exactions, and Paying for Growth in Hawaii," *University of Hawaii Law Review* 11: 295–333.

—— **(1989b)** "How Development Agreements Work in Hawaii," in Porter and Marsh (1989).

Callies, D.L. and Freilich, R.H. (1986) *Cases and Materials on Land Use*, St Paul, MN: West.

—— **(1988)** *1988 Supplement to Cases and Materials on Land Use*, St Paul, MN: West.

—— **(1991)** *1991 Supplement to Cases and Materials on Land Use*, St Paul, MN: West.

Callies, D.L. and Grant, M. (1991) "Paying for Growth and Planning Gain: An Anglo-American Comparison of Development Conditions, Impact Fees, and Development Agreements," *Urban Lawyer* 23: 221–48.

Callies, D.L. and Weaver, C.L. (1977) "The *Arlington Heights* Case: The Exclusion of Exclusionary Zoning Challenges," *Real Estate Issues* 2: 22–9.

—— **(1987)** "The *Arlington Heights* Case: A Reprise," *Real Estate Issues* 3: 37–40.

Callies, D.L., Neuffer, N.C., and Caliboso, C.P. (1991) "Ballot Box Zoning: Initiative, Referendum and the Law," *Journal of Urban and Contemporary Law* 39: 53–98.

Cameron, D.M. (1980) "Provincial Responsibilities for Municipal Government," *Canadian Public Administration* 23: 222–35.

Canada Year Book (annual), Ottawa: Statistics Canada.

***Canadian Encyclopedia* (1985)** Edmonton: Hurtig.

Canty, D. (ed.) (1969) *The New City* (National Committee on Urban Growth Policy), New York: Praeger.

Caplan, E.G. (1989) "Child Care Land Use Ordinances," in Dennison (1989).

Carnwath, R., Hart, G., and Williams, A. (1990) *Blundell and Dobry's Planning Appeals and Inquiries*, London: Sweet and Maxwell.

Carr, J.H. and Duensing, E.E. (1983) *Land Use Issues of the 1980s*, New Brunswick, NJ: Center for Urban Policy Research, Rutgers University.

Carstensen, V. (ed.) (1962) *The Public Lands: Studies in the History of the Public Domain*, Madison: University of Wisconsin Press.

Carter, L.J. (1974) *The Florida Experience: Land and Water Policy in a Growth State*, Baltimore: Johns Hopkins University Press.

Carter, T. and Herman, B.L (eds.) (1989) *Perspectives in Vernacular Architecture III*, Columbia: University of Missouri Press.

Carver, H. (1948) *Houses for Canadians*, Toronto: University of Toronto Press.

—— (1960) "Planning in Canada," *Habitat* 3, 5: 2–5.

—— (1962) *Cities in the Suburbs*, Toronto: University of Toronto Press.

—— (1975) *Compassionate Landscape*, Toronto: University of Toronto Press.

Case, F.E. and Gale, J. (1981) "Impact of Housing Costs From the California Coastal Zone Act," *Journal of the American Real Estate and Urban Economics Association* 9:345–66.

Castells, M. (1983) *The City and the Grass Roots*, Berkeley: University of California Press.

—— (ed.) (1985) *High Technology, Space and Society*, London: Sage.

Catanese, A.J. and Snyder, J.C. (eds.) (1988) *Urban Planning*, New York: McGraw-Hill.

Caves, R. (1991) *Land Use Planning: The Ballot Box Revolution*, Newbury Park, CA: Sage.

Cervero, R. (1986) *Suburban Gridlock*, New Brunswick, NJ: Center for Urban Policy Research, Rutgers University.

Chall, D.E. (1985) "Housing Reform in New Jersey: The Mount Laurel Decision," *Federal Reserve Bank of New York Quarterly Review* 10,4: 19–27 (Winter 1985).

Chapin, F.S. and Kaiser, E.J. (1979) *Urban Land Use Planning*, Champaign: University of Illinois Press.

Charles, Prince of Wales (1989) *A Vision of Britain: A Personal View of Architecture*, London: Doubleday.

Chavooshian, B.B. and Norman, T. (1973) "Transfer of Development Rights: A New Concept," *Urban Land* 32,11: 11–16.

Cheney, C.H. (1917) "Districting Progress and Procedure in California," *Proceedings of the Ninth National Conference on City Planning* (quoted in M. Scott, 1969: 161).

—— (1920a) "Zoning in Practice," *National Municipal Review* 9: 31–43.

—— (1920b) "Removing Social Barriers by Zoning," *The Survey* 44: 275–8 (May 22, 1920).

Chernoff, S.N. (1983) "Behind the Smokescreen: Exclusionary Zoning of Mobile Homes," *Journal of Urban and Contemporary Law*, 25: 235–68.

Cherry, G.E. (1980) *Shaping an Urban World*, London: Mansell.

—— (1981) *Pioneers in British Planning*, London: Architectural Press.

Chinitz, B. (1990) "Growth Management: Good for the Town, Bad for the Nation?," *Journal of the American Planning Association* 56: 3–8.

Chittenden, B. (1988) *Tax Incentives for Rehabilitating Historic Buildings: 1988*, Washington: National Trust for Historic Preservation.

Christeller, N.L. (1986) "Wrestling with Growth in Montgomery County, Maryland," in D.R. Porter, (1986).

Church, A. (1988) "Urban Regeneration in London Docklands: A Five Year Policy Review," *Environment and Planning C: Government and Policy* 6: 187–208.

Cibulskis, A. and Ritzdorf, M. (1989) *Zoning for Child Care*, Washington: American Planning Association, Planning Advisory Service Report 422.

Cirler, B.A. (1980) "Local Growth Management: Changing Assumptions About Land?," *Current Municipal Problems* 1980: 443–54.

Citizens' Commission on Civil Rights (1983) *A Decent Home: A Report on the Continuing Failure of the Federal Government to Provide Equal Housing Opportunity*, Washington: Center for National Policy Review.

City of . . . *see* name of city.

Clark, G.L. (1985) *Judges and the Cities: Interpreting Local Autonomy*, Chicago: University of Chicago Press.

Clark, G.L. and Dear, M. (1984) *State Apparatus*, London: Allen and Unwin.

Clarke, S.E. (1982) "Enterprise Zones: Seeking the Neighborhood Nexus," *Urban Affairs Quarterly* 18: 53–71.

—— **(1984)** "Neighborhood Policy Options: The Reagan Agenda," *Journal of the American Planning Association* 50: 493–501.

Clavel, P. (1982) *Opposition Planning in Appalachia and Wales*, Philadelphia: Temple University Press.

Clavel, P., Forester, J., and Goldsmith, W. (eds.) (1980) *Urban and Regional Planning in an Age of Austerity*, Oxford: Pergamon.

Clawson, M. (1981) *New Deal Planning: The National Resources Planning Board*, Baltimore: Resources for the Future, Johns Hopkins University Press.

—— **(1983)** *The Federal Lands Revisited*, Baltimore: Resources for the Future, Johns Hopkins University Press.

—— **(ed.) (1973)** *Modernizing Urban Land Policy*, Baltimore: Johns Hopkins University Press.

Clawson, M. and Hall, P. (1973) *Planning and Urban Growth: An Anglo-American Comparison*, Baltimore: Resources for the Future, Johns Hopkins University Press.

Cleaveland, F.N. (1969) "Legislating for Urban Areas: An Overview," in Cleaveland and Associates (1969).

Cleaveland, F.N. and Associates (1969) *Congress and Urban Problems*, Washington: Brookings Institution.

Cloke, P. and Little, J. (1990) *The Rural State? Limits to Planning in Rural Society*, Oxford: Clarendon Press.

Cohen, A. (1989) *Family Day Care Zoning*, Washington: National League of Cities.

Cohen, C.R. (1977) "The Equal Protection Clause and the Fair Housing Act: Judicial Alternatives for Exclusionary Zoning Challenges After *Arlington Heights*, *Environmental Affairs* 6: 63–99.

Cohen, N.M. (1983) "The Reagan Administration's Urban Policy," *Town Planning Review* 54: 304–15.

Coke, J.G. and Liebman, C.S. (1961) "Political Values and Population Density Control," *Land Economics* 37: 347–61.

Cole, R.L., Taebel, D.A., and Hissong, R.V. (1990) "America's Cities and the 1980s: The Legacy of the Reagan Years," *Journal of Urban Affairs* 12: 345–60.

Collier, M. (1975) *Contemporary Cathedrals*, Eugene, OR: Harvest House.

Collin, R. and Lytton, M. (1989) "Linkage; An Evaluation and Exploration," *Urban Lawyer* 21: 413–32.

Collins, B.R. and Russell, E.W.B. (eds.) (1988) *Protecting the New Jersey Pinelands*, New Brunswick, NJ: Rutgers University Press.

Collins, R.C., Waters, E.B., and Dotson, A.B. (1991) *America's Downtowns: Growth Politics and Preservation*, Washington: Preservation Press.

Colombo, J.R. (1976) *Colombo's Concise Canadian Quotations*, Edmonton: Hurtig.

Colwell, C. (1989) "Child Care Grows Up," *Planning* 55, 5: 12 (May 1989).

Comay Report (1977) *Report of the Ontario Planning Act Review Committee*, Toronto: Ontario Government Publications.

Commission on Civil Rights (1983) *A Sheltered Crisis: The State of Fair Housing in the Eighties*, Washington: Commission on Civil Rights.

Commission on Population Growth and the American Future (1971) "An Interim Report to the President and to Congress," printed in *Congressional Record* S 3460–66.

Conlan, T. (1988) *New Federalism: Intergovernmental Reform from Nixon to Reagan*, Washington: Brookings Institution.

Connerly, C.E. (1988) "The Social Implications of Impact Fees," *Journal of the American Planning Association* 54: 75–8.

Connors, D.L. and High, M.E. (1987) "The Expanding Circle of Exactions: From Dedication to Linkage," *Law and Contemporary Problems* 50: 69–83.

Conway, J. (1989) "An Adapted Organic Tradition," in Graubard (1989).

Cook, J. (1980) *Zoning for Downtown Urban Design*, Lexington, MA: Lexington.

Cooper, G. and Daws, G. (1985) *Land and Power in Hawaii*, Honolulu: Benchmark Books.

Cooper, S. (1986) "Growth Control in Boulder," in D.R. Porter (1986).

Corbet, J.G. (1978) "Canadian Cities: How 'American' Are They?," *Urban Affairs Quarterly* 13: 383-94.

Corden, C. (1977) *Planned Cities: New Towns in*

Britain and America, Newbury Park, CA: Sage.

Corke, S.E. (1983) *Land Use Control in British Columbia: A Contribution to a Comparative Study of Canadian Planning Systems*, Research Paper 138, Toronto: Centre for Urban and Community Studies, University of Toronto.

Costonis, J.J. (1972) "The Chicago Plan: Incentive Zoning and the Preservation of Urban Landmarks," *Harvard Law Review* 85: 574–634.

—— **(1973)** "Development Rights Transfer: An Exploratory Essay," *Yale Law Journal* 83: 75–128.

—— **(1974)** *Space Adrift: Saving Urban Landmarks Through the Chicago Plan*, Urbana-Champaign: University of Illinois Press.

—— **(1977)** "The Disparity Issue: A Context for the Grand Central Terminal Decision," *Harvard Law Review* 91: 402–26.

—— **(1982)** "Law and Aesthetics: A Critique and a Reformulation of the Dilemmas," *Michigan Law Review* 80: 355–461 (also in Deutsch 1983).

—— **(1989)** *Icons and Aliens: Law, Aesthetics, and Environmental Change*, Urbana-Champaign: University of Illinois.

Costonis, J.J., Berger, C.J., and Scott, S. (1977) *Regulation v Compensation in Land Use Control: A Recommendation, A Critique, and An Interpretation*, Berkeley: Institute of Governmental Studies, University of California.

Council of State Governments (1972) *The States' Role in Land Resource Management*, Lexington, KY: Council of State Governments.

—— **(1976)** *State Growth Management*, Lexington, KY: Council of State Governments.

Council of State Planning Agencies (1977) *History of State Planning: An Interpretive Commentary*, Washington: Council of State Planning Agencies.

Cox, W.M. (1988) *New Jersey Zoning and Land Use Administration*, Newark, NJ: Gann Law Books.

Crawford, C. (1969) *Strategy and Tactics in Municipal Zoning*, New York: Prentice-Hall.

Crawford, K.G. (1954) *Canadian Municipal Government*, Toronto: University of Toronto Press.

Crew, M.H. (1990) "Development Agreements After *Nollan v California Coastal Commission*," *Urban Lawyer* 22: 23–58.

Crosbie, J.C. (1956) "Local Government in Newfoundland," *Canadian Journal of Economic and Political Science* 22: 332–46.

Crosson, P. (ed.) (1982) *The Cropland Crisis*, Baltimore: Resources for the Future, Johns Hopkins University Press.

Crouch, C. and Marquand, D. (1989) *The New Centralism: Britain Out of Step in Europe?*, Oxford: Blackwell.

Crozier, M. (1984) *The Trouble with America: Why the System is Breaking Down*, Berkeley: University of California Press.

Crumplar, T. (1974) "Architectural Controls: Aesthetic Regulation of the Urban Environment," *Urban Lawyer* 6: 622–44.

Cullingworth, J.B. (1960) *Housing Needs and Planning Policy: A Restatement of the Problems of Housing Need and "Overspill" in England and Wales*, London: Routledge and Kegan Paul.

—— **(1966)** *Housing and Local Government in England and Wales*, London: Allen and Unwin.

—— **(1973)** *Problems of an Urban Society, Vol 3: Planning for Change*, London: Allen and Unwin.

—— **(1975)** *Environmental Planning 1939–1969, Vol I: Reconstruction and Land Use Planning 1939–1947*, London: HMSO.

—— **(1979a)** *Environmental Planning 1939–1969, Vol III: New Towns Policy*, London: HMSO.

—— **(1979b)** *Essays on Housing Policy: The British Scene*, London: Allen and Unwin.

—— **(1980)** *Environmental Planning 1939–1969, Vol IV: Land Values, Compensation, and Betterment*, London: HMSO.

—— **(1987)** *Urban and Regional Planning in Canada*, New Brunswick, NJ: Transaction Books.

—— **(1988)** *Town and Country Planning in England and Wales* (10th edn.), London: Unwin Hyman.

—— **(ed.) (1990)** *Energy, Land, and Public Policy*, Energy Policy Series 5, New

Brunswick, NJ: Transaction Books.

Cunliffe, M. (1991) *In Search of America: Transatlantic Essays, 1951–1990*, Westport, CT: Greenwood.

Curtin, D.J. (1989) "Status of Exactions After *First Lutheran Church* and *Nollan*," in Dennison (1989).

Curtin, D.J. and Jacobson, M.T. (1989) "Growth Management by the Initiative in California: Legal and Practical Issues," *Urban Lawyer* 21: 491–510.

Dahl, R. (1967) *Pluralist Democracy in the United States*, Skokie, IL: Rand McNally.

—— **(1971)** *Polyarchy*, New Haven: Yale University Press.

Dale, L. and Burton, T.L. (1984) "Regional Planning in Alberta: Performance and Prospects," *Alberta Journal of Planning Practice* 3: 17–41.

Dales, J.H. (1968) *Pollution, Property and Prices*, Toronto: University of Toronto Press.

Danels, P. and Magida, L. (1979) "Application of Transfer of Development Rights to Inner City Communities: A Proposed Municipal Land Use Rights Act," *Urban Lawyer* 11: 124–38.

Daniels, T.L. and Lapping, M.B. (1984) "Has Vermont's Land Use Control Failed?," *Journal of the American Planning Association* 50: 502–8.

Danielson, M.N. (1976) *The Politics of Exclusion*, New York: Columbia University Press.

Dart, A.K. and Costa, F.J. (1985) *Public Planning in the Netherlands: Perspectives and Change Since the Second World War*, Oxford: Oxford University Press.

David, S. and Kantor, P. (1983) "Urban Policy in the Federal System: Reconceptualization of Federalism," *Polity* 16: 284–303.

Davidoff, P. and Davidoff, L. (1971) "Opening the Suburbs: Toward Inclusionary Land Use Controls," *Syracuse Law Review* 22: 509–36.

Davidoff, P., Davidoff, L., and Gold, N.N. (1970) "Suburban Action: Advocate Planning for an Open Society," *Journal of the American Institute of Planners* 36: 12–21.

Davidson, D. and Usagawa, A. (eds.) (1988) *Paying for Growth in Hawaii: An Analysis of Impact Fees and Housing Exactions Programs,* Honolulu: Land Use Research Foundation of Hawaii.

Davidson, J.M. (1986) "Plan-Based Land Development and Infrastructure Controls: New Directions for Growth Management," *Journal of Land Use and Environment Law* 2: 151–75.

Davies, G.W. (1977) "A Model of the Urban Residential Land and Housing Markets," *Canadian Journal of Economics* 10: 393–410.

Davies, H.W.E., Edwards, D., Hooper, A.J., and Punter, J.V. (1989) *Planning Control in Western Europe*, London: HMSO.

Davies, P.J. (1958) *Real Estate in American History*, Washington: Public Affairs Press.

Davies, R.O. (1966) *Housing Reform under the Truman Administration*, Columbia: University of Missouri Press.

Davis, M. (1985) *State Tax Incentives for Historic Preservation*, State Legislation Project, Washington: National Trust for Historic Preservation.

—— **(1987)** *State Systems for Designating Historic Properties and the Results of Designation*, State Legislation Project, Washington: National Trust for Historic Preservation.

Dawson, G. (1977) *No Little Plans: Fairfax County's PLUS Program for Managing Growth*, Washington: Urban Institute.

Dawson, J. and Walker, C. (1990) "Mitigating the Social Costs of Private Development," *Town Planning Review* 61: 157–70.

Dear, M.J. and Taylor, S.M. (1982) *Not on Our Street: Community Attitudes to Mental Health Care*, London: Pion.

Dear, M.J. and Wolch, J.R. (1987) *Landscapes of Despair: From Deinstitutionalization to Homelessness*, Princeton: Princeton University Press.

Dearlove, J. and Saunders, P. (1984) *Introduction to British Politics*, Cambridge: Polity Press.

Decker, J.E. (1987) "Management and Organizational Capacities for Responding to Growth in Florida's Nonmetropolitan Counties," *Journal of Urban Affairs* 9: 47–61.

deCler, C. (1967) "Dutch Land Use Planning" in Eldredge, H.W. (ed.) *Taming Megalopolis* (vol. 2), New York: Praeger.

DeGrove, J.M. (1984) *Land Growth and Politics*, Chicago: Planners Press.
—— **(1988)** "The Battle Over Land Use: A Second Wave Emerges," *State Government News*, May 1988.
—— **(1989)** "Florida's Greatest Challenge: Managing Massive Growth," in Brumback and Marvin (1989).
—— **(ed.) (1991)** *Balanced Growth: A Planning Guide for Local Government*, Washington: International City Management Association.
DeGrove, J.M. and Juergensmeyer, J.C. (eds.) (1986) *Perspectives on Florida's Growth Management Act of 1985*, Cambridge, MA: Lincoln Institute of Land Policy, Monograph 86–5.
DeGrove, J.M. and Stroud, N.E. (1987) "State Land Planning and Regulation: Innovative Roles in the 1980s and Beyond," *Land Use Law and Zoning Digest* 3: 3–8.
—— **(1989)** "New Developments and Future Trends in Local Government Comprehensive Planning," in Dennison (1989).
deHaven-Smith, L. (1987) *Environmental Publics: Public Opinion on Environmental Protection and Growth Management*, Cambridge, MA: Lincoln Institute of Land Policy, Monograph 87–2.
deHaven-Smith, W.J. (ed.) (1988) *Growth Managment Innovations in Florida*, Miami: Florida Atlantic University and Florida International Joint Center for Environmental and Urban Problems.
Delafons, J. (1969) *Land Use Controls in the United States*, (2nd edn.), Cambridge, MA: MIT Press.
—— **(1990a)** *Development Impact Fees and Other Devices*, Monograph 40, Berkeley: Institute of Urban and Regional Development, University of California at Berkeley.
—— **(1990b)** *Aesthetic Control: A Report on Methods Used in the USA to Control the Design of Buildings*, Monograph 41, Berkeley: Institute of Urban and Regional Development, University of California at Berkeley.
Delaney, C.J. (1987) "Impact Fees, Housing Costs, and Housing Affordability: Who Bears the Impact of Impact Fees?," *University of Florida Journal of Law and Public Policy* 1: 87–101.

Delogu, O.E. (1980) "The Misuse of Land Use Control Powers Must End: Suggestions for Legislative and Judicial Responses," *Maine Law Review* 32: 19–82.
—— **(1981)** "The Dilemma of Local Land Use Control: Power Without Responsibility," *Maine Law Review* 33: 15–34.
—— **(1984)** "Local Land Use Control: An Idea Whose Time Has Passed," *Maine Law Review* 36: 261–310 (also in Deutsch 1985).
—— **(1986)** "A Comprehensive State and Local Government Land Use Control Strategy to Preserve the Nation's Farmland is Unnecessary and Unwise," *Kansas Law Review* 34: 519–38.
Denhez, M.C. (1980) "Protecting the Built Environment of Ontario," *Queen's Law Journal* 5: 73–115.
Dennison, M.S. (1989) *1989 Zoning and Planning Law Handbook*, New York: Clark Boardman.
—— **(1990)** *1990 Zoning and Planning Law Handbook*, New York: Clark Boardman.
Derry, A., Jandl, H.W., Shull, C.D., and Thorman, J. (1977) *Guidelines for Local Surveys: A Basis for Preservation Planning*, Washington: National Register of Historic Places, United States Department of the Interior.
Derthick, M. (1972) *New Towns In-Town: Why A Federal Program Failed*, Washington: Urban Institute.
de Smith, S. and Brazier, R. (1990) *Constitutional and Administrative Law* [British], Harmondsworth: Penguin.
Deutsch, S.L. (1983) *Land Use and Environment Law Review 1983*, New York: Clark Boardman.
—— **(1984)** *Land Use and Environment Law Review 1984*, New York: Clark Boardman.
—— **(1985)** *Land Use and Environment Law Review 1985*, New York: Clark Boardman.
—— **(1986)** *Land Use and Environment Law Review 1986*, New York: Clark Boardman.
—— **(1987)** *Land Use and Environment Law Review 1987*, New York: Clark Boardman.
—— **(1988)** *Land Use and Environment Law Review 1988*, New York: Clark Boardman.
Deutsch, S.L. and Tarlock, A.D. (1989) *Land Use and Environment Law Review 1989*, New York: Clark Boardman.

—— (1990) *Land Use and Environment Law Review 1990*, New York: Clark Boardman.

Dewey, J. (1934) *Art as Experience*, New York: Minton Balch (Putnam Capricorn 1959).

Dilger, R.J. (1982) *The Sunbelt/Snowbelt Controversy: The War Over Federal Funds*, New York: New York University Press.

—— (1989) *National Intergovernmental Programs*, Englewood Cliffs, NJ: Prentice Hall.

Dillon, D. (1991) "The Scoop on Houston: What's Behind the Turnabout in the Nation's Antizoning Stronghold?" *Planning* 57, 4: 13–16 (April 1981).

DiMento, J. (1990) *Wipeouts and Their Mitigation: The Changing Context for Land Use and Environment Law*, Cambridge, MA: Lincoln Institute of Land Policy.

Dion, L. (1989) "The Mystery of Quebec," in Graubard (1989).

Dixon, R.G. (1968) *Democratic Representation: Reapportionment in Law and Politics*, Oxford: Oxford University Press.

Dogan, M. and Pelassy, D. (1984) *How to Compare Nations: Strategies in Comparative Politics*, London: Chatham House.

Doig, A. (1984) *Corruption and Misconduct in Contemporary British Politics*, Harmondsworth: Penguin.

—— (1990) *Westminster Babylon*, London: Allison and Busby.

Dommel, P.R. (1974) *The Politics of Revenue Sharing*, Bloomington: Indiana University Press.

Dommel, P.R. and Rich, M.J. (1987) "The Rich Get Richer: The Attenuation of Targeting Effects of the Community Development Block Grant Program," *Urban Affairs Quarterly* 22: 552–79.

Dommel, P.R. and Rubinowitz, L. (1982) *Decentralizing Urban Policy: Case Studies in Community Development*, Washington: Brookings Institution.

Donovan, T.B. (1962) "Zoning: Variance Administration in Alameda County," *California Law Review* 50: 101–20.

Douglas Commission (1968) *Building the American City: Report of the National Commission on Urban Problems*, Washington: U.S. GPO (also published by Praeger 1969).

Dowall, D.E. (1980) "An Examination of Population-Growth-Managing Communities," *Policy Studies Journal* 9: 414–27.

—— (1981) "Reducing the Cost Effects of Land Use Controls," *Journal of the American Planning Association* 47: 145–53.

—— (1984) *The Suburban Squeeze: Land Conversion and Regulation in the San Francisco Bay Area*, Berkeley: University of California Press.

—— (1986) "Planners and Office Overbuilding," *Journal of the American Planning Association* 52: 131–2.

Downey, T.J. (1982) "Ontario's Local Governments in the 1980s: A Case for Policy Initiatives," *Canadian Journal of Regional Science* 5: 145–63.

Downs, A. (1967) *Inside Bureaucracy*, Boston: Little, Brown.

—— (1970) *Urban Problems and Prospects*, Chicago: Markham.

—— (1972) "Up and Down with Ecology – The 'Issue-Attention' Cycle," *Public Interest* 28: 38–50.

—— (1973) *Opening Up The Suburbs: An Urban Strategy for America*, New Haven, CT: Yale University Press.

—— (1981) *Neighborhoods and Urban Development*, Washington: Brookings Institution.

Doyle, F. (1990) "Concerns Aired During Negotiation Phase," *The Federation Planner* (New Jersey Federation of Planning Officials) 50,3: 1–4.

Dreier, P. and Ehrlich, B. (1991) "Downtown Development and Urban Reform: The Politics of Boston's Linkage Program," *Urban Affairs Quarterly* 26: 354–75.

Ducker, R. (1988) "Land Subdivision Regulations," in So and Getzels (1988).

Duddleson, W.J. (1978) "How the Citizens of California Secured Their Coastal Management Program," in Healey (1978).

Duerksen, C.J. (ed.) (1983) *A Handbook on Historic Preservation Law*, Washington: The Conservation Foundation and the National Center for Preservation Law.

—— (1986) *Aesthetics and Land Use Controls: Beyond Ecology and Economics*, Planning

Advisory Service Report 399, Washington: American Planning Association.

Dukeminier, J. (1955) "Zoning for Aesthetic Objectives: A Reappraisal," *Law and Contemporary Problems* 20: 218–37.

Dukeminier, J. and Stapleton, C.L. (1962) "The Zoning Board of Adjustment: A Case Study in Misrule," *Kentucky Law Journal* 50: 273–350.

Dunham, A. (1955) "Private Enforcement of City Planning," *Law and Contemporary Problems* 20: 463–80.

—— **(1958)** "A Legal and Economic Basis for City Planning: (Making Room for Robert Moses, William Zeckendorf, and a City Planner in the Same Community)," *Columbia Law Review* 58: 650–71.

Dunham, J.G. (1988) "Variances," Chapter 38 of Rathkopf (1988).

Du Page County (1991) *Impacts of Development on DuPage County*, Wheaton, IL: DuPage County Development Department, Planning Division.

Dupre, J.S. (1981) "Intergovernmental Relations and the Metropolitan Area" in L.D. Feldman (1981).

Dwievedi, O.P. (ed.) (1980) *Resources and the Environment: Policy Perspectives for Canada*, Toronto: McClelland and Stewart.

Dworsky, D., McVarish, V., Perry. K.M., and Robinson, S.M. (1983) "Federal Law" [Historic Preservation], in Duerksen (1983).

Dyckman, J.W. (1966) "The Public and Private Rationale for a National Urban Policy," in Warner (1966).

Dye, T.R. (1973) *Politics in States and Communities* (2nd edn.), London: Prentice-Hall.

Dyer, S. (1991) *Mobile Homes in England and Wales: Report of a Postal Survey of Local Authorities*, London: HMSO.

Eagles, P.F.J. (1981) "Environmentally Sensitive Planning in Ontario, Canada," *Journal of the American Planning Association* 47: 313–23.

ECO Northwest Inc (1986) "Growth Management Study of Boca Raton," in D.R. Porter (1986).

Edel, M. (1980) "People versus Places in Urban Impact Analysis," in Glickman (1980).

Edmiston, R.L. (1980) "Marshaling Preservation Law Resources," *Urban Lawyer* 12: 42–53.

Egan, T. (1991) "National Parks: An Endangered Species," *New York Times*, May 27, 1991.

Eichler, N. (1982) *The Merchant Builders*, Cambridge, MA: MIT Press.

Eisenstadt, L.J. (1987) Review of Haar and Fessler: *The Wrong Side of the Tracks*, *Harvard Civil Rights–Civil Liberties Law Review* 22: 278–83.

Eisinger, P.K. (1985) "The Search for a National Urban Policy," *Journal of Urban History* 12: 3–23.

Elazar, D.J. (1968) "Are We a Nation of Cities?," in Goldwin, R.A. (ed.), *A Nation of Cities: Essays on America's Urban Problems*, Skokie, IL: Rand McNally.

—— **(1984)** *American Federalism: A View from the States*, New York: Harper and Row.

—— **(1988)** "Introduction" and "Land and Liberty in American Civil Society," *Publius* 18: v–vii and 1–29.

Elkin, S.L. (1974) *Politics and Land Use: The London Experience*, Cambridge: Cambridge University Press.

Elkins, D.J. and Simeon, R. (1980) *Small Worlds: Provinces and Parties in Canadian Political Life*, Toronto: Methuen.

Ellickson, R.C. (1973) "Alternatives to Zoning: Covenants, Nuisance Rules, and Fines as Land Use Controls," *University of Chicago Law Review*, 40: 681–781 (also in Ackerman 1975).

—— **(1977)** "Suburban Growth Controls: An Economic and Legal Analysis," *Yale Law Journal* 86: 385–511.

—— **(1981)** "The Irony of 'Inclusionary' Zoning," *Southern California Law Review* 1167–1216 (also in Deutsch 1983).

—— **(1990)** "Three Systems of Land-Use Control," *Harvard Journal of Law and Public Policy* 13: 67–74.

Elliott, M. (1981) "The Impact of Growth Control Regulation on Housing Prices in California," *Journal of the American Real Estate and Urban Economics Association* 9: 115–33.

Ely, J.W. (1987) Review of Haar and Fessler: *The Wrong Side of the Tracks*, *Hastings Law Journal* 38: 1297–1303.

Emanuel, A. (1973) *Issues of Regional Politics*, Paris: Organization for Economic Cooperation and Development.

Environmental Board of Vermont (1981) *Act 250: A Performance Evaluation*, Montpelier: Environmental Board of Vermont.

Epling, J.W. (1987) "The New Jersey State Plan: Making the Future a Promising Reality," *New Jersey Municipalities*: 26, 41: 52–4.

Epstein, R.A. (1987) "Takings: Descent and Resurrection," *Supreme Court Review* 1: 3–47 (also in Deutsch and Tarlock 1989).

Erber, E. (1983) "The Road to *Mount Laurel*," *Planning*, November 1983: 4–9.

Essex County Council (1973) *A Design Guide for Residential Areas*, Tiptree, Essex: Anchor Press.

Eule, J.N. (1990) "Judicial Review of Direct Democracy," *Yale Law Journal* 99: 1503–90.

Evans, H. and Rodwin, L. (1979) "The New Towns Program and Why it Failed," *Public Interest* 56: 90–107.

Everett, R.O. and Johnston, J.D. (eds.) (1968) *Housing*, Dobbs Ferry, NY: Oceana.

Fainstein, S.S. and Fainstein, N.I. (1989) "The Ambivalent State: Economic Development Policy in the U.S. Federal System under the Reagan Administration," *Urban Affairs Quarterly* 25; 41–62.

Fainstein, S.S. Fainstein, N.I., Smith, M.P., Hill, R.C. and Judd, D.R. (1986) *Restructuring the City: The Political Economy of Urban Development*, (2nd edn.), New York: Longman.

Farr, C.A. (ed.) (1984) *Shaping the Local Economy*, Washington: International City Management Association.

Feagin, J.R. (1983) *The Urban Real Estate Game*, New York: Prentice Hall.

—— **(1988)** *Free Enterprise City: Houston in Political-Economic Perspective*, New Brunswick, NJ: Rutgers University Press.

—— **(1989)** "Arenas of Conflict: Zoning and Land Use Reform in Critical Political-Economic Perspective," in Haar and Kayden (1989a).

Feldman, E.J. and Goldberg, M.A. (1987) *Land Rites and Wrongs: The Management, Regulation, and Use of Land in Canada and the United States*, Cambridge, MA: Lincoln Institute of Land Policy.

Feldman, E.J. and Milch, J. (1981) "Coordination or Control? The Life and Death of the Ministry of State for Urban Affairs," in L.D. Feldman (1981).

Feldman, L.D. (ed.) (1981) *Politics and Government of Urban Canada: Selected Readings* (4th edn.), Toronto: Methuen.

Feldman, L.D. and Goldrick, M.D. (1976) *Politics and Government of Urban Canada: Selected Readings* (3rd edn), Toronto: Methuen.

Feldman, L.D. and Graham, K.A. (1979) *Bargaining for Cities: Municipalities and Intergovernmental Relations: An Assessment*, Montreal: Institute for Research on Public Policy and Butterworths (Toronto).

Fielding, R. (1975) "The Right to Travel: Another Constitutional Standard?," in R.W. Scott, (1975) volume 2.

Filion, P. (1988) "The Neighbourhood Improvement Plan: Montreal and Toronto – Contrasts Between a Participatory and a Centralized Approach to Urban Policy Making," *Urban History Review* 17: 16–27.

Finnell, G.L. (1978) "Coastal Land Management in California," *American Bar Foundation Research Journal* 4: 649–750.

Fischel, W.A. (1982) "The Urbanization of Agricultural Land: A Review of the National Agricultural Lands Study," *Land Economics* 58: 236–59.

—— **(1985)** *The Economics of Zoning Laws: A Property Rights Approach to American Land Use Controls*, Baltimore: Johns Hopkins University Press.

—— **(1989)** "Centralized Control: Do We Want a Double-Veto System?," *Journal of the American Association of Planners* 55: 205–6.

—— **(1990a)** *Do Growth Controls Matter? – A Review of Empirical Evidence on the Effectiveness and Efficiency of Local Government Land Use Regulations*, Cambridge, MA: Lincoln Institute of Land Policy.

—— **(1990b)** "Research on Land Use Controls," *Land Economics* 66: 229–36.

Fischer, M.L. (1985) "California's Coastal Program: Larger-than-Local Interests Built into Local Plans," *Journal of the American Planning Association* 51: 312–21.

Fisher, R.M. (1959) *Twenty Years of Public Housing*, New York: Harper.

Fishman, R. (1978) *Housing for All Under the Law*, Cambridge, MA: Ballinger.

—— **(1987)** *Bourgeois Utopias: The Rise and Fall of Suburbia*, New York: Basic Books.

Fitch, J.M. (1982) *Historic Preservation: Curatorial Management of the Built World* (1st edn.), New York: McGraw-Hill.

—— **(1990)** *Historic Preservation: Curatorial Management of the Built World* (2nd edn.), Charlottesville: University Press of Virginia.

Fitzgerald, R. and Peiser, R. (1989) "Santa Monica's Colorado Place Agreement," in Porter and Marsh (1989).

Flack, T.A. (1986) "*Euclid v Ambler*: A Retrospective," *Journal of the American Planning Association* 52: 326–37.

Flaherty, D.H. and McKercher, W.R. (1986) *Southern Exposure: Canadian Perspectives on the United States*, Toronto: McGraw-Hill Ryerson.

Fleischmann, A. (1989) "Politics, Administration, and Local Land Use Regulation: Analyzing Zoning as a Policy Process," *Public Administration Review* 49: 337–44.

Fleischmann, A. and Pierannunzi, C.A. (1990) "Citizens, Development Interests, and Local Land Use Regulation," *Journal of Politics* 52: 838–53.

Florida (1986) *The State Land Development Plan 1986-2000*, Tallahassee: Florida Department of Community Affairs.

—— **(1987)** *Keys to Florida's Future: Winning in a Competitive World; Final Report of the State Comprehensive Plan Committee to the State of Florida*, Tallahassee: State of Florida.

—— **(1989a)** *The State Land Development Plan*, Tallahassee: Florida Department of Community Affairs.

—— **(1989b)** *Local Government Comprehensive Planning Process*, Tallahassee: Florida Department of Community Affairs.

—— **(1991a)** "Techniques for Discouraging Sprawl," in DeGrove (1991).

—— **(1991b)** *Governor's Growth Management Task Force: Final Report*, Tallahassee: State of Florida.

Floyd, C.F. (1979a) "Billboard Control Under the Highway Beautification Act," *Real Estate Appraiser and Analyst*, July–August 1979: 19–26.

—— **(1979b)** "Billboard Control under the Highway Beautification Act – A Failure of Land Use Controls," *Journal of the American Institute of Planners* 45: 115–26.

—— **(1979c)** *Highway Beautification: The Environmental Movement's Greatest Failure*, Boulder, CO: Westview.

Flynn, K.M. (1983) "Impediments to the Increased Use of Manufactured Housing," *University of Detroit Journal of Urban Law* 60: 485–505.

Fogelson, R.M. (1967) *The Fragmented Metropolis: Los Angeles, 1850–1930*, Cambridge, MA: Harvard University Press.

Ford, G.B. (ed.) (1916a) *City Planning Progress 1917*, Washington: American Institute of Architects.

—— **(1916b)** "Fundamental Data for City Planning Work," in Nolen (1916).

Fossler, R.S. and Berger, R.A. (1982) *Public–Private Partnerships in American Cities*, Lexington, MA: Lexington.

Foster, M.S. (1981) *From Streetcar to Superhighway: American City Planners and Urban Transportation, 1900–1940*, Philadelphia: Temple University Press.

Fountaine, C.L. (1988) "Lousy Lawmaking: Questioning the Desirability and Constitutionality of Legislating by Initiative," *Southern California Law Review* 61: 733–76.

Fox, G.M. and Davis, B.R. (1976) "Density Bonus Zoning to Provide Low and Moderate Cost Housing," *Hastings Constitutional Law Quarterly* 3: 1015–71.

Fox, K. (1985) *Metropolitan America: Urban Life and Urban Policy in the United States 1940–1980*, New York: Macmillan.

Frank, J.E. (1989) *The Costs of Alternative Development Patterns*, Washington: Urban Land Institute.

Frank, J.E. and Downing, P.B. (1988) "Patterns of Impact Fee Usage," in Nelson (1988b).

Frank, J.E. and Rhodes R.M. (eds.) (1987) *Development Exactions*, Chicago: Planners Press.

Franklin, H.M. (1983a) "The Most Important Zoning Opinion Since *Euclid*," *Planning*, November 1983: 10–12.

—— (1983b) *Fundamental Fairness in Zoning: Mount Laurel Reaffirmed*, Washington: Potomac Institute.

Franklin, H.M., Falk, D., and Levin, A.J. (1974) *In-Zoning: A Guide for Policy-Makers on Inclusionary Land Use Programs*, Washington: Potomac Institute.

Franklin, H.M. and Levin, A.J. (1973) *Controlling Urban Growth – But For Whom?*, Washington: Potomac Institute.

Franklin, J.J. and Muller, T. (1977) "Environmental Impact Evaluation, Land Use Planning, and the Housing Consumer," *Journal of the American Real Estate and Urban Economics Association* 5: 279–301.

Franson, R.T. and Lucas, A.R. (1978) *Environmental Law Commentary and Case Digests*, Austin, TX: Butterworths.

Franzese, P.A. (1989) "*Mount Laurel III*: The New Jersey Supreme Court's Judicious Retreat," in Dennison (1989).

Freilich, R.H. (1975) "Awakening the Sleeping Giant: New Trends and Developments in Environmental and Land Use Controls," in *Southwestern Legal Foundation: Proceedings of the Institute on Planning, Zoning and Eminent Domain*, New York: Matthew Bender.

Freilich, R.H. and Chinn, S.P. (1987) "Transportation Corridors: Shaping and Financing Urbanization Through Integration of Eminent Domain, Zoning, and Growth Management Techniques," *University of Missouri–Kansas City Law Review* 55: 153–212 (also in Deutsch 1988).

Freilich, R.H. and Guemmer, D.B. (1989) "Removing Artificial Barriers to Public Participation in Land Use Policy: Effective Zoning and Planning by Initiative and Referenda," *Urban Lawyer* 21: 511–56.

Freilich, R.H. and Levi, P.S. (1975) *Model Subdivision Regulations: Text and Commentary*, Washington: American Society of Planning Officials.

Freilich, R.H. and Morgan, T.D. (1987) "Municipal Strategies for Imposing Valid Development Exactions: Responding to *Nollan*," *Zoning and Planning Law Report* 10: 169–75.

—— (1988) "Requiring Exactions After *Nollan*: Strategies and Advice for Municipalities," in Gordon (1988).

Freilich, R.H. and Stuhler, E.O. (1981) *The Land Use Awakening: Zoning Law in the Seventies*, Washington: American Bar Association.

French, S.P. and Isaacson, M.S. (1984) "Has Vermont's Land Use Control Program Failed? Evaluating Act 250," *Journal of the American Planning Association* 50: 502–8.

Freund, E. (1904) *The Police Power: Public Policy and Constitutional Rights*, (reprinted Arno Press, New York 1976).

Freyfogle, E.T. (1989) "Context and Accommodation in Modern Property Law," *Stanford Law Review* 41: 1529–56 (also in Deutsch and Tarlock 1990).

Frieden, B.J. (1979) *The Environmental Protection Hustle*, Cambridge, MA: MIT Press.

Frieden, B.J. and Kaplan, M. (1977) *The Politics of Neglect: Urban Aid from Model Cities to Revenue Sharing* (2nd edn.), Cambridge, MA: MIT Press.

Frieden, B.J. and Sagalyn, L.B. (1989) *Downtown Inc: How America Rebuilds Cities*, Cambridge, MA: MIT Press.

Friedenberg, E.Z. (1980) *Deference to Authority: The Case of Canada*, White Plains, NY: M.E. Sharpe (Random House).

Friedman, L.M. (1968) *Government and Slum Housing: A Century of Frustration*, Skokie, IL: Rand McNally.

Frisken, F. (1986) "Canadian Cities and the American Example: A Prologue to Urban Policy Analysis," *Canadian Public Administration* 29: 345–76.

—— (1992) "Planning and Servicing the Greater Toronto Metropolitan Area".

Frost, R.T. (1958) "The Trouble with Zoning," *National Municipal Review* 47: 275–8.

Frye, N. (1980) *Divisions on a Ground: Essays on Canadian Culture*, Toronto: Anansi.

Fulton, W. (1987) "Exactions Put to the Test," *Planning* 53, 12: 6–10 (December 1987).

—— (1989a) "In Land Use Planning, A Second Revolution Shifts Control to the States," *Governing* 2: 40–5.

—— (1989b) *Reaching Consensus in Land Use Negotiations*, Washington: American Planning

Association, Planning Advisory Service Report 417.

—— (1990) "The Trouble with Slow-Growth Politics: It Wins Elections, But the Subdivisions Keep Going Up," *Governing* 3,7: 27–33 (April 1990).

Gagnon, A.G. (ed.) (1984) *Quebec: State and Society*, Toronto: Methuen.

Gailey, J.B. (1984) *1984 Zoning and Planning Law Handbook*, New York: Clark Boardman.

—— (1985) *1985 Zoning and Planning Law Handbook*, New York: Clark Boardman.

—— (1986) *1986 Zoning and Planning Law Handbook*, New York: Clark Boardman.

Gall, G.L. (1990) *The Canadian Legal System*, Scarborough, Ontario: Carswell.

Gallagher, M.L. (1991) "Norman Williams," *Planning* 57, 3: 15 (March 1991).

Gans, H. (1967) *The Levittowners: Ways of Life and Politics in a Suburban Community*, New York: Pantheon.

Gardiner, J.A. and Lyman, T.R. (1978) *Decisions for Sale: Corruption and Reform in Land Use and Building Regulation*, New York: Praeger.

Garner, J.F. and Callies, D.L. (1972) "Planning Law in England and Wales and in the United States," *Anglo-American Law Review* 1: 292–334.

Garner, J.F. and Gravells, N.P. (eds.) (1986) *Planning Law in Western Europe*, Amsterdam: North-Holland.

Garreau, J. (1991) *Edge City: Life on the New Frontier*, New York: Doubleday.

GB [Great Britain] (1944) (White Paper) *The Control of Land Use*, Cmd 6537, London: HMSO.

—— (1947) (White Paper) *Town and Country Planning Bill 1947: Explanatory Memorandum*, Cmd 7006, London: HMSO.

—— (1982) *Town and Country Planning (Structure and Local Plans) Regulations 1982*, Statutory Instrument 1982, no. 555, London: HMSO.

—— (1983) (White Paper) *Streamlining the Cities*, Cmnd 9063, London: HMSO.

—— (1987) *An Evaluation of the Enterprise Zone Experiment*, London: HMSO.

—— (1988) *An Evaluation of the Urban Development Grant Programme*, London: HMSO.

—— (1990) *Enterprise Zone Information 1987–1988*, London: HMSO.

—— (1991) *Chief Planning Inspector's Report* (annual), London: HMSO.

Geisler, C.C. and Popper, F.J. (eds.) (1984) *Land Reform: American Style*, Totowa, NJ: Rowman and Allanheld.

Gelfand, M.I. (1975) *A Nation of Cities: The Federal Government and Urban America 1933–1965*, Oxford: Oxford University Press.

Gellen, M. (1985) *Accessory Apartments in Single-Family Housing*, New Brunswick, NJ: Center for Urban Policy Research, Rutgers University.

Georgia (1988) *Quality Growth Partnership: The Bridge to Georgia's Future*, Atlanta: Governor's Growth Strategies Commission.

Gerckens, L.C. (1988) "Historical Development of American City Planning," in So and Getzels (1988).

Germantown, City of (1989) *Design Review Manual*, City of Germantown, Tennessee.

Gertler, L.O. (ed.) (1982) "Public Policy: Urban and Regional Issues," *Canadian Journal of Regional Studies* 5: 1–224.

Getzels, J. and Jaffe, M. (1988) *Zoning Bonuses in Central Cities*, Washington: American Planning Association, Planning Advisory Service Report 410.

Gilbert, J.G. (1990) "Cross-Acceptance Gets New Jersey Moving Toward Strategic Planning," *New Jersey Bell Journal* 12: 32–43.

Gillette, C.P. (1987) Review of Haar and Fessler: *The Wrong Side of the Tracks*, *Harvard Law Review* 100: 946–68.

Gist, J.R. (1980) "Urban Development Action Grants: Design and Implementation," in Rosenthal (1980).

Gist, J.R. and Hill, R.C. (1984) "Political and Economic Influences on the Bureaucratic Allocation of Federal Funds: The Case of Urban Development Action Grants," *Journal of Urban Economics* 16: 158–72.

Gittel, M. (1985) "The American City: A National Priority or an Expendable Population?," *Urban Affairs Quarterly* 21: 13–19.

Glaab, C.N. (1963) *The American City: A Documentary History*, Homewood, IL: Dorsey.

Glaab, C.N. and Brown, A.T. (1983) *A History*

of Urban America (3rd edn.), New York: Macmillan.

Glassford, P. (1983) *Appearance Codes for Small Communities*, Washington: American Planning Association, Planning Advisory Service Report 379.

—— **(1988)** "Regulating Appearance," in Solnit *et al.* (1988)

Glazer, N. (1967) "Housing Problems and Housing Policies," *The Public Interest* 7: 21–51 (Spring 1967).

—— **(1988)** *The Limits of Social Policy*, Cambridge, MA: Harvard University Press.

Gleeson, M.E. *et al.* (1975) *Urban Growth Management Systems: An Evaluation of Policy-Oriented Research*, Washington: American Society of Planning Officials.

Glickfield, M., Graymer, L., and Morrison, K. (1987) "Trends in Local Growth Control Ballot Measures in California," *University of California Los Angeles Journal of Environmental Law* 6: 111–58.

Glickman, N.J. (1980) *The Urban Impacts of Federal Policies*, Baltimore: Johns Hopkins University Press.

—— **(1984)** "Economic Policy and the Cities: In Search of Reagan's Real Urban Policy," *Journal of the American Planning Association* 50: 471–8.

Godin, J. (1974) "Local Government Reform in the Province of Quebec," in ACIR (1974).

Godschalk, D.R., Brower, D.J., McBennett, L.D., Vestal, B.A. and Herr, D.C. (1979) *Constitutional Issues of Growth Management* (2nd edn.), Chicago: Planners Press.

Goetz, E. (1989) "Office–Housing Linkage in San Francisco," *Journal of the American Planning Association* 55: 66–77.

Goldberg, M.A. (1987) "Evaluating Urban Land Use and Development," in Feldman and Goldberg (1987).

Goldberg, M.A. and Chinloy, O. (1984) *Urban Land Economics*, New York: Wiley.

Goldberg, M.A. and Mark, J.H. (1985) "The Roles of Government in Housing Policy," *Journal of the American Planning Association* 51: 34–42.

Goldberg, M.A. and Mercer, J. (1980) "Canadian and U.S. Cities: Basic Differences, Possible Explanations, and their Meaning for Public Policy," *Regional Science Association Papers* 45: 159–83.

—— **(1986)** *The Myth of the North American City: Continentalism Challenged*, Vancouver: University of British Columbia Press.

Goldberger, P. (1981) *The Skyscraper*, New York: Knopf.

—— **(1990)** "A Commission that has Itself Become a Landmark: The Landmarks Commission, Born of the Loss of Penn Station, Celebrates 25 Years of Existence," *New York Times* April 15, 1990.

Goldfield, D.R. and Brownell, B.A. (1990) *Urban America: A History* (2nd edn.), Boston: Houghton Mifflin.

Goldman, S. and Jahnige, T.P. (1985) *The Federal Courts as a Political System*, New York: Harper and Row.

Goldsmith, W.W. (1982) "Enterprise Zones: If They Work We're in Trouble," *International Journal of Urban and Regional Research* 6: 435–42.

Goldsmith, W.W. and Derian, M.J. (1979) "Is There An Urban Policy?," *Journal of Regional Science* 19: 93–108.

Goldsmith, W.W. and Jacobs, H.M. (1982) "The Improbability of Urban Policy," *Journal of the American Planning Association* 48: 53–66.

Goldwich, D.S. (1989) "Development Agreements: A Critical Introduction," *Journal of Land Use and Environmental Law* 4: 249–69.

Goldwin, R.A. (ed.) (1968) *A Nation of Cities: Essays on America's Urban Problems*, Skokie, IL: Rand McNally.

Goode, V.L. (1986) Review of Haar and Fessler: *The Wrong Side of the Tracks*, *Thurgood Marshall Law Review* 12: 293–8.

Goodin, C.C. (1984) "The Honolulu Development Plan: An Analysis of Land Use Implications for Oahu," *University of Hawaii Law Review* 6: 33–108.

Goodman, R. (1971) *After the Planners*, New York: Simon and Schuster.

Goodwin, M. (1991) "Replacing a Surplus Population: The Employment and Housing Policies of the London Docklands Development Corporation," in Allen, J. and Hamnett, C., *Housing and Labour Markets*, London: Unwin Hyman.

Gordon, N.J. (1987) *1987 Zoning and Planning Law Handbook*, New York: Clark Boardman.
—— **(1988)** *1988 Zoning and Planning Law Handbook*, New York: Clark Boardman.
Gordon, R.J. and Gordon, L. (1990) "Neighborhood Responses to Stigmatized Urban Facilities: A Private Mental Hospital and Other Facilities in Phoenix, Arizona," *Journal of Urban Affairs* 12: 437–47.
Gorland, E. (1971) *Urban Renewal Administration: Practices, Procedures, Record Keeping*, Detroit: Wayne State University Press.
Gottdiener, M. (1977) *Planned Sprawl: Private and Public Interests in Suburbia*, Beverly Hills, CA: Sage.
—— **(1983)** "Some Theoretical Issues in Growth Control Analysis," *Urban Affairs Quarterly* 18: 565–9.
Gottdiener, M. and Neiman, M. (1981) "Characteristics of Support for Local Growth Control," *Urban Affairs Quarterly* 17: 55–73.
Gould, L.L. (1988) *Lady Bird Johnson and the Environment*, Lawrence: University Press of Kansas.
Governmental Advisory Associates Inc. (1988) *Resources Recovery Yearbook, Directory and Guide*, Washington: Governmental Advisory Associates.
Graham, O.L. (1976) *Toward A Planned Society: From Roosevelt to Nixon*, Oxford: Oxford University Press.
Grant, D.R. and Nixon, H.C. (1968) *State and Local Government in America* (2nd edn.), Needham Heights, MA: Allyn and Bacon.
Grant, J.A.C. (1931) "The 'Higher Law' Background of the Law of Eminent Domain," *Wisconsin Law Review* 6: 67–85.
Grant, M. (1982) *Urban Planning Law*, London: Sweet and Maxwell.
Gratz, R.B. (1989) *The Living City*, New York: Simon and Schuster.
Graubard, S.R. (ed.) (1989) *In Search of Canada*, New Brunswick, NJ: Transaction Books.
Green, P.P. (1955) "Is Zoning by Men Replacing Zoning by Law?," *Journal of the American Institute of Planners*, 21: 82–7.
—— **(1961)** "Rough Justice: Baby or Bathwater?," *Zoning Digest* 13: 161–6.
Green, R.E. (ed.) (1991) *Enterprise Zones: New Directions in Economic Development*, Newbury Park, CA: Sage.
Green, R.E. and Brintnall, M.A. (1987) "Reconnoitering State-Administered Enterprise Zones," *Journal of Urban Affairs* 9: 159–70.
—— **(1988)** "Comparing State Enterprise Zone Programs: Variations in Structure and Coverage," *Economic Development Quarterly* 2: 50–68.
—— **(1991)** "Conclusions and Lessons Learned" [Enterprise Zones], in R.E. Green (ed.) (1991).
Greenberg, M.R. and Popper, F.J. (1990) "Government Land Preservation and Communication Policies in Fast-Growing Counties of the United States of America," *Environment and Planning C; Government and Policy* 8: 417–26.
Greenhouse, L. (1990) "High Court Justices Face an Issue Close to Home," *New York Times* December 13, 1990.
Greenspan Report (1978) *Federal Provincial Task Force on the Supply and Price of Serviced Residential Land*, Ottawa: Canada Mortgage and Housing Corporation.
Greer, S.A. (1966) *Urban Renewal and American Cities: The Dilemma of Democratic Intervention*, Indianapolis: Bobbs-Merrill.
Grenell, P. (1988) "The Once and Future Experience of the California Coastal Conservancy," *Coastal Management* 16: 13–20.
Griffith, E.S. (1938a) *A History of American City Government: The Conspicuous Failure, 1970–1900*, New York: Da Capo (reprinted by University Press of America 1983).
—— **(1938b)** *A History of American City Government: The Progressive Years and Their Aftermath, 1900–1920*, New York: Da Capo (reprinted by University Press of America 1983).
Griffith, J.A.G. (1988) *Central Departments and Local Authorities*, London: Allen and Unwin.
Groberg, R.P. (1965) "Urban Renewal Realistically Reappraised," *Law and Contemporary Problems* 30: 212–29 (also in Wilson 1966).
Gruen, C. (1985) "The Economics of Requiring Office Space Development to Contribute to

the Production of Housing," in D.R. Porter (1985).

Gruen, N.J. (1985) "A Case History of the San Francisco Office–Housing Linkage Program," in D.R. Porter (1985).

Gurr, T.R. and King, D.S. (1987) *The State and the City*, Chicago: University of Chicago Press.

Gustafson, G.C., Daniles, T.L., and Shirack, R.P. (1982) "The Oregon Land Use Act: Implications for Farmland and Open Space Protection," *Journal of the American Planning Association*, 48: 365–73.

Haar, C.M. (1951) *Land Planning in a Free Society*, Cambridge, MA: Harvard University Press.

—— **(1953)** "Zoning for Minimum Standards: The *Wayne Township* Case," *Harvard Law Review* 66: 1051–63.

—— **(1954)** "Wayne Township: Zoning for Whom? – In Brief Reply," *Harvard Law Review* 67: 986–93.

—— **(1955a)** "In Accordance with a Comprehensive Plan," *Harvard Law Review*, 68: 1154–75.

—— **(1955b)** "The Content of the General Plan: A Glance at History," *Journal of the American Institute of Planners*, 21: 66–70.

—— **(1955c)** "The Master Plan: An Impermanent Constitution," *Law and Contemporary Problems* 20: 353–418.

—— **(1977)** *Land Use Planning: A Casebook on the Use, Misuse, and Re-use of Urban Land* (3rd edn.), Boston: Little, Brown.

—— **(ed.) (1984)** *Cities, Law, and Social Policy: Learning from the British*, Lexington, MA: Lexington.

Haar, C.M. and Fessler, D.W. (1986) *The Wrong Side of the Tracks*, New York: Simon and Schuster. (For reviews, see: Eisenstadt 1987; Ely 1987; Gillette 1987; Goode 1986; Kalsheur 1987; and Scurlock 1987).

Haar, C.M. and Iatridis, D.S. (1974) *Housing the Poor in Suburbia: Public Policy at the Grass Roots*, Cambridge, MA: Ballinger.

Haar, C.M. and Kayden, J.S. (1989a) *Zoning and the American Dream: Promises Still to Keep*, Chicago: Planners Press.

—— **(1989b)** *Landmark Justice: The Influence of William J. Brennan on America's Communities*, Washington: Preservation Press.

Haar, C.M. and Wolf, M.A. (1989) *Land Use Planning: A Casebook on the Use, Misuse, and Re-use of Urban Land* (4th edn.), Boston: Little, Brown.

Haar, C.M., Horowitz, S.G., and Katz, D.F. (1980) *Transfer of Development Rights: A Primer*, Cambridge, MA: Lincoln Institute of Land Policy.

Habe, R. (1989) "Public Design Control in American Communities," *Town Planning Review* 60: 195–219.

Hagman, D.G. (1971) *Urban Planning and Land Development Control Law*, St Paul, MN: West.

—— **(1975)** *Urban Planning and Land Development Control Law* (with additional chapters), St Paul, MN: West.

—— **(1979)** "Estoppel and Vesting in the Age of Multi-Land Use Permits," *Southwestern Law Review* 11: 545–96.

—— **(1980)** *Public Planning and Control of Urban and Land Development* (2nd edn.), St Paul, MN: West.

—— **(1982)** "Landowner-Developer Provision of Communal Goods Through Benefit-Based and Harm Avoidance 'Payments' (BHAPS)," *Zoning and Planning Law Report* 5: 17–23 and 25–32.

—— **(1990)** *Public Planning and Control of Urban and Land Development*, St Paul, MN: West.

Hagman, D.G. and Juergensmeyer, J.C. (1986) *Urban Planning and Land Development Control Law* (2nd edn.), St Paul, MN: West.

Hagman, D.G. and Misczynski, D.J. (1978) *Windfalls for Wipeouts: Land Value Capture and Compensation*, Washington: American Society of Planning Officials.

Hagman, D.G. and Pepe, S. (1974) "English Planning Law: A Summary of Recent Developments," *Harvard Journal on Legislation* 11: 557–93.

Haider, D.H. (1974) *When Governments Come to Washington: Governors, Mayors, and Intergovernmental Lobbying*, New York: Free Press.

Hall, P. (1981) *The Enterprise Zones Concept: British Origins, American Adaptations*, Institute of Urban and Regional Development, University of California, Berkeley, Working Paper 350.

—— **(1982a)** *Urban and Regional Planning*, London: Allen and Unwin.

—— **(1982b)** "Enterprise Zones: A Justification," *International Journal of Urban and Regional Research* 6: 416–21.

—— **(1982c)** "Enterprise Zones: A Justification," in "Urban Enterprise Zones: A Debate," *International Journal of Urban and Regional Research* 6: 415–21.

—— **(1989)** *Cities of Tomorrow: An Intellectual History of Urban Planning and Design in the Twentieth Century*, Oxford: Blackwell.

—— **(1991)** "The British Enterprise Zones," in R.E. Green (1991).

Hall, P., Gracey, H., Drewett, R., and Thomas, R. (1973) *The Containment of Urban England* (2 vols.), London: Allen and Unwin.

Hallett, G. (ed.) (1977) *Housing and Land Policies in West Germany and Britain: A Comparative Analysis*, London: Macmillan.

Hamilton, A., Jay, J., and Madison, J. (1788) *The Federalist: A Commentary on the American Constitution* (introduction by E.M. Earle), New York: Modern Library edn.

Hammer, Siler, George Associates (1975) *Ramapo, New York: Socio-Economic Profile* (quoted in Babcock 1980: 267).

Hamnett, S. (1978) "Leiden-Merenwikj: A Case Study of Dutch Local Planning," *Planning and Public Administration* 5,2: 28–42.

Hancock, J.L. (1967) "Planners in the Changing American City 1900–1940," *Journal of the American Institute of Planners* 33: 290–304.

—— **(1972)** "History and the American Planning Profession," *Journal of the American Institute of Planners*, 38: 274–5.

Hanna, R.W. (1982) "Right to Farm Statutes: The Newest Tool in Agricultural Land Preservation," *Florida State University Law Review* 10: 415–39.

Hanson, R. (1982a) *National Policy and the Post-Industrial City: An International Perspective*, Washington: National Academy Press.

—— **(1982b)** *The Evolution of National Urban Policy, 1970–1980*, Washington: National Academy Press.

—— **(1983)** *Rethinking Urban Policy: Urban Development in an Advanced Economy*, Washington: National Academy Press.

Hardin, H. (1974) *A Nation Unaware: The Canadian Economic Culture*, Vancouver: Douglas.

Harding, A. (1989) "Central Control in British Urban Economic Development Programmes," in Crouch and Marquand (1989).

Hare, K. (1989) "Canada – The Land," in Graubard (1989).

Harris, D.C. (1988) "Growth Management Reconsidered," *Journal of Planning Literature* 3: 466–81.

Hart, J. (1991) *Farming on the Edge: Saving Family Farms in Marin County, California*, Berkeley: University of California Press.

Hartke, V. (1973) "Toward a National Growth Policy," *Catholic University Law Review* 22: 231–78.

Hartman, C. (1974) *Yerba Buena: Land Grab and Community Resistance in San Francisco*, San Francisco: Glide.

—— **(1975)** *Housing and Social Policy*, Hemel Hempstead: Prentice-Hall.

—— **(1984)** *The Transformation of San Francisco*, Totowa, NJ: Rowman and Allanheld.

Hartman, C. and Kessler, R. (1978) "The Illusion and Reality of Urban Renewal: San Francisco's Yerba Buena Center," in Tabb and Sawers (1978).

Harvard Law Review Note (1969) "Administrative Discretion in Zoning," *Harvard Law Review* 82: 668–85.

—— **(1989)** "Municipal Development Exactions, the Rational Nexus Test, and the Federal Constitution," *Harvard Law Review* 102: 992–1012.

Harvey, D. (1973) *Social Justice and the City*, Baltimore: Johns Hopkins University Press.

Hausrath, L.L. (1988) "Economic Basis for Linking Jobs and Housing in San Francisco," *Journal of the American Planning Association* 54: 210–16.

Hawley, E. (1974) "Herbert Hoover, the Commerce Secretariat, and the Vision of an 'Associative State' 1921–28," *Journal of American History* 61: 116–40.

Hazell, R. (1989) "Freedom of Information in Australia, Canada, and New Zealand," *Public Administration* 67: 189–210.

Healey, R.G. (ed.) (1978) *Protecting the Golden Shore: Lessons from the California Coastal*

Commissions, Washington: Conservation Foundation.

Healy, M.R. (1974) "National Land Use Proposal: Land Use Legislation of Landmark Environmental Significance," *Environmental Affairs* 3: 355–95.

Healy, R.G. (1977) "Rural Land: Private Choices, Public Interests," *Conservation Foundation Letter* August 1977: 1–8.

Healy, R.G. and Rosenberg, J.S. (1979) *Land Use and the States*, (2nd edn.), Baltimore: Johns Hopkins University Press.

Hedman, R. (1977) *Stop Me Before I Plan Again*, Washington: American Society of Planning Officials.

Hedman, R. and Jaszewski, A. (1984) *Fundamentals of Urban Design*, Chicago: Planners Press.

Heeter, D. (1969) *Toward a More Effective Land-Use Guidance System: A Summary and Analysis of Five Major Reports*, Planning Advisory Service Report 250, Washington: American Society of Planning Officials.

Heeter, D.G. (1976) "Almost Getting it Together in Vermont," in Mandelker (1976a).

Heidenheimer, A.J., Heclo, H., and Adams, C.T. (1975) *Comparative Public Policy: The Politics of Social Choice in Europe and America* (2nd edn.), New York: St. Martin's Press.

—— **(1990)** *Comparative Public Policy: The Politics of Social Choice in Europe and America* (3rd edn.), New York: St. Martin's Press.

Heikoff, J.M. (1977) *Coastal Resources Management: Institutions and Programs*, Ann Arbor, MI: Ann Arbor Science.

Heilbroner, R.L. (1976) "Benign Neglect in the United States," in Tropman (1976).

Heiman, M. (1986) *Coastal Recreation in California: Policy, Management, Access*, Berkeley: Institute of Governmental Studies, University of California.

Heiman, M.K. (1988) *The Quiet Revolution: Power, Planning and Profits in New York State*, New York: Praeger.

Henderson, A. (1987) "Land Use Controls in Houston: What Protection for Owners of Restricted Property?," *South Texas Law Review* 29: 131–87.

Henig, J.R. (1985) *Public Policy and Federalism:*

Issues in State and Local Politics, New York: St. Martin's Press.

Hetzel, O. (undated) *Transplanting Housing Concepts: Some Potential Seedlings of British Origin*, Land Policy Rountable Case Studies Series 304, Cambridge, MA: Lincoln Institute of Land Policy.

Heyman, I.M. and Gilhool, T.K. (1964) "The Constitutionality of Imposing Increased Community Costs on New Subdivision Residents Through Subdivision Exactions," *Yale Law Journal* 73: 1119–57.

Hicks, D.A. (1982) "National Urban Land Policy: Facing the Inevitability of City and Regional Evolution," in Lefcoe (1982).

Hiemstra, H. and Bushwick, N. (eds.) (1989) *Plowing the Urban Fringe: An Assessment of Alternative Approaches to Farmland Preservation*, Miami: Florida Atlantic University and Florida International University Joint Center for Environmental and Urban Problems.

Higgins, D.J.H. (1986) *Local and Urban Politics in Canada*, Toronto: Gage.

Hill, G.R. (ed.) (1990) *Regulatory Taking: The Limits of Land Use Controls*, Section of Urban, State and Local Government Law, Washington: American Bar Association.

Hill, H.A (1984) "Government Manipulation of Land Values to Build Affordable Housing: The Issue of Compensating Benefits," *Real Estate Law Journal* 13: 3–27 (also in Deutsch 1985: 147–71).

Hill, R.C. (1983) "Market, State, and Community: National Urban Policy in the 1980s," *Urban Affairs Quarterly* 19: 5–20.

—— **(1986)** "Crisis in the Motor City: The Politics of Economic Development in Detroit," in Fainstein *et al.* (1986).

Hillman, J. (1990) *Planning for Beauty: The Case for Design Guidelines*, Royal Fine Art Commission, London: HMSO.

Hirschhorn, L. (1979) "The Urban Crisis: A Post-Industrial Perspective," *Journal of Regional Science* 19: 109–29.

Hodge, G. (1985) "The Roots of Canadian Planning," *Journal of the American Planning Association* 51: 8–22.

—— **(1986)** *Planning Canadian Communities*, Toronto: Methuen.

Hoffecker, C.E. (1983) *Corporate Capital: Wilmington in the Twentieth Century,* Philadelphia: Temple University Press.

Holford, W.G. (1953) "Design in City Centres," in *Design in Town and Village,* Ministry of Housing and Local Government (England), London: HMSO.

Holland, K.M. (1988) "The Courts in the United States," in Waltman and Holland (1988).

Holliman, W.G. (1981) "Development Agreements and Vested Rights in California," *Urban Lawyer* 13: 44–64.

Holloway, J.E. and Guy, D.C. (1990) "Rethinking Local and State Agricultural Land Use and Natural Resource Policies: Coordinating Programs to Address the Interdependency and Combined Losses of Farms, Soils and Farmland," in Dennison (1990).

Holman, B. (1990) "The Politics of Planning: Where is North Carolina Heading?," *Carolina Planning* 16,1: 40–7.

Home, R.K. (1991) "Deregulating UK Planning Control in the 1980s," *Cities* 8: 292–300.

Hopperton, R.J. (1980) "A State Legislative Strategy for Ending Exclusionary Zoning of Community Homes," *Urban Law Annual* 47: 47–85.

Horowitz, C.F. (1983) *The New Garden Apartment: Current Market Realities of an American Housing Form,* New Brunswick, NJ: Center for Urban Policy Research, Rutgers University.

Hosmer, C.B. (1965) *Presence of the Past: A History of the Preservation Movement in the United States Before Williamsburg,* New York: Putnam.

—— **(1981)** *Preservation Comes of Age: From Williamsburg to the National Trust 1926–1949,* Charlottesville: University Press of Virginia.

Hough, D.E. and Kratz, C.G. (1983) "Can 'Good' Architecture Meet the Market Test?," *Journal of Urban Economics* 14: 40–54.

Howarth, D. (1984) "A History of British Regional Policy in the 1970s," *Yale Law and Policy Review* 2: 215–55.

Howitt, A.M. (1984) *Managing Federalism: Studies in Intergovernmental Relations,* Washington: Congressional Quarterly.

Hubbard, T.K. and Hubbard, H.V. (1929) *Our Cities of To-day and To-morrow: A Survey of Planning and Zoning Progress in the United States,* Cambridge, MA: Harvard University Press.

Huffman, F.E. and Smith, M.T. (1988) "Market Effects of Office Development Linkage Fees," *Journal of the American Planning Association* 54: 217–24.

Hufford, M. (1986) *One Space, Many Places: Folklife and Land Use in New Jersey's Pinelands National Reserve,* Washington: American Wildlife Center.

Hulchanski, J.D. (1981) *The Origins of Urban Land Use Planning in Alberta 1900–1945,* Research Paper 119, Toronto: University of Toronto, Centre for Urban and Community Studies.

Hulten, C.R. and Sawhill, I.V. (1984) *The Legacy of Reaganomics: Prospects for Long Term Growth,* Washington: Urban Institute.

Hunt, E.L.R. (1988) *Managing Growth's Impact on the Mid-South's Historic and Cultural Resources,* Washington: National Trust for Historic Preservation.

Huthmacher, J.J. (1971) *Senator Robert F. Wagner and the Rise of Urban Liberalism,* New York: Atheneum.

Huxtable, A.L. (1986) *Goodbye History, Hello Hamburger,* Washington: Preservation Press.

Hyman, R. (1979) *National Land Use Planning: A Bibliography,* Monticello, IL: Vance Bibliographies, Public Administration Series 257.

Ince, J.G. (1977) *Land Use Law: A Study of Legislation Governing Land Use in British Columbia,* Vancouver: University of British Columbia.

—— **(1984)** *Land Use Law: British Columbia Handbook 1984,* Markham, Ontario: Butterworths.

Jackson, A. (1976) *A Place Called Home: A History of Low-Cost Housing in Manhattan,* Cambridge, MA: MIT Press.

Jackson, K.T. (1985) *Crabgrass Frontier: The Suburbanization of the United States,* Oxford: Oxford University Press.

Jacobs, A.B. (1980) *Making City Planning Work,* Chicago: Planners Press.

Jacobs, H.M. (1989) "Localism and Land Use

Planning," *Journal of Architectural and Planning Research* 6: 1–18.

Jacobs, J. (1961) *The Death and Life of Great American Cities*, New York: Random House.

Jaffe, M. (1983) "Mobile Homes in Single-Family Neighborhoods," *Land Use Law and Zoning Digest* 35, 6: 4–10.

James, F.J. (1977) *Zoning for Sale: A Critical Analysis of Transferable Development Rights Programs*, Washington: Urban Institute.

—— **(1991)** "The Evaluation of Enterprise Zone Programs," in Green (1991).

James, F.J. and Windsor, O.D. (1976) "Fiscal Zoning, Fiscal Reform, and Exclusionary Land Use Controls," *Journal of the American Institute of Planners* 42: 130–41.

James, H. (1926) *Land Planning in the United States for the City, State and Nation*, New York: Macmillan.

Jeglie, J. and Metzger, P. (1990) "Containing Urban Sprawl: Proceedings of the Third Annual Growth Management Conference," *Environmental and Urban Issues* 17, 4: 2–5 (Summer 1990), Boca Raton, FL: Joint Center for Environmental and Urban Problems, Florida State University.

Jencks, C. and Peterson, P. (1991) *The Urban Underclass*, Washington: Brookings Institution.

Jennings, E.T., Krane, D., Pattakos, A., and Reed, B. (1986) *From Nation to Cities: The Small Cities Community Development Block Grant Program*, Albany, NY: State University of New York Press.

Jennings, M.D. (1989) "The Weak Link in Land Use Planning," *Journal of the American Planning Association* 55: 206–8.

Johnson, M.B. (ed.) (1982) *Resolving the Housing Crisis*, San Francisco: Pacific Institute.

Johnson, R.R. (1970) "The Highway Beautification Act: Cosmetic for the City?," *Catholic University Law Review* 20: 69–90.

Johnson, R.W. and Schene, M.G. (eds.) (1987) *Cultural Resources Management*, Melbourne, FL: Krieger.

Johnston, R.A. (1980) "The Politics of Local Growth Control," *Policy Studies Journal* 9: 427–37.

—— **(1984)** *Residential Segregation, The State, and Constitutional Conflict in American Urban Areas*, London: Academic Press.

Johnston, R.A., Schwartz, S.I., Wandesforde-Smith, G.A., and Caplan, M. (1989) "Selling Zoning: Do Density Bonus Incentives for Moderate Cost Housing Work?," *Journal of Urban and Contemporary Law* 36: 45–61 (also in Dennison 1990).

Jones, G.W. (1988) "The Crisis in British Central–Local Government Relationships," *Governance* 1: 162–83.

Judd, D.R. (1988) *The Politics of American Cities: Private Power and Public Policy* (2nd edn.), Glenview, IL: Scott, Foresman.

—— **(ed.) (1990)** *Leadership and Urban Regeneration: Cities in North America and Europe*, Newbury Park, CA: Sage.

Judd, D.R. and Robertson, D.B. (1989) "Urban Revitalization in the United States: Prisoner of the Federal System," in Parkinson, Foley, and Judd (1989).

Juergensmeyer, J.C. and Blake, R.M. (1981) "Impact Fees: An Answer to Local Government's Capital Funding Dilemma," *Florida State University Law Review* 9: 415–45.

Kain, J.F. (1978) *Failure in Diagnosis: A Critique of Carter's National Urban Policy*, Cambridge, MA: Department of City and Regional Planning, Harvard University.

Kaiser Report (1968) *A Decent Home: The Report of the President's Committee on Urban Housing*, Washington: U.S. GPO.

Kalscheur, G.A. (1987) Review of Haar and Fessler: *The Wrong Side of the Tracks*, *Michigan Law Review* 85: 1124–29.

Kanner, G. (1973) "Condemnation Blight: Just How Just is Just Compensation?," *Notre Dame Lawyer* 48: 765–810.

Kantor, P. (1991) "A Case for a National Urban Policy," *Urban Affairs Quarterly* 26: 394–415.

Kantor, P. and David, S. (1988) *The Dependent City: The Changing Political Economy of Urban America*, Glenview, IL: Scott, Foresman.

Kaplan, B. (1983) "Houston: The Golden Buckle of the Sunbelt," in Bernard and Rice (1983).

Kaplan, H. (1963) *Urban Renewal Politics: Slum Clearance in Newark*, New York: Columbia University Press.

—— **(1967)** *Urban Political Systems: A Functional Analysis of Metro Toronto*, New York: Columbia University Press.

Kaplan, M. and Cuciti, P. (eds.) (1986) *The Great Society and its Legacy: Twenty Years of U.S. Social Policy*, Durham, NC: Duke University Press.

Kaplan, M. and James F. (eds.) (1990) *The Future of National Urban Policy*, Durham, NC: Duke University Press.

Kaplan, H., Gans, S.P., and Kahn, H.M. (1970) *The Model Cities Program: The Planning Process in Atlanta, Seattle, and Dayton*, New York: Praeger.

Karlin, N. (1981) "Zoning and Other Land Use Controls: From the Supply Side," *Southwestern University Law Review* 12: 561–79.

Karp, J.P. (1990) "The Evolving Meaning of Aesthetics in Land Use Regulation," *Columbia Journal of Environmental Law* 15: 307–28.

Kasowski, K. (1991) "Oregon: Fifteen Years of Land Use Planning," in DeGrove (1991).

Kayden, J.S. (1978) *Incentive Zoning in New York City: A Cost-Benefit Analysis*, Land Policy Roundtable Policy Analysis Series 201, Cambridge, MA: Lincoln Institute of Land Policy.

—— **(1990)** "Zoning for Dollars: New Rules for an Old Game? Comments on the *Municipal Art Society* and *Nollan* Cases," in Lassar (1990).

Kayden, J.S. and Zax, L.A. (1983) "*Mount Laurel II*: Landmark Decision on Zoning and Low Income Housing Holds Lessons for Nation," *Zoning and Planning Law Report* 6: 145–51.

Keating, M. (1989) "The Disintegration of Urban Policy: Glasgow and the New Britain," *Urban Affairs Quarterly* 24: 513–36.

Keating, W.D. (1986) "Linking Downtown Development to Broader Community Goals: An Analysis of Linkage Policy in Three Cities," *Journal of the American Planning Association* 52: 133–41.

—— **(1989)** "Linkage: Tying Downtown Development to Community Housing Needs," in Rosenberry, S. and Hartman, C., *Housing Issues of the 1990s*, New York: Praeger.

Keith, K.M. (1985) "The Hawaii State Plan Revisited," *University of Hawaii Law Review* 7: 29–61.

Keith, N.S. (1973) *Politics and the Housing Crisis*, New York: Universe Books.

Kelly, E.D. (1984) "Piping Growth: The Law, Economics, and Equity of Sewer and Water Connection Policies," *Land Use Law* 36, 7: 3–8 (July).

—— **(1988)** "Zoning," in So and Getzels (1988).

—— **(1989)** "The Quiet Revolution Continues," *Journal of Planning Literature* 4: 121–38 (July 1984).

Kelly, E.D. and Raso, G.J. (1989) *Sign Regulation for Small and Midsize Communities: A Planners Guide and A Model Ordinance*, Planning Advisory Service Report 419, Washington: American Planning Association.

Kemble, R. (1989) *The Canadian City, St John's to Victoria: A Critical Commentary*, Montreal: Harvest House.

Kendig, L. (1980) *Performance Zoning*, Chicago: Planners Press.

Kent, T.J. (1964) *The Urban General Plan*, San Francisco: Chandler.

Kent-Smith, H.L. (1987) "The Council on Affordable Housing and the *Mount Laurel* Doctrine: Will the Council Succeed?," *Rutgers Law Journal* 18: 929–60.

Kerner Report (1968) *Report of the National Commission on Civil Disorders*, Washington: U.S. GPO (also published by Bantam Books [New York]).

Kettl, D.F. (1980) *Managing Community Development in the New Federalism*, New York: Praeger.

—— **(1981)** "Regulating the Cities," *Publius* 11: 111–25.

Keune, R. (ed.) (1984) *The Historic Preservation Yearbook: A Documentary Record of Significant Policy Developments and Issues 1984/1985*, Washington: National Trust for Historic Preservation.

Kienz, G.C. (1990) *A Guide to Planning Boards and Zoning Boards of Adjustment*, Springfield, NJ: New Jersey Federation of Planning Officials.

King, A. (1973) "Ideas, Institutions and the Policies of Governments: A Comparative Analysis," *British Journal of Political Science* 3: 291–313 and 409–23.

King, J. (1990) "How the BRA [Boston Redevelopment Authority] Got Some Respect," *Planning*, 56, 5: 4–9 (May 1990).

Kirlin, J.J. (1985) "The Bargaining Process: Trends and Issues," in Levitt and Kirlin (1985).

Kirlin, J. J. and Kirlin, A. (1985) *Public Choices – Private Resources: Financing Capital Infrastructure for California's Growth Through Public–Private Bargaining*, California Tax Foundation.

Klein, M.S.G. (1984) *Law, Courts, and Social Policy*, New York: Prentice-Hall.

Kluger, R. (1976) *Simple Justice*, New York: Knopf.

Kmiec, D.W. (1981) "Deregulating Land Use: An Alternative Free Enterprise Development System," *University of Pennsylvania Law Review* 130: 28–130 (also in Deutsch 1983).

Knack, R.E. (1990) "Immobile Homes," *Planning* 56, 12: 4–9.

Knapp, G.J. (1985) "The Price Effects of Urban Growth Boundaries in Metropolitan Portland, Oregon," *Land Economics* 61: 26–35.

—— **(1990)** "State Land Use Planning and Inclusionary Zoning: Evidence from Oregon," *Journal of Planning Education and Research* 10: 39–46.

Koehler, C.T. (1989) *Federal and State Government Impacts on Local Land Use Policy*, Chicago: Council of Planning Librarians, Bibliography 241.

Koenig, J. (1990) "Down the Wire in Florida: Concurrency is the Byword in the Nation's Most Elaborate Statewide Growth Managment Scheme," *Planning* 56, 10: 4–11 (October 1990).

Kolis, A.B. (1979) "Architectural Expression: Police Power and the First Amendment," *Urban Law Annual* 16: 273–304.

Kozlowski, J.M. (1986) *Threshold Approach in Urban, Regional, and Environmental Planning: Theory and Practice*, Brisbane: University of Queensland Press.

Kramer, B. (1987) "*Nollan* and the Day of Judgment," *American Planning Association Planning and Law Division Newsletter*, 2–3.

Krasnowiecki, J.Z. (1972) "Zoning Litigation and the New Pennsylvania Procedures," *University of Pennsylvania Law Review* 120: 1029–165.

—— **(1980)** "Abolish Zoning," *Syracuse Law Review* 31: 719–53.

Krefetz, S. (1979) "Low and Moderate Income Housing in the Suburbs: The Massachusetts Anti-Snob Zoning Law Experience," *Policy Studies Journal* 8: 288–99.

Kristof, F.S. (1965) "Challenges Critical View of Urban Renewal," *Savings Bank Journal*, April 1965: 36–9.

Krueckeberg, D.A. (1983) *Introduction to Planning History in the United States*, New Brunswick, NJ: Center for Urban Policy Research, Rutgers University.

Krumholz, N. and Forester, J. (1990) *Making Equity Planning Work: Leadership in the Public Sector*, Philadelphia: Temple University Press.

Kyle, J.E. (ed.) (1987) *Children, Families and Cities*, Washington: National League of Cities.

Ladd, B. (1990) *Urban Planning and Civic Order in Germany, 1860–1914*, Cambridge, MA: Harvard University Press.

Lahey, K.E., Diskin, B.A., and Lahey, V.M. (1989) "Manufactured Housing: An Alternative to Site-Built Homes," *Real Estate Appraiser and Analyst* 55: 26–36.

Lai, R.T. (1988) *Law in Urban Design and Planning*, London: Van Nostrand Reinhold.

Lake, R.W. (1981) "The Fair Housing Act in a Discriminatory Market: The Persistent Dilemmas," *Journal of the American Planning Association* 47: 48–58.

Lamar, M. and Associates (1989a) *Affordable Housing in New Jersey: The Results of Mount Laurel II and the Fair Housing Act*, Springfield, NJ: New Jersey Federation of Planning Officials.

Lamar, M., Mallach, A., and Payne, J.M. (1989b) "*Mount Laurel* at Work: Affordable Housing in New Jersey 1983–88," *Rutgers Law Review* 41: 1197–277.

Lamb, C.M. (1981) "Housing Discrimination and Segregation in America: Problematic Dimensions and the Federal Legal Response," *Catholic University Law Review* 30: 363–430.

Lamm, R.D. and Davison, A.G. (1972) "The Legal Control of Population Growth and Distribution in a Quality Environment: The Land Use Alternatives," *Denver Law Journal* 49: 1–51.

Landis, J.D. (1986) "Land Regulation and the Price of New Housing," *Journal of the American Planning Association* 52: 9–21.

Langdon, P. (1990) *Urban Excellence*, New York: Van Nostrand Reinhold.

Lapping, M.B. (1987) "Peoples of Plenty: A Note on Agriculture as a Planning Metaphor and National Character in North America," *Journal of Canadian Studies* 22: 121–8.

Lapping, M.B. and Leutwiler, N.R. (1987) "Agriculture in Conflict: Right-to-Farm Laws and the Peri-Urban Milieu for Farming," in Lockeretz, W. (ed.) *Sustaining Agriculture Near Cities*, Ankenny, IA: Soil and Water Conservation Society (1987).

Large, D.W. (1973) "This Land is Whose Land? Changing Concepts of Land as Property," *Wisconsin Law Review* 1973: 1039–83.

Lasker, B. (1920) "The Issue Restated," *The Survey* 44: 278–9 (May 22, 1920).

—— **(1922)** "The Atlanta Zoning Plan," *The Survey* 48: 114–15 (April 22, 1922).

Laski, H.J. (1948) *The American Democracy: A Commentary and an Interpretation*, New York: Viking (reprinted 1977, Augustus M. Kelley [Fairfield, NJ]).

Lassar, T.J. (1989) *Carrots and Sticks: New Zoning Downtown*, Washington: Urban Land Institute.

—— **(1990)** *City Deal Making*, Washington: Urban Land Institute.

Latouche, D. (1986) *Canada and Quebec, Past and Future*, Toronto: University of Toronto Press.

Lauber, D. with Bangs, S.F. (1974) *Zoning for Family and Group Care Facilities*, Planning Advisory Service Report 300, Washington: American Society of Planning Officials.

Laux, F.A. (1971) "The Zoning Game: Alberta Style, Part II: Development Control," *Alberta Law Review* 10: 1–37.

—— **(1979)** *The Planning Act (Alberta)*, Markham, Ontario: Butterworths.

—— **(1990)** *Planning Law and Practice in Alberta*, Scarborough, Ontario: Carswell.

Lawless, P. (1981) *Britain's Inner Cities: Problems and Policies*, London: Harper and Row.

—— **(1986)** *The Evolution of Spatial Policy: A Case of Inner City Policy in the United Kingdom, 1968–1981*, London: Pion.

Lawrence, B.L. (1988) "New Jersey's Controversial Growth Plan," *Urban Land* 47: 18–21 (January 1988).

Lawrence, N.S. (1989) "Means, Motives, and Takings: The Nexus Test of *Nollan v*

California Coastal Commission," in Dennison (1989).

Lazarus, E. (1883) 'The New Colussus' (inscribed on the Statue of Liberty). Reprinted in *The Poem of Emma Lazarus*, Cambridge, MA: Riverside Press, 1888.

Leach, J. (1986) "A Homebuilder Looks at Boulder's Controls: Stonewalling Growth?," in D.R. Porter (1986).

Leach, R.H. (1985) *Whatever Happened to Urban Policy? A Comparative Study of Urban Policy in Australia, Canada, and the United States*, Centre for Research on Federal Financial Relations, Research Monograph 40, Canberra: Australian National University.

Leach, S. and Game, C. (1991) "English Metropolitan Government Since Abolition: An Evaluation of the Abolition of the English Metropolitan County Councils," *Public Administration* 69: 141–70.

Lefcoe, G. (1971) "The Right to Develop Land: The German and Dutch Experience," *Oregon Law Review* 56: 31–61.

—— **(1979)** *Land Development in Crowded Places: Lessons from Abroad*, Washington: Conservation Foundation.

—— **(1982)** *Urban Land Policies for the 1980s: The Message for State and Local Government*, Lexington, MA: Heath.

Leiss, W. (ed.) (1979) *Ecology versus Politics in Canada*, Toronto: University of Toronto Press.

Leitner, M.L. and Strauss, E.J. (1988) "Elements of Municipal Impact Fee Ordinance, with Commentary," *Journal of the American Planning Association* 54: 225–31.

Lemon, J. (1984) "Toronto Among North American Cities," in Russell, V.L. (ed.) *Forging a Consensus: Historical Essays on Toronto*, Toronto: University of Toronto Press (1984).

—— **(1986)** "Reflections on the American Experience," in Flaherty and McKercher (1986).

Lemon, S.J., Feinland, S.R. and Deihl, C.C. (1990) "The First Applications of the *Nollan* Nexus Test: Observations and Comments," in Dennison (1990).

Leo, C. (1977) *The Politics of Urban Development: Canadian Urban Expressway Disputes*,

Monographs on Canadian Urban Government 3, Toronto: Institute of Public Administration of Canada.

Leonard, H.J. (1983) *Managing Oregon's Growth: The Politics of Development and Planning,* Washington: Conservation Foundation.

Leung, H.L. (1979) *Redistribution of Land Values: A Re-Examination of the 1947 Scheme,* Cambridge: Department of Land Economy, University of Cambridge.

Levin, P.H. (1976) *Government and the Planning Process: An Analysis and Appraisal of Government Decision-Making Processes, with Special Reference to the Launching of New Towns and Town Development Schemes,* London: Allen and Unwin.

Levine, M.A. (1983) "The Reagan Urban Policy: Efficient National Economic Growth and Public Sector Minimization," *Journal of Urban Affairs* 5: 17–28.

Levitt, R.L. (1987) *Cities Reborn,* Washington: Urban Land Institute.

Levitt, R.L. and Kirlin, J.J. (1985) *Managing Development Through Public/Private Negotiations,* Washington: Urban Land Institute and American Bar Association.

Levy, L.A. (1989) "Impact Fees, Concurrency, and Reality: A Proposal for Financing Infrastructure," *Urban Lawyer* 21: 471–90.

Lewis, O. (1959) *Five Families: Mexican Case Studies in the Culture of Poverty,* New York: Basic Books.

—— (1966) "The Culture of Poverty," *Scientific American* 215, 16: 19–25 (October 1966).

Lewis, S. (1990a) "The Town that Said No to Sprawl," *Planner* 55, 4: 14–19 (April 1990).

—— (1990b) *Managing Urban Growth in the San Francisco Bay Region,* Hayward, CA: Center for Public Service Education and Research, California State University.

Leyton-Brown, D. (1986) "The Domestic Policy Making Process in the United States," in Flaherty, D.H. and McKercher, W.R., *Southern Exposure: Canadian Perspectives on the United States,* Toronto: McGraw-Hill Ryerson (1986).

Lieberman, J.K. (1981) *The Litigious Society,* New York: Basic Books.

Liebmann, G.W. (1991) "The Modernization of Zoning: Enabling Act Revision As a Means to

Reform," *Urban Lawyer* 23: 1–24.

Lillydahl, J.H. and Singell, L.D. (1987) "The Effects of Growth Management on the Housing Market: A Review of the Theoretical and Empirical Evidence," *Journal of Urban Affairs* 9: 63–77.

Lillydahl, J.H., Nelson, A.C., Ramis, T.V., Rivasplata, A., and Schell, S.R. (1988) "The Need for a Standard State Impact Fee Enabling Act," in A.C. Nelson, (1988b).

Lim, G.C. (ed.) (1983) *Regional Planning: Evolution, Crisis and Prospects,* Totowa, NJ: Allanheld, Osmun.

Liner, E.B. (1985) *Intergovernmental Land Management Innovations,* Monograph 85–7, Cambridge, MA: Lincoln Institute of Land Policy.

Linteau, P-A., Durocher, J-C., and Ricard, F. (1991) *Quebec Since 1930,* Toronto: Lorimer.

Lipset, S.M. (1963a) "The Value Patterns of Democracy: A Case Study in Comparative Analysis," *American Sociological Review* 28: 515–31.

—— (1963b) *The First Nation: The United States in Historical and Comparative Perspective,* New York: Basic Books.

—— (1988) *Revolution and Counterrevolution: Change and Persistence in Social Structures,* New Brunswick, NJ: Transaction Books.

—— (1990) *The United States and Canada: Two Polities in North America,* Washington: Canadian Embassy.

—— (1990) *Continental Divide: The Values and Institutions of the United States and Canada,* London: Routledge.

Listokin, D. (1976) *Fair Share Housing Allocation,* New Brunswick, NJ: Center for Urban Policy Research, Rutgers University.

Listokin, D. Neaigus, A., Winslow, J. and Nemeth, J. (1982) *Landmarks Preservation and the Property Tax: Assessing Landmark Buildings for Real Taxation Purposes,* New Brunswick, NJ: Center for Urban Policy Research, Rutgers University.

Listokin, D. and Walker, C. (1989) *The Subdivision and Site Plan Handbook,* New Brunswick, NJ: Center for Urban Policy Research, Rutgers University.

Lithwick, N.H. (1970) [Lithwick Report] *Urban Canada: Problems and Prospects,* Ottawa:

Canada Mortgage and Housing Corporation.
—— **(1972a)** "Political Intervention: A Case Study," *Plan Canada* 12: 45–56.
—— **(1972b)** "An Economic Interpretation of the Urban Crisis," *Journal of Canadian Studies* 7, 3: 36–49.
—— **(1972c)** "Urban Policy-Making: Shortcomings in Political Technology," *Canadian Public Administration* 15: 571–84.
—— **(1978)** *Regional Economic Policy: The Canadian Experience*, Toronto: McGraw-Hill Ryerson.
Little, C.E. (1974) *The New Oregon Trail: An Account of the Development and Passage of State Land Use Legislation in Oregon*, Washington: Conservation Foundation.
Lloyd, G. (1961) *Transferable Density in Connection with Zoning*, Technical Bulletin 40, Washington: Urban Land Institute.
Logan, J.R. and Molotch, H.L. (1987) *Urban Fortunes: The Political Economy of Place*, Berkeley: University of California Press.
Logan, J.R. and Zhou, M. (1989) "Do Suburban Growth Controls Control Growth?," *American Sociological Review* 54: 461–71.
Logan, T. (1976) "The Americanization of German Zoning," *Journal of the American Institute of Planners* 42: 377–85.
Logue, T.W. (1990) "Avoiding Takings Challenges While Protecting Historic Properties From Demolition," *Stetson Law Review* 19: 739–69.
Longtin, C. (1987) *Longtin's California Land Use* (2nd edn.), Malibu, CA: Local Government Publications.
Loon, R.J. Van, and Whittington, M.S. (1987) *The Canadian Political System: Environment, Structure and Process* (4th edn.), Toronto: McGraw-Hill Ryerson.
Lorimer, J. (1978) *The Developers*, Toronto: Lorimer.
Los Angeles Times (1981) "The California Coastal Commission at the Crossroads," *Los Angeles Times* (a series of articles reprinted by the California Coastal Commission).
Losness, C.W. (1989) "Growth Control Regulation: What Does it Take to Withstand Judicial Scrutiny?," *Western State University Law Review* 17: 157–70.
Lotz, J. (1977) *Understanding Canada: Regional and Community Development*, Toronto: N.C. Press.
—— **(1978)** "Community Development and Public Participation," in Sadler (1978).
Lowi, T.J. (1979) *The End of Liberalism: The Second Republic of the United States* (2nd edn.), New York: Norton.
Lubove, R. (1962) *The Progressives and the Slums: Tenement House Reform in New York City, 1890–1917*, Pittsburgh: University of Pittsburgh Press.
Lyday, N. (1976) *The Law of the Land: Debating National Land Use Legislation 1970–75*, Washington: Urban Institute.
Lynch, K. (1981) *Good City Form*, Cambridge, MA: MIT Press.
Lynch, K. and Hack, G. (1984) *Site Planning* (3rd edn.), Cambridge, MA: MIT Press.
McAuslan, P. (1980) *The Ideologies of Planning Law*, Oxford: Pergamon.
Macaulay, R.W. and Doumani, R.G. (1991) *Ontario Land Development: Legislation and Practice*, Scarborough, Ontario: Carswell.
McClaughry, J. (1974) "The Land Use Planning Act: An Idea We Can Do Without," *Environmental Affairs* 3: 595–626.
McClendon B.W. (1984) "Reforming the Rezoning Process," *Land Use Law* April 1984: 5–9.
McCloskey, R.G. (1960) *The American Supreme Court*, Chicago: University of Chicago Press.
McCormack, A. (1946) "A Law Clerk's Recollections," *Columbia Law Review* 46: 710–18.
Macdonald, D. (1984) *Alberta's Direct Control District: A Critical Examination*, Master of City Planning Thesis, University of Alberta.
MacDonald, D.C. (ed.) (1980) *Government and Politics of Ontario* (2nd edn.), New York: Van Nostrand Reinhold.
McDougall, H.A. (1984) "*Mount Laurel II* and the Revitalizing City," *Rutgers Law Journal* 15: 667–93.
—— **(1987)** "From Litigation to Legislation in Exclusionary Zoning Law," *Harvard Civil Rights–Civil Liberties Law Review* 22: 623–63 (also in Deutsch 1988).
McDowell, B.D. (1985) "The Evolution of American Planning," in So, Hand, and McDowell (1985).

—— (1985) "Regional Planning Today," in So, Hand, and McDowell (1985).

MacEwen, M. (1990) *Housing, Race and Law: The British Experience*, London: Routledge and Kegan Paul.

McFarland, C.M. (1978) *Federal Government and Urban Problems: HUD: Successes, Failures, and the Fate of Our Cities*, Boulder, CO: Westview.

Macfarlane, C.B. and Macauley, R.W. (1984) *Land Use Planning; Practice, Procedure, Policy* [Canada], Markham: Ontario: Butterworths.

McGeary, M.G.H. and Lynn, L.E. (eds.) (1988) *Urban Change and Poverty*, Washington: National Academy Press.

Mackenzie, R.M. (1978) "Land Use and Development Control in British Columbia," *Advocate* 36: 511–31.

—— (1985) *Canadian and U.S. Land Regulation: Due Process and Fundamental Fairness: Experiences in Canada in the Due Process*, Paper to the Planning and Law Division of the American Planning Association, 1985 International Planning Conference.

McKercher, W.R. (ed.) (1983) *The U.S. Bill of Rights and the Canadian Charter of Rights and Freedoms*, Toronto: Ontario Economic Council.

Mackintosh, J.P. (1977) *The Government and Politics of Britain*, London: Hutchinson.

McMahon, E.T. and Taylor, P.A. (1990) *Citizens' Action Handbook on Alcohol and Tobacco Billboard Advertising*, Washington: Center for Science in the Public Interest.

Macpherson, C.B. (ed.) (1978) *Property: Mainstream and Critical Positions*, Toronto: University of Toronto Press.

Maddex, D. (ed.) (1990) *Landmark Yellow Pages: Where to Find All the Names, Addresses, Facts, and Figures You Need*, Washington: Preservation Press.

Madison, J. (Hamilton, A. and Jay, J.) (1788) *The Federalist Papers* (No. 10), New York: Modern Library Edition.

Magnusson, W. and Sancton, A. (eds.) (1983) *City Politics in Canada*, Toronto: University of Toronto Press.

Maine (1988) *Guidelines for Maine's Growth Management Program*, Augusta, ME: Office of Comprehensive Planning, Maine Department of Economic and Community Development.

Maine Law Review (1972) "Contract Zoning: A Flexible Technique for Protecting Maine Municipalities," *Maine Law Review* 24: 263–79.

Major, L.D. (1986) "Linkage of Housing and Commercial Development: The Legal Issues," *Real Estate Law Journal* 15: 328–52.

Makielski, S.J. (1966) *The Politics of Zoning: The New York Experience*, New York: Columbia University Press.

Makuch, S.M. (1983) *Canadian Municipal and Planning Law*, Scarborough, Ontario: Carswell.

Malecki, E.J. (1982) "Federal R and D Spending in the United States of America: Some Impacts on Metropolitan Economies," *Regional Studies* 16: 19–35.

Mallach, A. (1984) *Inclusionary Housing Programs: Policies and Practices*, New Brunswick, NJ: Center for Urban Policy Research, Rutgers University.

Mandelker, D.R. (1962) *Green Belts and Urban Growth*, Madison: University of Wisconsin Press.

—— (1971a) *The Zoning Dilemma: A Legal Strategy for Urban Change*, Indianapolis: Bobbs-Merrill.

—— (1971b) "Judicial Review of Land Development Controls Under the ALI Model Code," *Land Controls Annual 1971*.

—— (1976a) *Environmental and Land Controls Legislation* (with 1982 Supplement), Indianapolis: Bobbs-Merrill.

—— (1976b) "The Role of the Local Comprehensive Plan in Land Use Regulation," *Michigan Law Review*, 74: 900–73.

—— (1977) "Racial Discrimination and Exclusionary Zoning," *Texas Law Review* 55: 1217–53.

—— (1982) *Land Use Law*, Charlottesville, VA: Michie.

—— (1985a) *1985 Supplement to Land Use Law*, Charlottesville, VA: Michie.

—— (1985b) *Planning and Control of Land Development: Cases and Materials* (2nd edn.), Charlotteville, VA: Michie.

—— (1988) *Land Use Law* (2nd edn.), Charlottesville, VA: Michie.

——— (1989a) "The Quiet Revolution – Success and Failure," *Journal of the American Planning Association* 55: 204–5.

——— (1989b) "The Shifting Presumption of Constitutionality in Land Use Law," *Journal of Planning Literature* 4: 383–93.

Mandelker, D.R. and Cunningham, R.A. (1985) *Planning and Control of Land Development: Cases and Materials* (2nd edn.), Charlottesville, VA: Michie.

——— (1987) *1987 Supplement: Planning and Control of Land Development: Cases and Materials*, Charlottesville, VA: Michie.

——— (1990) *Planning and Control of Land Development: Cases and Materials*, (3rd edn.), Charlottesville, VA: Michie.

Mandelker, D.R., Netsch, D.C., and Salsich, P.W. (1983) *State and Local Government in a Federal System: Cases and Materials*, Charlottesville, VA: Michie.

Mantell, M.A., Harper, S.F., and Propst, L. (1990) *Creating Successful Communities: A Guidebook to Growth Management Strategies*, Washington: Conservation Foundation/Island Press.

Marcus, N. (1974) "Mandatory Development Rights Transfers and the Taking Clause: The Case of Manhattan's Tudor City Parks," *Buffalo Law Review* 24: 77–110.

——— (1984) "Air Rights in New York City: TDR, Zoning Lot Merger and the Well-Considered Plan," *Brooklyn Law Review* 50: 867–911.

——— (1985) "A New Era of Zoning Exactions?," in Merriam, Brower, and Tegeler (1985).

Marcus, N. and Elliott, D.H. (1970) *The New Zoning: Legal, Administrative, and Economic Concepts and Techniques*, New York: Praeger.

Marcuse, P. (1980) "Housing Policy and City Planning: The Puzzling Split in the United States 1893–1931," in Cherry (1980).

Markusen, A.R. (1987) *Regions: The Economics and Politics of Territory*, Lanham, MD: Rowman and Littlefield.

Markusen, A.R. and Bloch, R. (1985) "Defensive Cities: Military Spending, High Technology, and Human Settlements," in Castells (1985).

Markusen, A.R. and Wilmoth, D. (1982) *The Political Economy of National Urban Policy in the U.S.A. 1978–81*, Berkeley: University of California, Berkeley, Institute of Urban and Regional Development, Working Paper 316.

Markusen, A.R., Hall, P., Campbell, S., and Deitrick, S. (1991) *The Rise of the Gunbelt: The Military Remapping of Industrial America*, New York: Oxford University Press.

Marris, P (1982) *Community Planning and Conceptions of Change*, London: Routledge and Kegan Paul.

Marris, P. and Rein, M. (1973) *Dilemmas of Social Reform: Poverty and Community Action in the United States*, Chicago: Aldine.

Marsh, B.C. (1909) *An Introduction to City Planning: Democracy's Challenge and the American City*, New York: privately published.

Marsh, J.H. (chief editor) (1985) *The Canadian Encyclopedia*, Edmonton: Hurtig.

Maryland (1991) *Protecting the Future: A Vision for Maryland*, Annapolis: Governor's Commission on Growth in the Chesapeake Bay Region.

Mason, A.T. (1968) *The Supreme Court From Taft to Warren*, Baton Rouge: Louisana State University Press.

Masser, I. (1984) "Cross National Comparative Planning Studies: A Review," *Town Planning Review* 55: 137–60.

Mather, G. (1988) *Pricing for Planning*, London: Institute of Economic Affairs.

Matuson, J.L. (1978) "A Legislative Approach to Solar Access: Transferable Development Rights," *New England Law Review* 13: 835–76.

Maurer, R.C. and Christenson, J.A. (1982) "Growth and Nongrowth Orientations of Urban, Suburban, and Rural Mayors: Reflections on the City as a Growth Machine," *Social Science Quarterly* 63: 350–8.

Mayer, M. (1990) *The Greatest-Ever Bank Robbery: The Collapse of the Savings and Loans Industry*, New York: Scribner.

Mazmanian, D.A. and Sabatier, P.A. (1983) *Implementation and Public Policy*, Glenview, IL: Scott, Foresman.

Meck, S. and Netter, E.M., (eds.) (1983) *A Planner's Guide to Land Use Law*, Chicago: Planners Press.

Meltzer, J. (1984) *Metropolis to Metroplex: The Social and Spatial Planning of Cities*, Baltimore: Johns Hopkins University Press.

Mercer, J. (1979) "On Continentalism, Distinctiveness, and Comparative Urban Geography: Canadian and American Cities," *Canadian Geographer* 23: 119–39.

—— **(1982)** *Comparing the Reform of Metropolitan Fragmentation, Fiscal Dependency, and Political Culture in Canada and the United States*, Syracuse, NY: Metropolitan Studies Program, Maxwell School of Citizenship and Public Affairs, Syracuse University.

Mercer, J. and Goldberg, M.A. (1982) *Value Differences and their Meaning for Urban Development in Canada and the USA*, Working Paper 12, Vancouver: Faculty of Commerce, University of British Columbia.

Mercer, L.J. and Morgan, W.D. (1985) "California City and County User Charges: A Post Proposition 13 Assessment," *Urban Law and Policy* 7: 187–205.

Merriam, D., Brower, D.J. and Tegeler, P.D. (1985) *Inclusionary Zoning Moves Downtown*, Chicago: Planners Press.

Merrill, T.W. (1988) "The Economics of Public Use," in Deutsch (1988) (also in Cornell Law Review 72: 61–116, 1986).

Merton, R.K. (1957) "The Latent Functions of the Machine," in *Social Theory and Social Structure*, New York: Free Press.

Meshenberg, M.J. (1976) *The Language of Zoning: A Glossary of Words and Phrases*, Planning Advisory Service Report 322, Washington: American Society of Planning Officials.

Metcalfe, G.R. (1988) *Fair Housing Comes of Age*, Westport, CT: Greenwood.

Metzenbaum, J. (1955) *The Law of Zoning*, (2nd edn.): Baker, Voorhis. (1st edn. 1930).

Meyer, M. (1990) *The Greatest-Ever Bank Robbery: The Collapse of the Savings and Loan Industry*, New York: Scribner.

Meyerson, M. and Banfield, E.C. (1955) *Politics, Planning and the Public Interest*, New York: Free Press.

Miami Valley Regional Planning Commission (1970) *A Housing Plan for the Miami Valley*, Dayton, OH: Miami Valley Regional Planning Commission.

Michelman, F.I. (1968) "Property, Utility, and Fairness: Comments on the Ethical Foundations of 'Just Compensation' Law," *Harvard Law Review* 80: 1165–258 (also in Ackerman 1975).

Mier, R. and Gelzer, S.E. (1982) "State Enterprise Zones: The New Frontier?," *Urban Affairs Quarterly* 18: 39–52.

Miller, B. (1987) "Department of Transportation's Section 4(f): Paving the Way Toward Preservation," *American University Law Review* 36: 633–67.

Miller, J.H. (1987) "Coordination of Historic Preservation and Land Use Controls: New Directions in Historic Preservation Regulation," *Preservation Law Reporter* 5: 2041–53.

Miller, T.I. (1986) "Must Growth Restrictions Eliminate Moderate-Priced Housing?," *Journal of the American Planning Association* 52: 319–25.

Miller, W. (1976) "The New Jersey Land Use Law Review: A Lesson for Other States," *Real Estate Law Journal* 5: 138–54.

Miller, Z.L. and Melvin, P.M. (1987) *The Urbanization of Modern America: A Brief History*, New York: Harcourt Brace Jovanovitch.

Mills, E.S. (1979) "Economic Analysis of Urban Land Use Controls," in Mieszkowsko, P. and Straszheim, M. *Current Issues in Urban Economics*, Baltimore: Johns Hopkins University Press.

Mills, E.S. and Hamilton, B.W. (1989) *Urban Economics* (4th edn.), Glenview, IL: Scott, Foresman.

Mills, S.F. (1987) "A Profile of the American City," in Welland (1987).

Minnesota (1979) *PUD: A Special Tool for Special Situations*, St. Paul: Minnesota State Planning Agency.

MIT Center (1984) *Spring Seminars 1984: An Examination of Public/Private Issues in Real Estate Development*, Cambridge, MA: MIT Center for Real Estate Development.

Mitchell, J.P. (ed.) (1985) *Federal Housing Policy and Programs*, New Brunswick, NJ: Center for Urban Policy Research, Rutgers University.

Mixon, J. (1967) "Jane Jacobs and the Law – Zoning for Diversity Examined," *Northwestern University Law Review* 62: 314–56.

—— **(1990)** "Neighborhood Zoning for Houston," *South Texas Law Review* 31: 1–41.

Mogulof, M.B. (1971) "Regional Planning, Clearance, and Evaluation: A Look at the A-95 Process," *Journal of the American Institute of Planners* 37: 418–22.

—— **(1975)** *Saving the Coast: California's Experiment in Intergovernmental Land Use Control*, Lexington, MA: Lexington.

Mohl, R.A. and Richardson, J.F. (eds.) (1973) *The Urban Experience: Themes in American History*, Belmont, CA: Wadsworth.

Mollenkopf, J.H. (1975) "The Post-War Politics of Urban Development," *Politics and Society* 5: 247–95 (also in Tabb and Sawers 1978).

—— **(1983)** *The Contested City*, Princeton: Princeton University Press.

Molotch, H. (1976) "The City as a Growth Machine," *American Journal of Sociology* 82: 309–32.

Moore, B.H. (1983) *The Entrepreneur in Local Government*, Washington: International City Management Association.

Moore, C.G. and Siskin, C. (1985) *PUDS in Practice*, Washington: Urban Land Institute.

Moore, M.M. (1988) "Legislative and Judicial Regulation of Junkyards," *Mercer Law Review* 39: 461-75.

Moore, V, (1990) *A Practical Guide to Planning Law* [England and Wales] (2nd edn.), London: Blackstone.

Morrill, R.L. (1989) "Regional Governance in the United States: For Whom?," *Environment and Planning C: Government and Policy* 7: 13–26.

Morris, R.S. (1980) *Bum Rap on America's Cities*, Hemel Hempstead: Prentice-Hall.

Morrison, D.C. (1991) "Gunbelt Diplomacy," *National Journal* 23, 39: 2341–3.

Morrison, J.H. (1972) *Supplement to Historic Preservation Law*, no publisher stated.

Morrison, P.A. (1977) *Migration and Rights of Access: New Public Concerns of the 1970s*, Santa Monica, CA: Rand Corporation, Report P-5785.

Morton, F.L. (1987) "The Political Impact of the Canadian Charter of Rights and Freedoms," *Canadian Journal of Political Science* 20: 31–55.

Moss, E. (ed.) (1977) *Land Use Controls in the United States: A Handbook on the Legal Rights of Citizens by Natural Resources Defence Council*, New York: Dial Press/James Wade.

Mowat, F. (1985) *My Discovery of America*, Toronto: McClelland and Stewart.

Moynihan, D.P. (1969) "Toward a National Urban Policy," *The Public Interest* 17: 3–20 (Fall 1969).

—— **(1970)** *Toward a National Urban Policy*, New York: Basic Books.

Mumford, L. (1945) *City Development: Studies in Disintegration and Renewal*, New York: Harcourt Brace.

—— **(1961)** *The City in History: Its Origins, Its Transformations, and Its Prospects*, New York: Harcourt Brace.

—— **(1986)** *The Lewis Mumford Reader*, (edited by Miller, D.L.), New York: Pantheon.

Munby, D.L. (1954) "Development Charge and the Compensation-Betterment Problem," *Economic Journal* 64: 87–97.

Murtagh, W.J. (1988) *Keeping Time: The History and Theory of Preservation in America*, Pittstown, NJ: Main Street Press.

Muth, R. and Wetzler, E. (1976) "The Effect of Constraints on Housing Costs," *Journal of Urban Economics* 3: 57–67.

Myers, P. (1974) *So Goes Vermont*, Washington: Conservation Foundation.

—— **(1976)** *Zoning Hawaii: An Analysis of the Passage and Implementation of Hawaii's Land Classification Law*, Washington: Conservation Foundation.

Myerson, M. and Banfield, E.C. (1955) *Politics, Planning, and the Public Interest: The Case of Public Housing in Chicago*, New York: Free Press.

Nathan, R.P., Manvell, A.D., and Calkins, S.E. (1975) *Monitoring Revenue Sharing*, Washington: Brookings Institution.

National Advisory Commission on Criminal Justice Standards and Goals (1973) *Community Crime Prevention*, Washington: U.S. GPO.

National Commission on Urban Problems (1968) *Report* [Douglas Report], Washington: U.S. GPO.

National Committee Against Discrimination in Housing and Urban Land Institute (1974) *Fair Housing and Exclusionary Land Use*, Research Report 23, Washington: Urban Land Institute.

National Committee on Urban Growth *see* Canty, D.

National Research Council (1982) *Critical Issues for National Urban Policy: A Reconnaissance and Agenda for Further Study*, Committee on National Urban Policy, Washington: National Academy Press.

—— **(1986)** *Emerging Issues in National Urban Policy: Report of a Seminar*, Washington: National Academy Press.

—— **(1988)** *Committee on National Urban Policy Report*, in McGeary and Lynn (1988).

Natural Resources Defense Council (1977) *Land Use Controls in the United States*, Washington: Natural Resources Defence Council.

Needham, B. (1988) "An Approach to Land Policy: Ideas from the Dutch Experience," *Urban Law and Policy* 9: 439–51.

Nelles, H.V. (1974) *The Politics of Development: Forests, Mines and Hydro-Electric Power in Ontario 1849–1941*, Toronto: Macmillan.

Nelson, A.C. (1988a) "Downtown Office Development and Housing Linkage Fees," *Journal of the American Planning Association* 54: 197–8.

—— **(ed.) (1988b)** *Development Impact Fees: Policy Rationale, Practice, Theory, and Issues*, Chicago: Planners Press.

—— **(1991)** "Blazing New Planning Trails in Oregon," in DeGrove (1991).

Nelson, R.H. (1977) *Zoning and Property Rights: An Analysis of the American System of Land-Use Regulation*, Cambridge, MA: MIT Press.

—— **(1984)** "Private Neighborhoods: A New Direction for the Neighborhood Movement," in Geisler and Popper (1984).

—— **(1985)** "Marketable Zoning: A Cure for the Zoning System," *Land Use Law and Zoning Digest* 37, 11: 3–8 (November 1985).

—— **(1989)** "Zoning Myth and Practice – From *Euclid* into the Future," in Haar and Kayden (1989a).

Nelson, R.R. (1977) *The Moon and the Ghetto*, New York: Norton.

Nenno, M.K. (1986) *New Money and New Methods: A Catalog of State and Local Initiatives in Housing and Community Development*, Washington: National Association of Housing and Redevelopment Officials.

—— **(1987)** "Reagan's '88 Budget: Dismantling HUD," *Journal of Housing* 44,3: 103–8.

Ness, S.C. Van (1984) "On the Public Advocate's Involvement in *Mount Laurel*," *Seton Hall Law Review* 14: 832–50.

Netter, E. and Vranicar, J. (1981) *Linking Plans and Regulations: Local Responses to Consistency Laws in California and Florida*, Planning Advisory Service Report 363, Washington: American Planning Association.

Neutz, M. (1968) *The Suburban Apartment Boom*, Baltimore: Resources for the Future/Johns Hopkins Press.

Nevins, D. (ed.) (1982) *Grand Central Terminal: City Within the City*, New York: Municipal Art Society of New York.

New Jersey (1977) *The Princeton Housing Proposal: A Strategy to Achieve Balanced Housing Without Government Subsidy*, Trenton: New Jersey Department of Community Affairs.

—— **(1988a)** *The Draft Preliminary State Development and Redevelopment Plan, Vol I: "Building a Legacy"*, Trenton: New Jersey State Planning Commission.

—— **(1988b)** *The Draft Preliminary State Development and Redevelopment Plan, Vol II: "Strategies, Policies and Standards,"* Trenton: New Jersey State Planning Commission.

—— **(1988c)** *Communities of Place: The Preliminary State Development and Redevelopment Plan*, Trenton: New Jersey State Planning Commission.

Newton, K. (1975) "'American Democracy Reconsidered' – Some Comments" [on Sharpe 1973], *British Journal of Political Science* 5: 123–8.

New York (1982a) *Public–Private Partnerships: An Opportunity for Urban Communities*, New York Committee for Economic Development Research and Policy Committee.

—— **(1982b)** *Midtown Zoning*, New York City Planning Commission.

—— **(1987)** *Midtown Development Review* New York Department of City Planning.

—— **(1988a)** *The Role of Amenities in the Land Use Process*, Association of the Bar of the City of New York.

—— **(1988b)** *Buildings Approved under the Special Midtown Zoning District, May 1982 to May 1988*, New York Department of City Planning.

—— (1989a) *New York City Zoning: The Need for Reform*, Natural Resources Defense Council and Women's City Club of New York.

—— (1989b) *Regulating Residential Towers and Plazas: Issues and Options*, New York Department of City Planning.

—— (1990) *Zoning Handbook: A Guide to New York City's Zoning Resolution*, New York Department of City Planning.

Nicholas, H.G. (1986) *The Nature of American Politics* (2nd edn.), Oxford: Oxford University Press.

Nicholas, J.C. (1981) "Housing Costs and Prices Under Growth Control Regulation," *Journal of the American Real Estate and Urban Economics Association* 9: 384-96.

—— (1988) *The Calculation of Proportionate-Share Impact Fees*, Planning Advisory Service Report 408, Washington: American Planning Association.

Nicholas, J.C. (ed.) (1985) *The Changing Structure of Infrastructure Finance*, Monograph 85–5, Cambridge, MA: Lincoln Institute of Land Policy.

Nicholas, J.C. and Nelson, A.C. (1988) "The Rational Nexus Test and Appropriate Development Impact Fees," in Nelson (1988b).

Nitikman, M.A. (1988) "Instant Planning – Land Use Regulation by Initiative in California," *Southern California Law Review* 61: 497–539.

NLC [National League of Cities] (1979) *City Economic Development*, Washington: NLC.

—— (1988) *Your City's Kids*, Washington: NLC.

Nolan, Van and Horack, F.E. (1954) "How Small a House – Zoning for Minimum Space Requirements" *Harvard Law Review* 67: 967–86.

Nolen, J. (1909) "City Making," *The American City* 1: 15–19.

Nolen, J. (ed.) (1916) *City Planning: A Series of Papers Presenting the Essential Elements of a City Plan*, New York: Appleton.

North Carolina Department of Natural Resources and Community Development (1985) *Striking a Balance: Reflections on The Years of Managing the North Carolina Coast*, Raleigh: The Department.

Noyelle, T. and Stanback, J. (1983) *Economic Transformation of American Cities*, Totowa,

NJ: Allanheld and Rowman.

O'Keefe, T.C. (1972) "Time Controls on Land Use: Prophylactic Law for Planners," *Cornell Law Review* 57: 827–49.

Ong, P. (1981) "An Ethnic Trade: The Chinese Laundries in Early California," *Journal of Ethnic Studies* 8: 95–113.

Ontario, Province of (1988) *Financing Growth-Related Capital Needs*, Toronto: Ontario Ministry of Treasury and Economics.

Oregon (1990) *Oregon's Statewide Planning Goals*, Salem: Oregon Land Conservation and Development Commission.

—— (1991a) *Urban Growth Management Study: Summary Report*, Salem: Oregon Department of Land Conservation and Development.

—— (1991b) *Managing Growth to Promote Affordable Housing*, Portland: 1000 Friends of Oregon and Home Builders Assocation of Metropolitan Portland.

Oregon Coastal Management Program (1987) *Oregon's Coastal Management Program: A Citizen's Guide*, Salem: Oregon Department of Land Conservation and Development.

Orlebeke, C.J. (1990) "Chasing Urban Policy: A Critical Retrospect," in Kaplan and James (1990).

Orosz, L. (1991) "New Governor Reviews Florida Growth Law," *Planning* 57, 1: 30 (January 1991).

Osborn, J.E. (1977) "New York's Urban Development Corporation: A Study on Unchecked Power of a Public Authority," *Brooklyn Law Review* 43: 237–382.

Palmer, J.L. and Sawhill, I.V. (eds.) (1982) *The Reagan Experiment*, Washington: Urban Institute.

—— (1984) *The Reagan Record: An Assessment of America's Changing Domestic Priorities*, Cambridge, MA: Ballinger.

Parizeau Report (1986) *Report of the Municipal Study Commission*, Sainte-Foy, Quebec: Union of Quebec Municipalities.

Parker, C.E.T. (1982) "Inclusionary Zoning – A Proper Police Power Function or a Constitutional Anathema?," *Western State University Law Review* 9: 175–97.

Parkinson, M. (1989) "The Thatcher Government's Urban Policy, 1979–1989," *Town Planning Review* 60: 421–40.

**Parkinson, M., Foley, B. and Judd, D.R.
(1989)** *Regenerating the Cities: The UK Crisis
and the US Experience*, Glenview, IL: Scott,
Foresman.

Parris, J.H. (1969) "Congress Rejects the
President's Urban Department, 1961–62," in
Cleaveland and Associates (1969).

Passow, S.S. (1970) "Land Reserves and
Teamwork in Planning Stockholm," *Journal of
the American Institute of Planners* 36: 179–88.

Patterson, T.W. (1979) *Land Use Planning:
Techniques of Implementation*, London: Van
Nostrand Reinhold.

Paul, E.F., (1987) *Property Rights and Eminent
Domain*, New Brunswick, NJ: Transaction
Books.

Payne, J.M. (1976) "Delegation Doctrine in the
Reform of Local Government Law: The Case
of Exclusionary Zoning," *Rutgers Law Review*
29: 803–66.

—— **(1987)** "Rethinking Fair Share: The
Judicial Enforcement of Affordable Housing
Policies," *Real Estate Law Journal* 16: 20–44.

Peiser, R. (1981) "Land Development and
Regulation: A Case Study of Dallas and
Houston, Texas," *Journal of the American Real
Estate and Urban Economics Association* 9:
397–417.

—— **(1988)** "Calculating Equity-Neutral Water
and Sewer Impact Fees," in Nelson (1988b).

**Pelham, T.G., Hyde, W.L., and Banks, R.P.
(1985)** "Managing Florida's Growth: Toward
an Integrated State, Regional, and Local
Comprehensive Planning Process," *Florida
State University Law Review* 13: 515–98.

Peltason, J.W. (1971) *Fifty-eight Lonely Men:
Southern Federal Judges and School
Desegregation*, Champaign: University of
Illinois Press.

**Penfold, G., Glenn, J.M., Fleming, D. and
Brown, D. (1989)** "Right-to-Farm in
Canada," *Plan Canada* 29, 2: 47–55 (March
1989).

Pennsylvania Economy League (1989)
*Prospects for Linking Commercial and
Community Development in Philadelphia*,
Philadelphia: Pennsylvania Economy League.

Perin, C. (1977) *Everything in its Place: Social
Order and Land Use in America*, Princeton:
Princeton University Press.

Perks, W.T. and Jamieson, W. (1991)
"Planning and Development in Canadian
Cities," in Bunting and Filion (1991).

Perks, W.T. and Robinson, I.M. (eds.) (1979)
*Urban and Regional Planning in a Federal State:
The Canadian Experience*, Strasburg, PA:
Dowden, Hutchinson and Ross/McGraw-Hill.

Perks, W.T. and Smith, P.J. (1985) "Urban and
Regional Planning," in *The Canadian
Encyclopedia*, Edmonton: Hurtig.

Perloff, H.S. and Sandberg, N.C. (1972) *New
Towns: Why – And For Whom?*, New York:
Praeger.

Perry, D.C. and Keller, L.F. (1991) "The
Structures of [Local] Government," in
Bingham, R.D. *et al.*, *Managing Local
Government*, Newbury Park, CA: Sage.

Peter, L.J. (1985) *Why Things Go Wrong, or The
Peter Principle Revisited*, New York: Morrow.

Peters, J.E. (1990) "Saving Farmland: How Well
Have We Done?," *Planning* 56, 9: 12–17
(September 1990).

Peterson, G.E. and Lewis, C.W. (eds.) (1986)
Reagan and the Cities, Washington: Urban
Institute.

**Peterson, G.E. Bovbjerg, R.R., Davis, B.A.,
Davis, W.G., Durman, E.C. and Gullo,
T.A. (1986)** *The Reagan Block Grants: What
Have We Learned?*, Washington: Urban
Institute.

Peterson, J.A. (1976) "The City Beautiful
Movement: Forgotten Origins and Lost
Meanings," *Journal of Urban History* 2:
415–34; also in Krueckeberg (1983).

—— **(1979)** "The Impact of Sanitary Reform
upon American Urban Planning," *Journal of
Social History* 13: 83–103; also in Krueckeberg
(1983).

Peterson, P.E. (1981) *City Limits*, Chicago:
University of Chicago Press.

Peterson, P.E. and Kantor, P. (1977) "Political
Parties and Citizen Participation in English
City Politics," *Comparative Politics* 9: 197–217.

Petrillo, J.E. and Grenell, P. (1985) *The Urban
Edge: Where the City Meets the Sea*,
Sacramento: California State Coastal
Commission and William Kaufman Inc.

Philbrick, F.S. (1938) "Changing Concepts of
Property in Law," *University of Pennsylvania
Law Review* 86: 691–732.

Phillips, K. (1969) *The Emerging Republican Majority*, London: Arlington House.

Pierce, N.R. and Guskind, R. (1985) "Reagan Budget Cutters Eye Community Block Grant Program on its 10th Birthday," *National Journal* January 5, 1985: 12–15 (reprinted in Callies and Freilich 1986: 702–7).

Piven, F.F. and Cloward, R.A. (1971) *Regulating the Poor: The Functions of Public Welfare*, New York: Pantheon.

Pizor, P.J. (1986) "Making TDR Work: A Study of Program Implementation," *Journal of the American Planning Association* 52: 203–11.

Pizzorno, A. (1971) "Amoral Familism and Historical Marginality," in Dogan, M. and Rose, R. *European Politics: A Reader*, Boston: Little, Brown.

Planning (1990) "Washington Joins the Crowd," *Planning* 56, 5: 31 (May 1990).

Platt, R.H. (1991) *Land Use Control: Geography, Law, and Public Policy*, New York: Prentice-Hall.

Plaut, T.R. (1980) "Urban Expansion and the Loss of Farmland in the United States: Implications for the Future," *American Journal of Agricultural Economics* 62: 537–42.

Plotkin, S. (1987) *Keep Out: The Struggle for Land Use Control*, Berkeley: University of California Press.

Plunkett, T.J. (1972) *The Financial Structure and the Decision-Making Process of Canadian Municipal Government*, Ottawa: Central Mortgage and Housing Corporation.

Plunkett, T.J. and Betts, G.M. (1978) *The Management of Canadian Urban Government*, Kingston, Ontario: Institute of Local Government, Queen's University.

Plunkett, T.J. and Hooson, W. (1975) "Municipal Structure and Services," *Canadian Public Policy* 1: 367–75.

Pollakowski, H.O. and Wachter, S.M. (1990) "The Effects of Land Use Constraints on Housing Prices," *Land Economics* 66: 315–24.

Pollard, O.A. (1987) "Billboard Removal: What Amount of Compensation is Just?" *Virginia Journal of Natural Resources Law* 6: 323–74.

Poole, S.E. (1987) "Architectural Appearance Review Regulations and the First Amendment: The Good, The Bad, and The Consensus Ugly," *Urban Lawyer* 19: 287–344.

Pooley, B.J. (1961) *Planning and Zoning in the United States*, Ann Arbor: University of Michigan Law School (reprinted by Hein [Buffalo, NY] 1982).

Popp, T.E. (1988) "A Survey of Governmental Response to the Farmland Crisis: States' Application of Agricultural Zoning," *University of Arkansas at Little Rock Law Journal* 11: 515–52.

Popper, F.J. (1981) *The Politics of Land-Use Reform*, Madison: University of Wisconsin Press.

—— **(1986)** "The Strange Case of the Contemporary American Frontier," *Yale Review* 76: 101–21.

—— **(1988)** "Understanding American Land Use Regulation Since 1970," *Journal of the American Planning Association* 54: 291–301.

Porter, D.R. (1985) *Downtown Linkages*, Washington: Urban Land Institute.

—— **(1986)** *Growth Management: Keeping on Target?*, Washington: Urban Land Institute.

—— **(1989a)** "Foreword" to Brower, Godschalk, and Porter (1989).

—— **(1989b)** *Development Agreements: Practice, Policy, and Prospects*, Washington: Urban Land Institute.

Porter, D.R., Lin, B.C., and Peiser, R.B. (1987) *Special Districts: A Useful Technique for Financing Infrastructure*, Washington: Urban Land Institute.

Porter, D.R. and Marsh, L.L. (eds.) (1989) *Development Agreements: Practice, Policy and Prospects*, Washington: Urban Land Institute.

Porter, D.R. and Peiser, R.B. (1984) *Financing Infrastructure to Support Community Growth*, Washington: Urban Land Institute.

Porter, D.R., Phillips, P.L., and Lassar, T.J. (1988) *Flexible Zoning: How it Works*, Washington: Urban Land Institute.

Porter, J. (1965) *The Vertical Mosiac*, Toronto: University of Toronto Press.

Potter, D. (1954) *People of Plenty*, Chicago: University of Chicago Press.

Power, G. (1983) "Apartheid Baltimore Style: The Residential Segregation Ordinances of 1910–1913," *Maryland Law Review* 42: 289–328.

Prescott, J.R. (1970) "The Planning for Experimental City," *Land Economics* 56: 68–75.

President *see also* White House.

President's Commission on Housing (1982) *Report*, Washington: U.S. GPO.

President's Commission on a National Agenda for the Eighties, (1980) *A National Agenda for the Eighties*, Washington: U.S. GPO.

President's Interagency Coordinating Council (1979) *Urban Action: A New Partnership to Conserve America's Communities: A Report After One Year*, Washington: U.S. GPO.

President's National Advisory Commission on Rural Poverty (1967) *The People Left Behind*, Washington: U.S. GPO.

—— **(1968)** *Rural Poverty in the United States*, Washington: U.S. GPO.

President's Urban and Regional Policy Group (1977) *Cities and People in Distress: National Urban Policy Discussion Draft*, Washington: U.S. HUD.

—— **(1978)** *A New Partnership to Conserve America's Communities: A National Urban Policy*, Washington: U.S. GPO (This report is paginated in four parts: P – for Preface – and Sections I, II, and III).

Pressman, J.L. (1975) *Federal Programs and City Politics: The Dynamics of the Aid Process in Oakland*, Berkeley: University of California Press.

Pressman, J.L. and Wildavsky, A.B. (1984) *Implementation: How Great Expectations in Washington Are Dashed in Oakland; Or Why It's Amazing that Federal Programs Work at all, This Being the Saga of the Economic Development Administration as Told by Two Sympathetic Observers Who Seek to Build Morals on a Foundation of Ruined Hopes* (3rd edn.), Berkeley: University of California Press.

Presthus, R. (ed.) (1977) *Cross-National Perspectives: United States and Canada*, Leiden, Netherlands: Brill.

Public Lands Law Review Commission (1970) *One Third of the Nation's Land: A Report to the President and to the Congress*, Washington: U.S. GPO.

Puget Sound Council of Governments (1990) *Growth Management Techniques: A Report to the Puget Sound Council of Governments*, Seattle: Puget Sound Council of Governments.

Punter, J.V. (1986) "A History of Aesthetic Control: The Control of the External Appearance of Development in England and Wales, Part 1, 1909–1953," *Town Planning Review* 57: 351–81.

—— **(1987)** "A History of Aesthetic Control: The Control of the External Appearance of Development in England and Wales, Part 2, 1953–1985," *Town Planning Review* 58: 29–62.

—— **(1990)** *Design Control in Bristol 1940–1990*, Bristol, UK: Redcliffe.

Pye, L.W. (1972) "Culture and Political Science: Problems in the Evaluation of the Concept of Political Culture," *Social Science Quarterly* 53: 285–96.

Quimby, I.M.G. (ed.) (1978) *Material Culture and the Study of American Life*, New York: Norton.

Rabin, Y. (1989) "Exclusive Zoning: The Inequitable Legacy of *Euclid*" in Haar and Kayden (1989a).

Rabinovitz, F. (1969) *City Politics and Planning*, New York: Atherton.

—— **(1989)** "Rezoning Los Angeles: The Administration of Comprehensive Planning," *Public Administration Review* 49: 330–6.

Rabinovitz, F. and Smookler, H.V. (1972) "Rhetoric versus Performance: The National Politics and Administration of U.S. New Community Development Legislation," in Perloff and Sandberg (1972).

Rapoport, A. (1982) *The Meaning of the Built Environment: A Non-Verbal Communication Approach*, Beverly Hills, CA: Sage.

Rapp, D. (1988) *How the US got into Agriculture: And Why it Can't Get Out*, Washington: Congressional Quarterly.

Rathkopf, A.H. and Rathkopf, D.A. (1988) *The Law of Planning and Zoning*, New York: Clark Boardman (updated by regular releases).

Raup, P.M. (1975) "Urban Threats to Rural Lands: Background and Beginnings," *American Institute of Planners Journal* 41: 371–8.

—— **(1980)** *The Federal Dynamic in Land Use*,

Washington: National Planning Association.

—— (1982) "An Agricultural Critique of the National Agricultural Lands Study," *Land Economics* 58: 260–74.

Ravis, D. (1973) *Advance Land Acquisition by Local Government: The Saskatoon Experience*, Community Planning Association of Canada.

Ravo, N. (1988) "In Greenwich, Housing Problem is Huge," *New York Times*, August 25, 1988.

Raymond, V. (1988) *Surviving Proposition Thirteen: Fiscal Crisis in California Counties*, Berkeley: Institute of Governmental Studies, University of California, Berkeley.

Reagan, M.D. (1972) *The New Federalism* (1st edn.), Oxford: Oxford University Press.

Reagan, M.D. and Sanzone, J.G. (1981) *The New Federalism* (2nd edn.), Oxford: Oxford University Press

Real Estate Research Corporation (1974a) *The Costs of Sprawl: Literature Review and Bibliography*, Washington: U.S. GPO.

—— (1974b) *The Costs of Sprawl: Detailed Cost Analysis*, Washington: U.S. GPO.

Redcliffe-Maud Report (1974) *Report of the Prime Minister's Committee on Local Government Rules of Conduct*, Cmnd 5636, London: HMSO.

Redman, M. (1991) "Planning Gain and Obligations," *Journal of Planning and Environment Law* 1991: 203–18.

Rehfuss, J. and Stowe, E. (1975) "A Symposium: The Governance of New Towns," *Public Administration Review* 35: 221–62.

Reich, C.A. (1964) "The New Property," *Yale Law Journal* 73: 733–87.

—— (1966) "The Law of the Planned Society," *Yale Law Journal* 75: 1227–70.

Reidy, M.W. (1977) "H.B. 2876: Providing Cities with Flexibility in Land Use Decisionmaking," *Oregon Law Review* 56: 270–84.

Reilly, W.K. (1973) *The Use of Land: A Citizen's Policy Guide to Urban Growth: A Task Force Report Sponsored by the Rockefeller Brothers Fund*, New York: Thomas Y. Crowell.

Reilly, W.K. and Schulman, S.J. (1969) "The State Urban Development Corporation: New York's Innovation," *Urban Lawyer* 1: 129–46.

Reiner, T.A. (1979) "Towards a National Urban Policy: Critical Reviews: Introduction,"

Journal of Regional Science 19: 67–8.

Reps, J.W. (1965) *The Making of Urban America: A History of City Planning in the United States*, Princeton: Princeton University Press.

—— (1967) "The Future of American Planning: Requiem or Renascence?," *Planning 1967: Selected Papers from the ASPO National Planning Conference*, Washington: American Society of Planning Officials 47–65.

—— (1979) *Cities of the American West: A History of Frontier Urban Planning*, Princeton: Princeton University Press.

Reuss, H.S. (1977) *To Save Our Cities: What Needs To Be Done*, Washington: Public Affairs Press.

Reynolds, L. (1990) "Local Subdivision Regulation: Formulaic Constraints in an Age of Discretion," *Georgia Law Review* 24: 525–82.

Rhode Island (1988) *Rhode Island State Enabling Acts Relating to Land Use Planning*, Providence: Rhode Island Division of Planning, Department of Administration.

—— (1989) *State Guide Plan Element 121: Land Use 2010, State Land Use Policies and Plan*, Providence: Rhode Island Division of Planning, Department of Administration.

Rhodes, R. (1988) *Beyond Westminster and Whitehall*, London: Allen and Unwin.

Rhodes, R.M. (1988) "The Florida Local Government Development Agreement Act," *Florida Bar Journal* October 1988: 81–4.

Richardson, H.W. and Aldcroft, D.H. (1968) *Building in the British Economy Between the Wars*, London: Allen and Unwin.

Ripley, R.B. (1972) *The Politics of Economic and Human Resource Development*, Indianpolis: Bobbs-Merrill.

Riznik, B. (1989) "Hanalei Bridge: A Catalyst for Rural Preservation," *The Public Historian* 11: 45–66.

Roanoke City Planning Commission (1986) *Roanoke Vision – Zoning: A Process for Balancing Preservation and Change, 1986*, Roanoke, VA: City of Roanoke, Office of Community Planning.

Roark, K. (ed.) (1985) *Land for Housing: Developing a Research Agenda*, Monograph 85–3, Cambridge, MA: Lincoln Institute of Land Policy.

Roberts, N.A. (1975) "Land Storage – The Swedish Example," *Modern Law Review* 38: 121–38.
—— (1977) *The Government Land Developers*, Lexington, MA: Lexington.
Robertson, D.B. and Judd, D.R. (1989) *The Development of American Public Policy: The Structure of Policy Constraint*, Glenview, IL: Scott, Foresman.
Robinson, N.A. (1979) *Historic Preservation Law*, New York: Practising Law Institute.
Rocher, G. (1988) *Comments on Seymour Martin Lipset's "Canada and the United States: The Cultural Dimension"*, Faculty of Law, University of Quebec at Montreal.
Rockefeller Report (1973): see Reilly, W.K.
Roddewig, R.J. (1983) "Preservation Law and Economics: Government Incentives to Encourage For-Profit Preservation," in Duerksen (1983).
—— (1990) "Recent Developments in Land Use, Planning and Zoning Law," *Urban Lawyer* 22: 719–831.
Roddewig, R.J. and Duerksen, C.J. (1989) *Responding to the Takings Challenge: A Guide for Officials and Planners*, Planning Advisory Service Report 416, Washington: American Planning Association.
Rodwin, L. (1970) *Nations and Cities A Comparison of Strategies for Urban Growth*, Boston: Houghton Mifflin.
Roeseler, W.G. (1987) "Regulating Adult Entertainment Establishments under Conventional Zoning," *Urban Lawyer* 19: 125–40.
Rogers, I.M. (1991) *Canadian Law of Planning and Zoning*, Scarborough, Ontario: Carswell (loose-leaf; updated by periodic supplements).
Rohse, M. (1987) *Land Use Planning in Oregon: A No-Nonsense Handbook in Plain English*, Corvallis: Oregon State University Press.
Romanow, R.J., Ryan, C., and Stanfield, R.L. (1984) *Ottawa and the Provinces*, Toronto: Ontario Economic Council.
Rose, A. (1972) *Governing Metropolitan Toronto: A Social and Political Analysis*, Berkeley: University of California Press.
—— (1980) *Canadian Housing Policy, 1935–1980*, Markham, Ontario: Butterworths.

Rose, C.M. (1981) "Preservation and Community: New Directions in the Law of Historic Preservation," *Stanford Law Review* 33: 473–534.
—— (1983) "Planning and Dealing: Piecemeal Land Controls as a Problem of Local Legitimacy," *California Law Review* 71: 837–912.
Rose, J.B. (1984) "Landmarks Preservation in New York," *The Public Interest* 74: 132–45 (Winter 1984).
Rose, J.G. (1979) "Myths and Misconceptions of Exclusionary Zoning Litigation," *Real Estate Law Journal* 8: 99–124.
—— (1983) "The *Mount Laurel II* Decision: Is it Based on Wishful Thinking?," *Real Estate Law Journal* 12: 115–37.
—— (1985) "New Jersey Enacts a Fair Housing Law," *Real Estate Law Journal* 14: 195–217.
—— (1987) "Historic Preservation Law: A New Hybrid Statute with New Legal Problems," *Real Estate Law Journal* 15: 195–222.
—— (1988) "Waning Judicial Legitimacy: The Price of Judicial Promulgation on Urban Policy," *Urban Lawyer* 20: 801–39 (also in Deutsch and Tarlock 1989).
Rose, J.G. and Rothman, R.E. (eds.) (1977) *After Mount Laurel*, New Brunswick, NJ: Center for Urban Policy Research, Rutgers University.
Rose, J.L. and Duncan, M.J. (1985) "Local Regulation of Manufactured Housing: Current Issues," *Planning and Public Policy* (Bureau of Urban and Regional Planning Research, University of Illinois at Urbana-Champaign) 11, 1:1 (February 1985).
Rose, M. (1979) *Interstate: Express Highway Politics 1941–1956*, Knoxville: Regents Press of Kansas.
Rosen, K.T. and Katz, L. (1981) "Growth Management and Land Use Controls: The San Francisco Bay Area Experience," *American Real Estate and Urban Economics Association Journal* 9: 321–44.
Rosenbaum, N. (1976) *Land Use and the Legislatures: The Politics of State Innovation*, Washington: Urban Institute.
Rosenberg, G.N. (1991) *The Hollow Hope: Can Courts Bring About Social Change?*, Chicago: University of Chicago Press.

Rosenthal, D.B. (1980) *Urban Revitalization*, Beverly Hills, CA: Sage.
—— (1988) *Urban Housing and Neighborhood Revitalization: Turning a Federal Program into Local Projects*, New York: Greenwood Press.
Roth, L.M. (ed.) (1983) *America Builds: Source Documents in American Architecture and Planning*, New York: Harper and Row.
Rothenberg, J. (1967) *Economic Evaluation of Urban Renewal*, Washington: Brookings Institution.
Rowan-Robinson, J. and Young, E. (1989) *Planning By Agreement in Scotland*, Glasgow, UK: The Planning Exchange, Glasgow.
Rubin, B.M. and Wilder, M.G. (1989) "Urban Enterprise Zones: Employment Incentives and Fiscal Impacts," *Journal of the American Planning Association* 55: 418–31.
Rubin, B.M. and Zorn (1985) "Sensible State and Local Economic Development," *Public Administration Review* 45: 333–9.
Rubino, R.G. and Wagner, W.R. (1972) *The States' Role in Land Resource Management*, Lexington, KY: Council of State Governments.
Rubinowitz, L.S. (1974) *Low Income Housing: Suburban Strategies*, Cambridge, MA: Ballinger.
Sabatier, P.A. and Mazmanian, D.A. (1983) *Can Regulation Work? The Implementation of the 1972 California Coastal Initiative*, New York: Plenum.
Sabatier, P.A. and Pelkey, N.W. (1990) *Land Development at Lake Tahoe, 1960–84*, Florida Atlantic University and Florida International University Joint Center for Environmental and Urban Problems, and Institute of Ecology, University of California, Davis.
Sadler, B. (ed.) (1978) *Involvement and Environment: Proceedings of the Canadian Conference on Public Participation, Vol 1: A Review of Issues and Approaches*, Edmonton: Environment Council of Alberta.
—— (1979) *Involvement and Environment Proceedings of the Canadian Conference on Public Participation, Vol 2: Working Papers and Case Studies*, Edmonton: Environment Council of Alberta.
Sagalyn, L.B. and Sternlieb, G. (1973) *Zoning and Housing Costs: The Impact of Land-Use Controls on Housing Price*, New Brunswick, NJ: Center for Urban Policy Research, Rutgers University.
Sager, L. (1969) "Tight Little Islands: Exclusionary Zoning, Equal Protection, and the Indigent," *Stanford Law Review* 21: 767–800.
—— (1978) "Insular Majorities Unabated: *Warth v Seldin* and *City of Eastlake v Forest City Enterprises*," *Harvard Law Review* 91: 1373–1425.
—— (1979) "Questions I Wish I Had Never Asked: The Burger Court on Exclusionary Zoning," *Southwestern University Law Review* 11: 509–44.
Salmon Report (1976) *Report of the Royal Commission on Standards of Conduct in Public Life*, Cmnd 6524, London: HMSO.
Salmons, W. (1986) "Petaluma's Experiment," in D.R. Porter (1986).
Sancton, A. (1991) "The Municipal Role in the Governance of Canadian Cities," in Bunting and Filion (1991).
Sancton. A. and Woolner, P. (1990) "Full-Time Municipal Councillors: A Strategic Challenge for Canadian Urban Government," *Canadian Public Administration* 33: 482–505.
Sanders, H.T. (1980) "Urban Renewal and the Revitalized City: A Reconsideration of Recent History," in Rosenthal (1980).
Sanderson, F.H. (1990) *Agricultural Protectionism in the Industrialized World*, Washington: Resources for the Future.
San Francisco (1991) *Bay Vision 2020 – The Commission Report*, Bay Vision 2020 Commission.
Sanger, John M. Associates (1982) *A Preservation Strategy for Downtown San Francisco*, San Francisco: Foundation for San Francisco's Architectural Heritage.
Saunders, J.H. (1981) "State and Federal Housing Policy vs Local Land Use Regulation: The New Conflict Between State and Municipal Powers in California," *University of San Francisco Law Review* 15: 509–36.
Savas, E.S. (1982) *Privatizing the Public Sector*, London: Chatham House.
—— (1983) "A Positive Urban Policy for the Future," *Urban Affairs Quarterly* 18: 447–53.
—— (1987) *Privatization: The Key to Better*

Government, London: Chatham House.

Sax, J.L. (1964) "Takings and the Police Power," *Yale Law Journal* 74: 36–76.

—— **(1971)** "Takings, Private Property and Public Rights," *Yale Law Journal* 81: 149–86 (also in Ackerman 1975).

Saywell, J.T. (1975) *Housing Canadians: Essays on the History of Residential Construction in Canada*, Working Paper 24, Ottawa: Economic Council of Canada.

Scarrow, H.A. (1971) "Policy Pressures by Local Government: The Case of Regulation in the Public Interest," *Comparative Politics* 4: 1–28.

Schafer, R. (1974) *The Suburbanization of Multifamily Housing*, Lexington, MA: Heath.

Schaffer, D. (1982) *Garden Cities for America: The Radburn Experience*, Philadelphia: Temple University Press.

Schaffer, D. (ed.) (1988) *Two Centuries of American Planning*, London: Mansell.

—— **(1990)** *Housing Discrimination: Law and Litigation*, New York: Clark Boardman.

Schiffman, I. (1990) *Alternative Techniques for Managing Growth*, Berkeley: Institute of Governmental Studies, University of California at Berkeley.

Schlereth, T. (1980) *Material Culture Studies in America*, Nashville, TN: American Association for State and Local History.

Schnare, A.B., and Struyk, R.J. (1976) "Segmentation in Urban Housing Markets," *Journal of Urban Economics* 3: 146–66.

Schnidman, F. (1986) "The Evolving Federal Role in Agricultural Land Preservation," *Urban Lawyer* 18: 425–44.

Schnidman, F. and Silverman, J. (1980) *Management and Control of Growth, Volume 5: Updating the Law*, Washington: Urban Land Institute.

Schnidman, F., Silverman, J.A., and Young, R.C. (1978) *Management and Control of Growth, Volume 4: Techniques in Application*, Washington: Urban Land Institute.

Schnidman, F., Smiley, M., and Woodbury, E.G. (1990) *Retention of Land for Agriculture: Policy, Practice and Potential in New England*, Cambridge, MA: Lincoln Institute of Land Policy.

Schon, D.A. (1971) *Beyond the Stable State*, New York: Norton.

Schultz, C. (1977) *The Public Use of Private Interest*, Washington: Brookings Institution.

Schultz, S.K. (1989) *Constructing Urban Culture: American Cities and City Planning 1800–1920*, Philadelphia: Temple University Press.

Schwartz, B. (1977) *The Great Rights of Mankind: A History of the American Bill of Rights*, Oxford: Oxford University Press.

Schwartz, S.I., Hansen, D.E., Green, R., Moss, W.G., and Belzer, R. (1979) *The Effect of Growth Management on New Housing Prices: Petaluma, California*, Environmental Quality Series 32, Davis, CA: University of California-Davis, Institute of Governmental Affairs.

Schwartz, S.I., Hansen, D.E., and Green, R. (1981) "Suburban Growth Controls and the Price of New Housing," *Journal of Environmental Economics and Management* 8: 303–20.

—— **(1984)** "The Effect of Growth Control on the Production of Moderate-Priced Houisng," *Land Economics* 60: 110–14.

Schwartz, S.I. and Johnston, R.A. (1981) *Local Government Initiatives for Affordable Housing: An Evaluation of Inclusionary Housing Programs in California*, Davis, CA: Institute of Governmental Affairs, University of California, Davis.

—— **(1983)** "Inclusionary Housing Programs," *Journal of the American Planning Association* 49, 1: 3–21.

Schwemm, R.G. (ed.) (1989) *The Fair Housing Act after Twenty years: A conference at the Yale Law School*, New Haven: Yale Law School.

—— **(1990)** *Housing Discrimination: Law and Litigation*, New York: Clark Boardman.

Scott, J. (1969) "Corruption, Machine Politics and Political Change," *American Political Science Review* 43: 1142–58.

Scott, J.N. (1973) "Toward a Strategy for Utilization of Contract and Conditional Zoning," *Journal of Urban Law* 51: 94–111.

Scott, M. (1969) *American City Planning Since 1890*, Berkeley: University of California Press.

Scott, R.W. (1975) *Management and Control of Growth: Issues, Techniques, Problems, Trends* (3 vols.), Washington: Urban Land Institute.

Scott, S. (1975) *Governing California's Coast*, Berkeley: Institute of Governmental Studies,

University of California, Berkeley.
—— **(1981)** *Coastal Conservation: Essays on Experiments in Governance*, Berkeley: Institute of Governmental Studies, University of California, Berkeley.

Scott, S. and Bollens, J. (1968) *Governing a Metropolitan Region: The San Francisco Bay Area*, Berkeley: Institute of Governmental Studies, University of California, Berkeley.

Scurlock, J. (1987) Review of Haar and Fessler: *The Wrong Side of the Tracks*, *Urban Lawyer* 19: 176–80.

Sedway, P.H. (1985) "The San Francisco Downtown Plan: Office Development Boom Brings Housing Boom," in Merriam, Brower, and Tegeler (1985).

Segal, D. and Srinivasan, P. (1985) "The Impact of Suburban Growth Restrictions on U.S. Housing Price Inflation, 1975–1978," *Urban Geography* 6: 14–26.

Seidel, S.R. (1978) *Housing Costs and Government Regulations: Confronting the Regulatory Maze*, New Brunswick, NJ: Center for Urban Policy Research, Rutgers University.

Seidman, H. (1970) *Politics, Position and Power: The Dynamics of Federal Organization* (1st edn.), Oxford: Oxford University Press; (3rd edn. 1980).

Self, P. (1982) *Planning the Urban Region: A Comparative Study of Policies and Organizations*, London: Allen and Unwin.

Semer, M.P., Zimmerman, J.H., Foard, A., and Frantz, J.M. (1985) "Evolution of Federal Legislative Policy in Housing: Housing Credits," in Mitchell (1985).

Sewell, J. (1983) as reported in the Toronto *Globe and Mail*, 28 January 1983.

Shafer, B.E. (1991) *Is America Different? A New Look at American Exceptionalism*, Oxford: Clarendon.

Shalala, D.E. and Vitullo-Martin, J. (1989) "Rethinking the Urban Crisis: Proposals for a National Urban Agenda," *Journal of the American Planning Association* 55: 3–13.

Shanahan, N.C. (1984) "Downtown Preservation Strategies and Techniques: Summary of Transcript of an Invitational Workshop of the National Trust for Historic Preservation and the Foundation for San Francisco's Architectural Heritage," *Preservation Law Reporter* 3: 2031–43.

Shapiro, R.M. (1968) "The Case for Conditional Zoning," *Temple Law Quarterly* 41: 267–87.
—— **(1969)** "The Zoning Variance Power – Constructive in Theory, Destructive in Practice," *Maryland Law Review* 29: 3–23.

Sharpe, L.J. (1973) "American Democracy Reconsidered," *British Journal of Political Science* 3: 1–28 and 129–67.
—— **(1981)** "The Failure of Local Government Modernization in Britain: A Critique of Functionalism," in L.D. Feldman (1981).
—— **(1988)** "The Growth and Decentralization of the Modern Democratic State," *European Journal of Political Research* 16: 365–80.

Sherer, S.A. (1969) "Snob Zoning in New Jersey and Massachusetts," *Harvard Journal on Legislation* 7: 246–70.

Shirvani, H. (1981) *Urban Design Review: A Guide for Planners*, Washington: American Planning Association.
—— **(1985)** *The Urban Design Process*, London: Van Nostrand Reinhold.

Shlay, A.B. and Rossi, P.H. (1981) "Keeping Up the Neighborhood: Estimating Net Effects of Zoning," *American Sociological Review* 46: 703–19.

Shonfield, A. (1965) *Modern Capitalism: The Changing Balance of Public and Private Power*, Oxford: Oxford University Press.

Shumiatcher, M.C. (1988) "Property and the Canadian Charter of Rights and Freedoms," *Canadian Journal of Law and Jurisprudence* 1: 189–208.

Sibley, J. (1989) *How Georgia Responded to Growth Issues*, Atlanta, GA: Governor's Development Council.

Siegan, B.H. (1970) "Non-Zoning in Houston," *Journal of Law and Economics* 13: 71–147.
—— **(1972)** *Land Use Without Zoning*, Lexington, MA: Heath.
—— **(1976)** *Other People's Property*, Lexington, MA: Heath.

Siegan, B.H. (ed.) (1977) *Planning Without Prices: The Taking Clause As it Relates to Land Use Regulation*, Lexington, MA: Heath.

Siegel, D. (1980) "Provincial–Municipal Relations in Canada: An Overview," *Canadian Public Administration* 23: 281–317.

Siemon, C.L. (1987) "Exactions – Who Bears

The Cost?," *Law and Contemporary Problems* 50: 115–26.

Siemon, C.L., Larsen, W.V., and Porter, D.R. (1982) *Vested Rights: Balancing Public and Private Development Expectations*, Washington: Urban Land Institute.

Silvern, P.J. (1985) "Negotiating the Public Interest: California's Development Agreement Statute," *Land Use Law* 51, 10: 3–9 (October 1985).

Simon, J. (1981) *The Ultimate Resource*, Princeton: Princeton University Press.

Simpson, M. (1981) "Thomas Adams 1871–1940" in Cherry (1981).

Sinclair, S.M. (1988) *Expectations and Opportunities: Growth Management in the Late Eighties*, Washington: National Governors Association.

Sipress, A. (1990) "Despite Ruling, Affordable Homes Still Scarce in N.J.," *New York Times*, November 25, 1990.

Slater, D.C. (1984) *Management of Local Planning*, Washington: International City Management Association.

Slitor, R.E. (1985) "Rationale of the Present Tax Benefits for Homeowners," in Mitchell 1985).

Smerk, G.M. (1965) *Urban Transportation: The Federal Role*, Bloomington: Indiana University Press.

Smiley, D.V. (1980) *Canada in Question: Federalism in the Seventies*, Toronto: McGraw-Hill Ryerson.

—— **(1987)** *The Federal Condition in Canada*, Toronto: McGraw-Hill Ryerson.

Smith, B. (1988) "California Development Agreements and British Planning Agreements," *Town Planning Review* 59: 277–87.

Smith, M.E.H. (1989) *Guide to Housing* [British], London: Housing Centre Trust.

Smith, M.P. (1984) *Cities in Transformation*, Beverly Hills, CA: Sage.

Smith, P.J. (1986) "American Influences and Local Needs: Adaptations to the Alberta Planning System in 1928–1929," in Stelter and Artibise (1986).

Smith, R.A. (1985) "A Critical Review of HUD and Fair Housing," *Journal of the American Planning Association* 51: 478–81.

Smith, R.M. (1987) "From Subdivision Improvement Requirements to Community Benefit Assessments and Linkage Payments: A Brief History of Land Development Exactions," *Law and Contemporary Problems* 50: 5–30.

Smith, W.F. (1965) "The Federal Bulldozer: A Review," *Journal of the American Institute of Planners* 31: 179–80 (also in Wilson 1966).

Snedcof, H.R. (1985) *Cultural Facilities in Mixed-Use Development*, Washington: Urban Land Institute.

Snyder, T.P. and Stegman, M.A. (1986) *Paying for Growth: Using Development Fees to Finance Infrastructure*, Washington: Urban Land Institute.

So, F.S. and Getzels, J. (1988) *The Practice of Local Government Planning* (2nd edn.), Washington: International City Management Association.

So, F.S., Hand, I., and McDowell, B.D. (1985) *The Practice of State and Regional Planning*, Washington: American Planning Association in cooperation with the International City Management Association.

Sobel, L.A. (ed.) (1976) *New York and the Urban Dilemma*, New York: Facts on File.

Solnit, A., with Reed, C., Glassford, P., and Erley, D. (1988) *The Job of the Practicing Planner*, Chicago: Planners Press.

Solomon, A.P. (1980) *The Prospective City: Economic, Population, Energy, and Environmental Developments*, Cambridge, MA: MIT Press.

Souter, D.V. (1961) "Zoning Appeals: How a Board of Zoning Appeals Works," *Michigan State Bar Journal* 40: 26–30.

Southern Environmental Law Center (1987) *Visual Pollution and Sign Control: A Legal Handbook on Billboard Reform*, Charlottesville, VA: Southern Environmental Law Center.

Southworth, M. (1989) "Theory and Practice of Contemporary Urban Design Plans in the United States," *Town Planning Review* 60: 369–402.

Spence-Sales, H. (1949) *Planning Legislation in Canada*, Ottawa: Central Mortgage and Housing Corporation.

Spurr, P. (1976) *Land and Urban Development: A Preliminary Study*, Toronto: Lorimer.

Squire, P. and Scott, S. (1984) *The Politics of California Coastal Legislation: The Crucial*

Year, 1976, Berkeley: Institute of Governmental Studies, University of California, Berkeley.

Stach, P.B. (1988) "Zoning – To Plan or to Protect?," *Journal of Planning Literature* 2: 472–81.

Stanfield, R.L. (1978) "The Development of Carter's Urban Policy: One Small Step for Federalism," *Publius* 8: 39–53.

—— **(1991)** "Strains in the Family," *National Journal* 23, 39: 2316–33.

Stapleford, J.E. (1985) *Mobile Home Park Impacts in New Castle County*, Newark, DE: Bureau of Economic and Business Research, University of Delaware.

Statistics Canada (1982) *The Municipal Structure in Canada: Problems It Creates for Users of Statistics*, Geography Staff Working Paper 1, Ottawa: Statistics Canada.

Stegman, J.E. (1986) "Development Fees for Infrastructure," *Urban Land* 45, 2: 2–5 (May 1986).

Stegman, M.A. and Holden, J.D. (1987) *Nonfederal Housing Programs: How States and Localities are Responding to Federal Cutbacks in Low-Income Housing*, Washington: Urban Land Institute.

Stein, C.S. (1951) *Toward New Towns for America*, Cambridge, MA: MIT Press.

Steinman, L.D. (1988) "The Impact of Zoning on Group Homes for the Mentally Disabled: A National Survey," in Gordon (1988), (also in *Urban Lawyer* 1987, vol. 19).

Steiss, A.W. (1968) *A Framework for Planning in State Government*, Lexington, KY: Council of State Governments.

Stelter, G.A. and Artibise, A.F.J. (eds.) (1977) *The Canadian City: Essays in Urban History*, New York: Macmillan.

—— **(1982)** *Shaping the Urban Landscape: Aspects of the Canadian City Building Process*, Ottawa: Carleton University Press.

—— **(1986)** *Power and Place: Canadian Urban Development in the North American Context*, Vancouver: University of British Columbia Press.

Sternlieb, G. (1986) *Patterns of Development*, New Brunswick, NJ: Center for Urban Policy Research, Rutgers University.

Sternlieb, G., Hughes, J.W., and Hughes, C.O.

(1983) *Demographic Trends and Economic Reality: Planning and Markets in the '80s*, New Brunswick, NJ: Center for Urban Policy Research, Rutgers University.

Stevens, J.P. (1985) "Judicial Restraint," *San Diego Law Review* 22: 437–52.

Stewart, M. (1987) "Ten Years of Inner Cities Policy," *Town Planning Review* 58: 129–45.

Stipe, R. (1972) *Legal Techniques in Historic Preservation*, Washington: National Trust for Historic Preservation.

—— **(1987)** *The American Mosaic: Preserving a National Heritage*, Washington: U.S. Committee, International Council on Monuments and Sites.

Stockman, G.R. (1986) "The Art of the Possible," *Seton Hall Legislative Journal* 9: 581–4.

Stoebuck, W.B. (1972) "A General Theory of Eminent Domain," *Washington Law Review* 47: 553–608.

Stokes, S.N. (1989) *Saving America's Countryside: A Guide to Rural Conservation*, Baltimore: Johns Hopkins University Press.

Stone, C.N. (1976) *Economic Growth and Neighborhood Discontent: System Bias in the Urban Renewal Program of Atlanta*, Chapel Hill: University of North Carolina Press.

Storey, B.A. (1987) "The Advisory Council on Historic Preservation: Its Role in the Developing American Preservation Program," in Johnson and Schene (1987).

Strong, A.L. (1971) *Planned Urban Environments: Sweden, Finland, Israel, The Netherlands, France*, Baltimore: Johns Hopkins University Press.

—— **(1975)** *Private Property and the Public Interest: The Brandywine Experience*, Baltimore: Johns Hopkins University Press.

Stroud, N. (1988) "Legal Considerations of Development Impact Fees," in Nelson (1988b).

Stroud, N. and DeGrove, J.M. (1985) "State and Federal Innovations in Land Acquisition and Regulation in Florida," in Liner (1985).

Suagee, D.B. (1983) "American Indian Religious Freedom and Cultural Resources Management: Protecting Mother Earth's Caretakers," *American Indian Review* 10: 1–58.

Sundquist, J.L. (1969) *Making Federalism Work: A Study of Program Coordination at the Community*

Level, Washington: Brookings Institution.

——— (1975) *Dispersing Population: What America Can Learn from Europe*, Washington: Brookings Institution.

——— (1978) "Needed: A National Growth Policy," *Brookings Bulletin*, 14, 4: 1–5.

Surenian, J.R. (1987) *Mount Laurel II and the Fair Housing Act*, New Jersey Institute for Continuing Legal Education.

Sutcliffe, A. (1981) *The History of Urban and Regional Planning: An Annotated Bibliography*, New York: Facts on File.

Sutcliffe, A. (ed.) (1980) *The Rise of Modern Urban Planning 1800–1914*, London: Mansell.

——— (1981a) *British Town Planning: The Formative Years*, Leicester, UK: Leicester University Press.

——— (1981b) *Towards the Planned City: Germany, Britain, the United States, and France, 1780–1914*, New York: St Martin's Press.

——— (1984) *Metropolis 1890–1940*, Chicago: University of Chicago Press.

Suttles, G.D. (1990) *The Man-Made City: The Land-Use Confidence Game in Chicago*, Chicago: University of Chicago Press.

Swan, H.S. (1920) "Does Your City Keep its Gas Range in the Parlor and its Piano in the Kitchen?," *American City* 12, 4: 339–44.

Tabb, W. and Sawers, L. (1978) *Marxism and the Metropolis: New Perspectives in Urban Political Economy*, Oxford: Oxford University Press.

Taeuber, K. (1988) "The Contemporary Context of Housing Discrimination," *Yale Law and Policy Review* 6: 339–47.

Tant, A.P. (1990) "The Campaign for Freedom of Information: A Participatory Challenge to Elitist British Government," *Public Administration* 68: 477–91.

Taraska Report (1976) *Committee of Review, City of Winnipeg Act: Report and Recommendations*, Winnipeg: Queen's Printer.

Tarlock, A.D. (1973) "Toward a Revised Theory of Zoning," in Bangs, F.S. (ed.) *Land Use Controls Annual 1972*, Washington: American Society of Planning Officials.

Tarn, J.N. (1980) "Housing Reform and the Emergence of Town Planning in Britain Before 1914," in Sutcliffe (1980).

Taub, T.C. (1990) "The Future of Affordable Housing," *Urban Lawyer* 22: 659–92.

Taub, T.C. and Rhodes, R.M. (1989) "The Florida Local Government Development Agreement Act," in Porter and Marsh (1989).

Tennant, P. and Zirnhelt, D. (1973) "Metropolitan Government in Vancouver: The Strategy of Gentle Imposition," *Canadian Public Administration* 16: 124–38.

Thomas, D., Minett, J., Hopkins, S., Hamnett, S., Faludi, A. and Barrell, D. (1983) *Flexibility and Commitment in Planning: A Comparative Study of Local Planning and Development in the Netherlands and England*, Dordrecht: Martinus Nijhoff.

Thomas, M.H. (1982) "Municipal Land Use Control in British Columbia," *Urban Lawyer* 14: 847–54.

Thompson, E. (1982) "Defining and Protecting the Right to Farm," *Zoning and Planning Law Report* 5: 57.

——— (1990) "Purchase of Development Rights: Ultimate Tool for Farmland Preservation," in Dennison (1990).

Thompson, W.R. (1972) "The National System of Cities as an Object of Public Policy," *Urban Studies* 9: 99–116.

Thornley, A. (1991) *Urban Planning Under Thatcherism: The Challenge of the Market*, London: Routledge.

Thurber, P. (1985) *Controversies in Historic Preservation: Understanding the Preservation Movement Today*, Washington: National Trust for Historic Preservation.

Thurber, P. and Moyer, R. (1984) *State Enabling Legislation for Local Preservation Commissions*, State Legislation Project, Washington: National Trust for Historic Preservation.

Tiebout, C. (1956) "A Pure Theory of Expenditure," *Journal of Political Economy* 64: 416–35.

Tindal, C.R. and Tindal, S.N. (1990) *Local Government in Canada* (3rd edn.), Toronto: McGraw-Hill Ryerson.

Titmuss, R.M. (1958) "War and Social Policy," in his *Essays on "The Welfare State"*, London: Allen and Unwin (1958).

Tocqueville, A de (1848), *Democracy in America*, (1969 edition edited by J.P. Mayer) New York: Doubleday Anchor Books.

Toll, S.I. (1969) *Zoned American*, New York: Grossman.

Tomioka, S. and Tomioka, E.M. (1984) *Planned Unit Developments: Design and Regional Impact*, New York: Wiley.

Torma, C. (1987) "Assessing the Work to Date," in *Proceedings of the Workshop on Historic Mining Resources*, Pierre, SD: State Historical Preservation Center, South Dakota State Historical Preservations Society.

Toronto, City (1983) *Social Change in Toronto: A Context for Human Services*, Toronto: City of Toronto Planning and Development Department.

——— **(1987)** *Assisted Housing: Options for Private Sector Involvement and Section 36 Guidelines*, Toronto: City of Toronto Planning and Development Department.

——— **(1988)** *Section 36 Guidelines: Further Report on Guidelines for Bonusing Pursuant to Section 36 of the Planning Act*, Toronto: City of Toronto Planning and Development Department.

Tribe, L.H. (1985) *God Save This Honorable Court: How The Choice of Supreme Court Justices Shapes Our History*, New York: New American Library.

Tripp, J.T.B. and Dudek, D.J. (1990) "Institutional Guidelines for Designing Successful Transferable Rights Programs," in Dennison (1990); also in *Yale Journal on Regulation*, vol. 6, 1989.

Tropman, J.E., Dluhy, M.J. and Lind, R.M. (1976) *Strategic Perspectives on Social Policy*, Oxford: Pergamon.

Truman, T. (1971) "A Critique of Seymour M. Lipset's Article 'Value Differences, Absolute or Relative: The English-Speaking Democracies'," *Canadian Journal of Political Science* 4: 497–525.

Turner, R.S. (1990a) "New Rules for the Growth Game: The Use of Rational State Standards in Land Use Policy," *Journal of Urban Affairs* 12: 35–47.

——— **(1990b)** "Intergovernmental Growth Management: A Partnership Framework for State–Local Relations," *Publius* 20: 79–95.

Tustian, R.E. (1984) "TDRs in Practice: A Case Study of Agriculture Preservation in Montgomery County, Maryland," in Gailey (1984).

Tyer, C.B. (1977) *Substate Planning and Development Districts: An Analysis of Regional Planning Agency Behavior*, Knoxville, TN: Bureau of Public Administration, University of Tennessee.

Tym, R. and Partners *et al.* (1982) *Enterprise Zones: Year One Report*, London: Roger Tym Associates.

——— **(1983)** *Enterprise Zones: Year Two Report*, London: Roger Tym Associates.

——— **(1984)** *Monitoring Enterprise Zones: Year Three Report*, London: Roger Tym Associates.

UK (1989) *Britain 1989: An Official Handbook*, London: Central Office of Information, HMSO.

——— **(1991)** *Britain 1991: An Official Handbook*, London: Central Office of Information, HMSO.

University of Pennsylvania Law Review (1955) "Zoning Variances and Exceptions: The Philadelphia Experience," *University of Pennsylvania Law Review* 103: 516–55.

Unwin, R. (1909) *Town Planning in Practice: An Introduction to the Art of Designing Cities and Suburbs*, London: T. Fisher Unwin.

Upton, D. and Vlach, J.M. (eds.) (1986) *Common Places: Readings in American Vernacular Architecture*, Athens, GA: University of Georgia Press.

Urban Land Institute (1989) *Development Trends 1989*, Washington: Urban Land Institute.

——— **(1990)** *Market Profiles 1989*, Washington: Urban Land Institute.

U.S. Bureau of the Budget (1969) *Circular A-95*, (July 24, 1969).

U.S. Bureau of the Census (1976) *Historical Statistics of the United States, Colonial Times to 1970*, Washington: U.S. GPO.

——— **(1988)** *1987 Census of Governments: Vol 1, No.1: Government Organization*, Washington: U.S. GPO.

U.S. Conference of Mayors (1966) *With Heritage So Rich: A Report of a Special Committee on Historic Preservation under the Auspices of the United States Conference of Mayors with a Grant from the Ford Foundation*, New York: Random House.

U.S. Department of Agriculture (1981) *National Agricultural Lands Study*, Washington: U.S. GPO.

U.S. Department of Justice (1979) *Corruption in*

Land Use and Building Regulation, 6 volumes: vol. 1: *An Integrated Report of Conclusions*, Washington: U.S. GPO.

U.S. Department of Transportation, Federal Highway Administration (1981) *Final Report of the National Advisory Committee on Outdoor Advertising and Motorist Information*, Washington: U.S. Department of Transportation.

—— **Office of the Secretary of Transportation, Office of Inspector General (1984)** *Report on Highway Beautification Program, Federal Highway Administration*, Washington: U.S. Department of Transportation.

U.S. GAO [General Accounting Office] (1978) *Why Are New Housing Prices So High , How Are They Influenced By Government Regulation, and Can Prices be Reduced?*, Washington: U.S. GAO.

—— **(1979)** *Preserving America's Farmland: A Goal the Federal Government Should Support*, Washington: U.S. GAO.

—— **(1981)** *The Community Development Block Grant Can Be More Effective in Revitalizing the Nation's Cities*, Washington: U.S. GAO.

—— **(1983)** *An Analysis of Zoning and Other Problems Affecting the Establishment of Group Homes for the Mentally Disturbed*, Washington: U.S. GAO.

—— **(1984a)** *Information on Historic Preservation Tax Incentives*, Washington: U.S. GAO.

—— **(1984b)** *Insights into Major Urban Development Action Grant Issues*, Washington: U.S. GAO.

—— **(1988a)** *Cultural Resources: Implementation of Federal Historic Preservation Program Can Be Improved*, Washington: U.S. GAO.

—— **(1988b)** *Federal Estate Tax on Historic Properties*, Washington: U.S. GAO.

—— **(1988c)** *Enterprise Zones: Lessons from the Maryland Experience*, Washington: U.S. GAO.

U.S. HUD [Department of Housing and Urban Development] (1970) *The Model Cities Program: A Comparative Analysis of the Planning Process in Eleven Cities*, Washington: U.S. GPO.

—— **(1972)** *Report on National Growth 1972*, Washington: U.S. GPO.

—— **(1978a)** *Final Report of the Task Force on Housing Costs*, Washington: U.S. GPO.

—— **(1978b)** *The President's National Urban Policy Report 1978*, Washington: U.S. GPO.

—— **(1980a)** *The President's National Urban Policy Report 1980*, Washington: U.S. GPO.

—— **(1980b)** *Urban Development Action Program: Second Annual Report*, Washington: U.S. GPO.

—— **(1982a)** *An Impact Evaluation of the Urban Development Action Grant Program*, Washington: U.S. GPO.

—— **(1982b)** *The President's National Urban Policy Report 1982*, Washington: U.S. GPO.

—— **(1982c)** *Implementing Community Development: A Study of the Community Development Block Grant*, Washington: U.S. GPO.

—— **(1984a)** *The President's National Urban Policy Report 1984*, Washington: U.S. GPO.

—— **(1984b)** *An Evaluation of the Federal New Communities Program*, Washington: U.S. GPO.

—— **(1986a)** *The President's National Urban Policy Report 1986*, Washington: U.S. GPO.

—— **(1986b)** *State-Designated Enterprise Zones: Ten Case Studies*, Washington: U.S. GPO.

—— **(1988a)** *The President's National Urban Policy Report 1988*, Washington: U.S. GPO.

—— **(1988b)** *Consolidated Annual Report to Congress on Community Development Programs 1988*, Washington: U.S. GPO.

—— **(1989)** *Annual Report to Congress on Community Development Programs 1989*, Washington: U.S. GPO.

—— **(1991)** *Fair Housing Amendments Act of 1988: A Selected Resource Guide*, Washington: U.S. GPO.

U.S. Senate (1966) Senate Operations Committee, Subcommittee on Intergovernmental Relations, *Creative Federalism Hearings*, Washington: 89th Congress, 2nd Session.

Uthwatt Report (1942) *Report of the Expert Committee on Compensation and Betterment*, Cmnd 6386, London: HMSO.

Vaillancourt, F. and Monty, L. (1985) "The Effect of Agricultural Zoning on Land Prices, Quebec, 1975–1981," *Land Economics* 61: 36–42.

Vancouver, City of (1981) *Eight Years After: Case Studies Under Discretionary Zoning in Vancouver*, City of Vancouver Planning Department.

—— **(1984)** "Eyes on Urban Design: Vancouver's Urban Design Panel," *Quarterly Review*, October 1984.

Van Nus, W. (1977) "The Fate of the City Beautiful Thought in Canada 1893–1930," in Stelter and Artibise (1977).

—— **(1979)** "Towards the City Efficient: The Theory and Practice of Zoning 1919–1939," in Artibise and Stelter (1979).

Van Til, J. (1982) *Living with Energy Shortfall: A Future for American Towns and Cities*, Boulder, CO: Westview.

Vaughan, R.J. (1977) *The Urban Impacts of Federal Policies: Vol. 2: Economic Development*, Santa Monica, CA: Rand Corporation.

—— **(1980)** "The Impact of Federal Policies on Urban Economic Development," in Solomon (1980).

Vaughan, R.J. and Vogel, M.E. (1979) *The Urban Impacts of Federal Policies: Vol.4, Population and Residential Location*, Santa Monica, CA: Rand Corporation.

Vaughan, R.J., Pascal, A.H., and Vaiana, M.E. (1980) *The Urban Impacts of Federal Policies: Vol. 1, Overview*, Santa Monica, CA: Rand Corporation.

Vaughn, P.M. (1974) "The Massachusetts Zoning Appeals Law; First Breach in the Exclusionary Wall," *Boston University Law Review* 54: 37–77.

Vermont (1981) *Act 250: A Performance Evaluation*, Montpelier: State of Vermont Environmental Board.

—— **(1988a)** *Report of the Governor's Commission on Vermont's Future: Guidelines for Growth*, Montpelier: Office of Policy Research and Coordination, State of Vermont.

—— **(1988b)** *Vermont Municipal and Regional Planning and Development Act, including 1988 Growth Management (Act 200) Amendments*, Montpelier: Department of Housing and Community Affairs, State of Vermont.

—— **(1990)** *Act 250: Vermont's Land Use and Development Act, Title 10, Chapter 151, Including All Legislative Amendments Effective July 1, 1990*, Montpelier: State of Vermont.

—— **(1991)** *Shaping Vermont's Future: The Citizen's Guide to Open State Agency Planning*, Montpelier: Vermont State Agency Planning Implementation Committee.

Vettel, S.L. (1985) "San Francisco's Downtown Plan: Environmental and Urban Design Values in Central Business District Regulation," *Ecology Law Quarterly* 12: 511–66.

Vining, D.R. (1979) "The President's National Urban Policy Report: Issues Skirted and Statistics Omitted," *Journal of Regional Science* 19: 69–77.

Vining, D.R., Plaut, T., and Bieri, K. (1977) "Urban Encroachment on Prime Agricultural Land in the United States," *International Regional Science Review* 2: 143–56.

Vlear, J.E. Van (1988) "Land Use Aesthetics: A Citizen Survey Approach to Decision Making," *Pepperdine Law Review* 15: 207–66.

Vogel, D. (1986) *National Styles of Regulation: Environmental Policy in Great Britain and the United States*, Ithaca, NY: Cornell University Press.

Vogel, R.K. and Swanson, B.E. (1989) "The Growth Machine Versus The Antigrowth Coalition: The Battle for Our Communities," *Urban Affairs Quarterly* 25: 63–85.

Vranicar, J., Sanders, W., and Mosena, D. (1980) *Streamlining Land Use Regulation: A Guidebook for Local Governments*, Washington: Urban Land Institute.

Wakeford, R. (1990) *American Development Control: Parallels and Paradoxes from an English Perspective*, London: HMSO.

Walker, R.A. (1950) *The Planning Function in Urban Government* (2nd edn.), Chicago: University of Chicago Press.

Waller, H., Sabetti, F., and Elazar, D.J. (eds.) (1988) *Canadian Federalism: From Crisis to Constitution*, Lanham, MD: University Press of America.

Wallis, A.D. (1991) *Wheel Estate: The Rise and Decline of Mobile Homes*, Oxford: Oxford University Press.

Waltman, J.L. and Holland, K.M. (eds.) (1988) *The Political Role of Law Courts in Modern Democracies*, New York: St Martin's Press.

Waltman, J.L. and Studlar, D.T. (eds.) (1987) *Political Economy: Public Policies in the United*

States and Britain, Jackson, MS: University Press of Mississippi.

Walton, J. (1982) "Cities and Jobs and Politics," *Urban Affairs Quarterly* 18: 5–17.

Warner, S.B. (1962) *Streetcar Suburbs: The Process of Growth in Boston 1870–1900*, Cambridge, MA: Harvard University Press and MIT Press.

—— **(1966)** *Planning for a Nation of Cities*, Cambridge, MA: MIT Press.

—— **(1972)** *The Urban Wilderness: A History of the American City*, New York: Harper and Row.

Warren, C.R. (ed.) (1985) *Urban Policy in a Changing Federal System: Proceedings of a Symposium*, Washington: National Academy Press.

Washington State (1990) *A Growth Strategy for Washington State*, Olympia: Washington State Growth Strategies Commission.

—— **(1990)** *An Act Relating to Growth* (Substitute House Bill 2929), Olympia: Office of the Governor.

Watson, K.F. (ed.) (1973) *Landbanking: Investment in the Future, Symposium, Vancouver*, Vancouver: Centre for Continuing Education, University of British Columbia.

—— **(1974)** *Landbanking in Red Deer*, Vancouver: Centre for Continuing Education, University of British Columbia.

Watts, R.L. (1987) "The American Constitution in Comparative Perspective: A Comparison of Federalism in the United States and Canada," in Thelen, D. *The Constitution and American Life*, Ithaca, NY: Cornell University Press (1987).

Weaver, C.L. and Babcock, R.F. (1979) *City Zoning: The Once and Future Frontier*, Chicago: Planners Press.

Webman, J.A. (1981) "UDAG: Targeting Urban Economic Development," *Political Science Quarterly* 96: 189–207.

Wegner, J.W. (1987) "Moving Toward the Bargaining Table: Contract Zoning, Development Agreements, and the Theoretical Foundations of Government Land Use Deals," *Maryland Law Review* 65: 957–1038 (also in Deutsch 1988).

Weicher, J.C. (1972) *Urban Renewal: National Program for Local Problems*, Washington:

American Enterprise Institute for Public Policy Research.

Weiher, G.R. (1989) "Rumors of the Demise of the Urban Crisis Are Greatly Exaggerated," *Journal of Urban Affairs* 11: 225–42.

Weinberg, S. (1982) "Lobbying Congress. . . The Inside Story," *Historic Preservation* 34: 17–24.

Weinstein, B.L., Gross, H.T., and Rees, J. (1985) *Regional Growth and Decline in the United States* (2nd edn.), New York: Praeger.

Weir, M., Orloff, A.S., amd Skocpol, T. (eds.) (1988) *The Politics of Social Policy in the United States*, Princeton: Princeton University Press.

Weiss, M.A. (1980) "The Origins and Legacy of Urban Renewal," in Clavel, Forester, and Goldsmith (1980).

—— **(1987)** *The Rise of the Community Builders: The American Real Estate Industry and Urban Land Planning*, New York: Columbia University Press.

Weitz, S. (1985) "Who Pays Infrastructure Benefit Charges – The Builder or the Home Buyer?," in J.C. Nicholas (1985).

Welfeld, I. (1976) "The Courts and Desegregated Housing: The Meaning (if any) of the Gautreaux Case," *The Public Interest* 45: 123–35.

Welland, D. (1987) *The United States: A Companion to American Studies* (2nd edn.) London: Methuen.

Wells, C. (ed.) (1982) *Perspectives in Vernacular Architecture*, Annapolis, MD: Vernacular Architecture Forum.

—— **(1986)** *Perspectives in Vernacular Architecture II*, Columbia: University of Missouri Press.

Westin, A.F. (1983) "The United States Bill of Rights and the Canadian Charter: A Socio-Political Analysis," in McKercher (1983).

Wheaton, W.L.C. (1949) "The Housing Act of 1949," *Journal of the American Institute of Planners* 15, 3: 36–41.

Whitaker's Almanac, annual, London: Whitaker.

White, A.G. (1988) *Transferable Development Rights in the 1980s: A Selected Bibliography*, Monticello, IL: Vance Bibliographies.

White, M. (1978) "Self-Interest in the Suburbs:

The Trend Toward No-Growth Zoning," *Policy Analysis* 4: 185–203.

White House (1978) *President's Message to Congress on the National Urban Policy*, Washington: Office of the President; also in U.S. HUD (1978b).

—— **(1981)** *America's New Beginning: A Program For Economic Recovery*, Washington: Office of the President.

White Paper *see* GB.

Whitten, R. (1922) "Social Aspects of Zoning," *Survey* 48: 418–19 (June 15, 1922), (quoted in Toll 1969: 262).

Whittington, M. and Williams, G. (eds.) (1984) *Canadian Politics in the 1980s* (2nd edn.), Toronto: Methuen.

Whorton, J.W. (1989) "Innovative Strategic Planning: The Georgia Model," *Environmental and Urban Issues* Fall 1989: 15–23.

Whyte, W.H. (1964) *Cluster Development*, New York: American Conservation Association.

—— **(1968)** *The Last Landscape*, New York: Doubleday.

—— **(1980)** *The Social Life of Small Places*, Washington: Conservation Foundation.

—— **(1988)** *City: Rediscovering the Center*, New York: Doubleday.

Widdicombe Committee (1986) *The Conduct of Local Authority Business*, Cmnd 9797, London: HMSO.

Wilburn, G.W. (1983) "Transportation Projects and Historic Preservation: Recent Developments under Section 4(f) of the Department of Transportation Act," *Preservation Law Reporter* 2: 2017–29.

Wildavsky, A.B. (1973) "If Planning is Everything, Maybe It's Nothing," *Policy Sciences* 4: 127–53.

Wilder, M.G. and Rubin, B.M. (1988) "Targeted Redevelopment Through Urban Enterprise Zones," *Journal of Urban Affairs* 10: 1–17.

Wilensky, H.L., Luebert, G.M., Hahn, S.R., and Jamieson, A.M. (1985) *Comparative Social Policy: Theories, Methods, Findings*, Berkeley: Institute of International Studies, University of California, Berkeley.

Williams, N. (1955) "Planning Law and Democratic Living," *Law and Contemporary Problems* 20: 317–50.

—— **(1970)** "The Three Systems of Land Use Control," *Rutgers Law Review* 25: 81–101.

—— **(1984a)** "A Look at Implementation," *Environmental Law* 14: 831–41.

—— **(1984b)** "The Background and Significance of *Mount Laurel II*," *Washington University Journal of Urban and Contemporary Law* 26: 3–23.

—— **(1985–90)** *American Land Planning Law*, vols. 1, 2, 3, 3a, 4, 4a, and 5, with supplements, Deerfield, IL: Callaghan. (The individual volumes have been revised at different dates from 1985 to 1988; at the time of writing the latest supplement was dated 1990.)

—— **(1990)** "Scenic Protection as a Legitimate Goal of Public Regulation," *Journal of Urban and Contemporary Law* 38: 3–24.

Williams, N. and Ernst, H. (1989) "And Now We Are On The Darkling Plain," *Vermont Law Review* 13: 635–73.

Williams, N. and Kellogg, E.H. (eds.) (1983) *Readings in Historic Preservation: Why? What? How?*, New Brunswick, NJ: Center for Urban Policy Research, Rutgers University.

Williams, N. and Norman, T. (1971) "Exclusionary Land Use Control: the Case of Northeastern New Jersey," *Syracuse Law Review* 22: 475–507.

Williams, N. and Wacks, E. (1969) "Segregation of Residential Areas Along Economic Lines: Lionshead Lake Revisited," *Wisconsin Law Review* 27: 827–47.

Williams, N., Kellogg, E.H., and Lavigne, P.M. (1987) *Vermont Landscape*, New Brunswick, NJ: Center for Urban Policy Research, Rutgers University.

Williams, R.A. (1985) *On The Inclination of Developers to Help the Poor: Designing Affirmative Measures to Induce the Construction of Lower Income Housing after Mount Laurel II*, Land Policy Roundtable Policy Analysis Series 211, Cambridge, MA: Lincoln Institute of Land Policy.

—— **(1988)** "Legal Discourse, Social Vision and the Supreme Court's Land Use Planning Law: The Genealogy of the *Lochnerian* Recurrence in *First English Lutheran* and *Nollan*," *University of Colorado Law Review* 59: 427–74; also in Deutsch and Tarlock (1989).

—— (1989) "Euclid's *Lochnerian* Legacy," in Haar and Kayden (1989a).

Williams, S.F. (1977) "Subjectivity, Expression, and Privacy: Problems of Aesthetic Regulation," *Minnesota Law Review* 62: 1–58.

Wilson, J.Q. (ed.) (1966) *Urban Renewal: The Record and the Controversy*, Cambridge, MA: MIT Press.

Wilson, J.W. and Pierce, J.T. (1982) "The Agricultural Land Commission of British Columbia," *Environments* 14: 11–20.

Wilson, M. (1989) *Land Use Planning Forever Altered? The 1987 Supreme Courts's Takings Trilogy Decisions: A Partially Annotated Bibliography of Periodical Articles*, Chicago: Council of Planning Librarians, Bibliography 243.

Wilson, W.H. (1980) "The Ideology, Aesthetics and Politics of the City Beautiful Movement," in Sutcliffe (1980).

—— (1989) *The City Beautiful Movement*, Baltimore: Johns Hopkins University Press.

Windsor, D. (1979) "A Critique of *The Costs of Sprawl*," *Journal of the American Planning Association* 45: 279–92.

Wingo, L. (1972) "Issues in a National Urban Development Strategy for the United States," *Urban Studies* 9: 3–27.

Winter, W.O. (1969) *The Urban Polity*, New York: Dodd, Mead.

Winters, J. (1991) "Approaches to Growth Management in Palm Beach," in DeGrove (1991).

Wise, H.F. (1977) *History of State Planning: An Interpretive Commentary*, Washington: Council of State Planning Agencies.

Witt, J.W. and Sammartino, J. (1990) "Facility Planning in the 1990s: The Second Step in Urban Growth Management," *Journal of Urban and Contemporary Law* 38: 115–55.

Wolf, M.A. (1989) "The Presence and Centrality of *Euclid v Ambler*," in Haar and Kayden (1989a).

Wolf, P. (1981) *Land in America: Its Value, Use and Control*, New York: Pantheon.

Wolfinger, R.E. (1974) *The Politics of Progress*, New York: Prentice-Hall.

Wolman, H. (1980) "The Presidency and Policy Formulation: President Carter and the Urban Policy," *Presidential Studies* 10: 402–15.

—— (1985) "National–Urban Relations in Foreign Federal Systems: Lessons for the United States," in C.R. Warren (1985).

—— (1986) "The Reagan Urban Policy and its Impacts," *Urban Affairs Quarterly* 21: 311–35.

—— (1988) "Understanding Recent Trends in Central–Local Relations: Centralization in Great Britain and Decentralization in the United States," *European Journal of Political Research* 16: 425–35.

Wolman, H. and Goldsmith (1990) "Local Autonomy as a Meaningful Analytical Concept: Comparing Local Government in the United States and the United Kingdom," *Urban Affairs Quarterly* 26: 3–27.

Wood, E.E. (1934) "A Century of the Housing Problem," *Law and Contemporary Problems* 1: 137–47.

Wood, P. (1986) "London: The Emerging Docklands City," *Built Environment* 12: 117–27.

World Almanac (annual), New York: Pharos Books.

Wright, R.R. and Gitelman, M. (1982) *Land Use: Cases and Materials* (3rd edn.) St. Paul, MN: West.

Wright, R.R. and Wright, S.W. (1985) *Land Use in a Nutshell* (2nd edn.) St. Paul, MN: West.

Wrong, D.H. (1955) *American and Canadian Viewpoints*, Washington: American Council on Education.

Wyatt, B. (1986) "The Challenge of Addressing Vernacular Architecture in a State Historic Preservation Survey Program," in Wells (1986).

Wylie, J. (1989) *Poletown: Community Betrayed*, Urbana-Champaign: University of Illinois Press.

Yale Law Journal Note (1962a) "Toward an Equitable and Workable Program of Mobile Home Taxation," *Yale Law Journal* 71: 702–19.

—— (1962b) "Zoning Against the Public Welfare: Judicial Limitations on Municipal Parochialism," *Yale Law Journal* 71: 720–35.

Yale Law and Policy Review (1988) "The Fair Housing Act After Twenty Years," Conference Papers, *Yale Law and Policy Review* 6: 331–92.

Yannacone, V.J., Rahenkamp, J., and Cerchione, A.J. (1976) "Impact Zoning: Alternative to Exclusion in the Suburbs," *Urban Lawyer* 8: 417–48.

Yearwood, R.M. (1971) *Land Subdivision Regulations: Policy and Legal Considerations for Urban Planning*, New York: Praeger.

Yelin, A.B. (1990) "Routes to Landmark 'De-Designation': An Analysis of Selected Cases," *Urban Lawyer* 22: 307–29.

Zick, S.J. (1984) *Preservation Easements: The Legislative Framework*, Washington: National Trust for Historic Preservation.

Ziegler, E.H. (1982) "The Twilight of Single Family Zoning," *University of California Los Angeles Journal of Environmental Law* 3: 161–217.

—— **(1986)** "Visual Environment Regulation and Derivative Human Values – The Engineering Rational Basis for Modern Aesthetic Doctrine," *Zoning and Planning Law Report* 9: 17–24.

Zimmerman, J.F. (1983) *State–Local Relations: A Partnership Approach*, London: Praeger.

Zingale, J.A. and Davies, T.R. (1986) "Why Florida's Tax Revenues Go Boom or Bust, and Why We Can't Afford it Anymore," *Florida State University Law Review* 14: 433–61.

Zorn, P.M., Hansen, D.E., and Schwartz, S.I. (1986) "Mitigating the Price Effects of Growth Control: A Case Study of Davis, California," *Land Economics* 62: 46–57.

Zotti, E. (1987) "Design by Committee," *Planning* 53, 5: 22–7 (May 1987).

List of main cases

Des Plaines, City of, v Trottner: 216 N.E.2d 116 (1966)
Euclid, Village of, v Ambler Reality Co.: 272 U.S. 365 (1926)
Fairfax County: Board of Supervisors of Fairfax County *et al.* v DeGroff Enterprises Inc *et al.*: 198 S.E.2d 600 (1973)
Fasano v Board of County Commissioners: 507 P.2d 23 (1973)
First English Evangelical Lutheran Church of Glendale v County of Los Angeles: 482 U.S. 304 (1987)
Fisher v Bedminster Township: 93 A.2d 378 (1952)
Ginzburg v U.S.: 383 U.S. 463 (1966)
Golden v Planning Board of the Town of Ramapo: 285 N.E.2d 291 (1972)
Green v Lima Township: 199 N.W.2d 243 (1972)
Group House of Port Washington v Board of Zoning and Appeals: 380 N.E.2d 207 (1978)
Hadacheck v Sebastian: 239 U.S. 394 (1915)
Haveis v Far Hills: No.L-73360-80-PW, letter opinion, App. Div, October 9, 1986
Hawaii Housing Authority v Midkiff: 467 U.S. 229 (1984)
Hawkins v Town of Shaw: 437 F.2d 1286 (1971)
Hills Development Company v Township of Bernards: 510 A.2d 621 (1986)
Home Builders League of South Jersey v Township of Berlin: 405 A.2d 381 (1979)
Home Builders and Contractors Association of Palm Beach v Board of County Commisioners: 446 So.2d 140 (1983)
John Donelly and Sons v Outdoor Advertising Board: 339 N.E.2d 709 (1975)
Jordan v Village of Menomonee Falls: 137 N.W.2d 442 (1965)
Kavanewsky v Zoning Board of Appeals: 279 A.2d 567 (1971)
Keystone Bituminous Coal Association v DeBenedictis: 480 U.S. 470 (1987)
Kuehne v Town of East Hartford: 72 A.2d 474 (1950)
Ladue, City of, v Horn: 720 S.W.2d 745 (1986)
Lincoln Trust Co. v Williams Building Corporation: 128 N.E. 209 (1920)
Linmark Associates v Township of Willingboro: 431 U.S. 85 (1977)
Lionshead Lake Inc. v Wayne Township: 89 A.2d 693 (N.J. 1952); 344 U.S. 919 (1953)
Members of City Council v Taxpayers for Vincent: 466 U.S. 789 (1984)
Metromedia Inc. v City of San Diego: 453 U.S. 490 (1981)
Miss Porter's School Inc. v Town Plan and Zoning Commission: 151 Conn, 425, 198 A.2d 707 (1964)
Montgomery County v Woodward and Lothrop: 376 A.2d 483 (1977)
Moore v City of East Cleveland: 431 U.S. 494 (1977)
Moriarty v Planning Board of Village of Sloatsburg: 506 N.Y.S.2d 184 (1986)
Mount Laurel I: Southern Burlington County NAACP v Township of Mount Laurel: 336 A.2d 713 (1975)
Mount Laurel II: Southern Burlington County NAACP v Township of Mount Laurel: 456 A.2d 390 (1983)
Mount Laurel III: *see* Hills Development Company
Mugler v Kansas: 123 U.S. 623 (1887)
Municipal Art Society of New York v City of New York: 522 N.Y.S. 800 (1987)
National Land and Investment Co. v Kohn: 215 A.2d 597 (1965)
Nectow v City of Cambridge: 277 U.S. 183 (1928)
Neuberger v City of Portland: 603 P.2d (1979)
Nollan v California Coastal Commission: 483 U.S. 825 (1987)
Oakwood at Madison Inc. v Township of Madison: 283 A.2d 353 (1971)
Oakwood at Madison Inc. v Township of Madison: 371 A.2d 1192 (1977)
Ogo Associates v City of Torrance: 112 Cal. Rptr. 761 (1974)
Oka v Cole: 145 So.2d 233 (1962)
Otto v Steinhilber: 24 N.E.2d 851 (1939)
Parkes v Watson: 716 F.2d 646 (1983)
Passaic, City of, v Paterson Bill Posting, Advertising and Sign Painting Company: 62 A.267 (1905)

Index